BASIC QUESTIONS IN
Paleontology

BASIC QUESTIONS IN
Paleontology

Geologic Time, Organic Evolution,
and Biological Systematics

Otto H. Schindewolf

Translated by
Judith Schaefer

Edited and with an Afterword by
Wolf-Ernst Reif

With a Foreword by
Stephen Jay Gould

The University of Chicago Press/Chicago and London

OTTO H. SCHINDEWOLF was professor of geology and paleontology at the University of
Tübingen. WOLF-ERNST REIF is associate professor of paleontology at the University of
Tübingen. JUDITH SCHAEFER is an interpretive writer at the Denver Museum of Natural History.

The University of Chicago Press, Chicago 60637
The University of Chicago Press, Ltd., London
© 1993 by The University of Chicago
Foreword © 1993 by Stephen Jay Gould
All rights reserved. Published 1993
Printed in the United States of America
02 01 00 99 98 97 96 95 94 93   1 2 3 4 5

ISBN (cloth): 0-226-73834-5
ISBN (paper): 0-226-73835-3

Originally published as *Grundfragen der Paläontologie*, © 1950 E. Schweizerbart'sche
Verlagsbuchhandlung, Erwin Nägele, Stuttgart, Germany.

Library of Congress Cataloging-in-Publication Data

Schindewolf, Otto H. (Otto Heinrich), 1896–1971.
    [Grundfragen der Paläontologie. English]
    Basic questions in paleontology : geologic time, organic evolution, and biological systematics
/ Otto H. Schindewolf ; translated by Judith Schaefer ; edited and with an afterword by Wolf-Ernst
Reif ; with a foreword by Stephen Jay Gould.
        p.   cm.
    Includes bibliographical references and index.
    ISBN 0-226-73834-5. — ISBN 0-226-73835-3 (pbk.)
    1. Paleontology.   I. Reif, Wolf-Ernst.   II. Title.
QE761.S3413   1993
560—dc20   93-10189
                                                                                        CIP

∞ The paper used in this publication meets the minimum requirements of the American National
Standard for Information Sciences—Permanence of Paper for Printed Library Materials, ANSI
Z39.48-1984.

Contents

Foreword

STEPHEN JAY GOULD

Obviously, a persona need not match the etymology of a name: Ms. Petit may be a six-footer; Mr. Gross may be thin; and Mr. Fletcher need not make arrows for a living. Still, perhaps we should not be surprised that so tough and influential a man as Otto Schindewolf should be the "wolf skinner."

When I was a graduate student in 1965, I asked my advisor, Dr. Norman Newell: "who, in your opinion, is the world's greatest living paleontologist." He replied, without hesitation, Otto H. Schindewolf. Newell admired Schindewolf for his voluminous work on the morphology and phylogeny of Mesozoic ammonites, but Newell also honored his own highly developed sense of fairness in giving Schindewolf the nod. For Newell had been the most important American invertebrate paleontologist in an interdisciplinary movement, beginning in the 1930s and solidifying in the 1950s, known as the Modern Synthesis—the establishment of Darwinian natural selection, in conjunction with Mendelian principles in genetics, as a paradigm for evolutionary change at all scales of magnitude and time. Schindewolf had remained steadfastly apart from this movement. His thoughts, best expressed and summarized in this volume (originally published in 1950), represent the purest expression of a complete and thoroughgoing anti-Darwinian evolutionism based on traditions that had run deep in continental Europe ever since the late eighteenth-century works of Goethe and the *Naturphilosophen*. In having this landmark work available for the first time in English, anglophone evolutionists can finally understand the full force of the alternative against which their own orthodoxy was originally defined.

Schindewolf was a powerful man, and his position as professor of the most prestigious chair at Tübingen gave him virtual hegemony over evolutionary paleontology in postwar Germany. The hierarchical structure of European universities gave such power to leading professors; in addition, he had little competition, for most paleontologists were practically or geologically oriented and few others studied evolutionary questions. I met Schindewolf once, during a meeting in 1970 sponsored in Tübingen by Adolf Seilacher, his successor as professor. Schindewolf was kind, even courtly, to me and the few other foreign visitors, but I well remember the hushed awe that attended his few interventions, and I became decidedly uncomfortable when not a single younger German paleontologist dared to question anything he said during the public forum.

The frustration imposed by such customary obeisance emerges in several passages of Wolf Reif's superb afterword to this volume—an essay all the more excellent for its clarity and fairness in the face of Schindewolf's status as an oppressive "old guard" against the modernistic Darwinism represented by Reif and paleontological colleagues of his (I should say our) generation. Reif speaks of "a strong anti-Darwinian movement" that "still today has not been completely overcome." And, at the very end of his essay, Reif lists the impacts of Schindewolf's volume and general prestige, mostly baleful for a younger scientist with divergent views. Reif ends: "Finally, as late as the 1970s young authors risked censure by their superiors if they discussed typostrophism [Schindewolf's main concept] critically. Under the influence of Schindewolf's authority, evolution was no topic for the would-be paleontologist." I believe that I sense some legitimate bitterness in Reif's words.

But if younger German Darwinians felt oppressed by a weighty orthodoxy of contrary commitments from an older and powerful generation, their English and American counterparts were involved in an almost precisely opposite professional experience—one that makes the English translation of Schindewolf's major volume an event of great interest, note, and utility in our linguistic community.

Darwin had convinced the thinking world of evolution's factuality, but his theory for the mechanism of change, natural selection, enjoyed little popularity at the time of his death in 1882. Ironically, the Mendelian rediscovery, and the rise of genetics in the early twentieth century, only decreased the prestige of natural selection, as macromutational ideas flourished. But, beginning in the 1930s, micromutational and population genetics built a style of Mendelism that could be synthesized with notions of gradual and adaptive change favored by many systematists and field naturalists. This emerging Darwinian orthodoxy gained major expression in a series of books, classics of the so-called Modern Synthesis, that began with Theodosius Dobzhansky's *Genetics and the Origin of Species* in 1937. By 1950, year of publication for Schindewolf's *Gundfragen der Paläontologie*, this orthodoxy had swept the field in England and America. Schindewolf's ideas may have been a powerful force in Germany; in America, they had already become, by the time of this book's first publication, a curiosity and a monument to reaction.

It is a sad commonplace of orthodoxies that they spawn derisive dismissals of contrary views, even (or especially) when the ideas of dissenters are not read, or known only from biased accounts written by high priests—a point so well understood by Orwell in his novel, *1984*. All orthodoxies must designate whipping boys, and the Modern Synthesis was no exception. Richard Goldschmidt (cited favorably by Schindewolf) became the chief focus of Darwinian deprecation, particularly for his *Material Basis of Evolution,* published in 1940. Goldschmidt, though German to the core (as a Jew, he had left for obvious reasons during the 1930s), was then teaching in California, and he wrote his book in English—so he was eminently available as a foil. The other focus of opposition to the Modern Synthesis was Otto Schindewolf—but in night-and-day contrast with his former colleague in California. Goldschmidt was present and palpable; Schindewolf hidden, inaccessible, and rather mysterious. He was the man who said that the first bird hatched from a reptile's egg, the man who spoke of phyletic life cycles independent of environment. But hardly anyone—given the lamentable linguistic parochialism of most anglophones—had read a word of Schindewolf's actual writing. Nothing was ever translated, while the war delayed Schindewolf's composition and publication until 1950 when, for most anglophone evolutionists, the battle had already been fought and won. Schindewolf therefore stands as a primary anti-Darwinian icon almost entirely by reputation, rather than by explicit study or confrontation with his actual words. This translation, so long overdue, finally allows us direct access to a document that many of us rejected, explicitly and vociferously, only because our teachers had so instructed us. Perhaps we shall find something of value herein?

Schindewolf's ideas are particularly fascinating for historical reasons, because his anti-Darwinism is so fully formulated. This work is no compromise or amalgam of acceptable notions welded together from opposing sides, but a trenchant and utterly consistent (if spectacularly flawed) account of how an uncompromisingly anti-Darwinian, but fully evolutionary, worldview can operate. Schindewolf follows a conventional scientific conceit in claiming that he worked objectively, moving upward and outward from an empirical base in the groups he knew best—ammonites and corals. In fact, this book represents the imposition of a worldview upon selected components of the empirical record. Opposing views are scarcely discussed or even cited. Schindewolf's *Grundfragen* is a credo or manifesto for a saltational, internally driven evolutionism in the Continental formalist tradition versus the longstanding English preference for functionalist theories based on continuous adaptation to a changing external environment.

Such opposition has unusual potential to disturb Darwinian orthodoxy when it claims a paleontological basis, for data of macroevolution pose a special challenge. Darwinism may work perfectly well, and to near exclusivity, in our observable world of lifetimes and generations. But suppose this lovely and proven mechanism cannot encompass, by extrapolation, the full scale of vastly more extensive changes that unfold during millions of years. Suppose that one (or both) of the two anti-uniformitarian ghosts reign over geological time—steady changes too slow to notice on observable times scales, or cataclysmic events occurring so rarely that little probability exists for direct observation during human historical time. These ghosts had been faced by the Modern Synthesis, and laid to rest by paleontology's champion within this movement: George Gaylord Simpson, who, in two books (*Tempo and Mode in Evolution* [1944], and *The*

Major Features of Evolution [1953]), persuaded the anglophone world that macroevolution could be fully rendered as Darwinian microevolution extended. But
could Schindewolf shatter, or even dent, this consensus?

Darwinian natural selection, in its essence, is a theory of trial-and-error externalism. Organisms propose (but do not direct or constrain) by presenting copious
and random variation to a selecting environment. The external environment
(both physical and biotic) then disposes by granting differential reproductive
success to those variations fortuitously better adapted to changing local conditions. The application of this notion to all scales of macroevolution requires
smooth, uniformitarian extrapolation. Gradual and adaptive changes, accumulating step by step through countless generations of competition, underlie all
scales of morphological transformation and the waxing and waning of large
clades.

Schindewolf's anti-Darwinism resides primarily in his denial of both these
essential postulates—externalism (gradualism and adaptation) and extrapola-
tionism. He examines Darwinian mechanisms and proclaims them capable of
explaining small-scale adaptation at generational scales. But can this mode then
be extrapolated to the history of life? Schindewolf sets up the issue as any Darwinian would: "The question now arises whether those minuscule mutational
steps and morphological changes within an existing framework are also sufficient to explain macroevolution, the comprehensive remodeling of forms of
higher taxonomic categories, which means the elaboration of new organs and
structural designs." And Schindewolf answers his own question firmly in the
negative. Fundamentally new features arise saltationally, "discontinuously, suddenly, between one individual and the next, during an embryonic developmental
stage." Following this saltational origin, new taxa unfold their history along internally forced, orthogenetic pathways, virtually independent of the external environment: "Many attempts have been made to disavow orthogenesis, but it is a
fact, and there is no way around it. . . . We see [evolution] follow its course as
if from inner compulsion, automatically, like clockwork, no matter whether it
leads to ascendary or decline." Thus, both ghosts of anti-uniformity rule macroevolutionary history: rare and sudden events of origin, followed by internally
driven unfolding, too slow to observe on generational scales but able to overcome any external adaptational modling, thus imparting direction to the history
of clades.

Darwinian adaptation exists in this world, Schindewolf continues, but it is
entirely secondary, superficial, and not the cause of anything macroevolutionary—a mere epiphenomenal molding upon an internally unfolding *Bauplan*.
"Selection concerns and affects only the most superficial layer of characters."
"Selection is only a negative principle, an eliminator, and as such is trivial."
Environment, the source of selective forces, is therefore similarly unimportant
in macroevolution: "Tempo, extent, and diversity of modification are conse-

quently grounded in internal factors of organic lineages and are not determined by the environment and its changes."

I shall not discuss Schindewolf's particular anti-Darwinian mechanisms in this essay, for Reif has given such a fine account in his afterword. I shall only mention in passing that Schindewolf relied upon two major theories: (1) The cyclical theory of typostrophism, according to which new groups arise by multiple, saltational origin of many vigorous sublineages [typogenesis], proceed then to unfold in a gradual but orthogenetic and therefore fundamentally nonadaptive way [typostasis], and finally suffer a period of degeneration before extinction [typolysis] (note that the first phase of the threefold cycle is saltational, while all are nonadaptational for different reasons—thus contravening all essential Darwinian postulates of gradualism, adaptationism, and selection set by environmental pressures); (2) The ontogenetic theory of proterogenesis for the saltational origin of new types. By this mechanism, new features arise all at once [first bird in a reptile's egg] by transformation of the earliest ontogenetic stages. These changes then work their way forward by neoteny until they finally appear in the adult stages of descendant forms.

In this book, Schindewolf says nothing at all about a proposal that would win him much notoriety during the 1960s—mass extinction provoked by explosions of supernovae. Such an idea, at first glance, sounds contrary to Schindewolf's emphatic denial of external environmental change as a spur to macroevolution. But as Reif points out so well, Schindewolf actually used this speculation to defend his internalist theory against accumulating evidence for the importance of mass extinction. Schindewolf argued that supernovae produced their effects by cosmic-ray zapping, thus raising mutational rates, and sealing the fates of lineages according to their own internal status at the time—degeneration to extinction for typolytic lineages, rejuvenation by saltation for typogenetic lineages.

Schindewolf's work offers much to scholars in other fields as well. How can we, for example, understand any product of Germany in the 1940s without reference to the surrounding greatest cataclysm for our own historical age—Hitler and the Third Reich. I am intrigued, for example, that the two most important paleontologists who became active Nazis—Othenio Abel and Karl Beurlen—both accepted evolutionary internalism, but with a decidedly vitalistic and romantic interpretation (rooted in such Nazi-friendly notions as "will to power"), whereas Schindewolf maintained a naturalistic and mechanistic perspective more congenial to the usual interpretations of modern Western science (he expected that internal laws of change would one day be rendered in mechanical, rather than mystical or vitalistic, terms). Historians have studied Nazi scientists intensely, but perhaps we should devote more attention to honorable men like Schindewolf (and Bernhard Rensch, who shared his mechanical view, but preferred Darwinian explanations)—scientists who remained in Germany, trying to continue their work without collaborating for political gain or belief.

Schinden, in modern German, is a largely derogatory word, used (as in English) for such metaphorical forms of "skinning" as traveling without a ticket, gatecrashing at a concert, or sitting in a restaurant without ordering anything. Too often, we try to take such self-serving ways out in our intellectual lives—to live with our certainties by avoiding the responsibility of seeking out and struggling with alternatives. I don't think that the typostrophic theory offers much to our search for mechanisms (though it describes a common pattern that we will have to render in other causal ways). I doubt that proterogenesis is more than a quite occasional mode of heterochronic change. But any probe from the surface of appearances to the core of things requires that we confront all the great wolf-skinners on opposing sides of our own certainties.

Preface

Paleontology, like its elder sister or progenitor, geology, is for most people not exactly a familiar science. Not until recently, during the disastrous war years, when we were required as never before to explore thoroughly the available mineral resources of our own country, has the significance of geology become evident to a wider circle. Paleontology, however, remains modestly in the background, its existence almost unknown to the vast majority. People are probably aware that this science has something to do with fossils, that is, "things that have turned to stone," and occasionally, in museums, may even have looked with astonishment upon the skeletons of huge dinosaurs or the remains of extinct giant mammals; beyond that, however, almost nothing is known of the actual work of paleontology—of its task, goals, and methods, its vast implications for our world view, and its practical application.

The layman, in general, is inclined to regard the work of paleontologists as being scholarly and esoteric and may even disapprove of it for those qualities. What is the use of collecting and describing antediluvian monsters, petrified snails, and other "figurines" of stone? Do they, like objects in old curio cabinets, merely serve to satisfy the public's curiosity and their curator's fancy?

Unfortunately, newspapers and the very popular illustrated weeklies provide no answers. Instead, what they usually publish is alarmingly uninformed—a level of reporting that would not be acceptable for other branches of science that fall within the scope of "general education." It is absolutely essential that this situation be remedied with explanation and clarification to provide a broader public with insight into one of the most intriguing, many-faceted areas in all of natural history.

But even in the related disciplines of natural science, which stand to gain much from paleontology, the dominant view reflects suspicion, lack of understanding, and unfounded prejudice against the findings of paleontologists. Most botanists, zoologists, geneticists, and other scientists know too little about fossils and the logical methods by which they are studied to appraise properly the importance of the conclusions arrived at and to fully exploit for their own discipline the rich trove of experience available to them. The cliché about "the gaps in the fossil record" makes them think that the paleontologist deals only with random, isolated, accidental finds, leaving any statement made about them without compelling proof. This view, widespread among biologists today, has more serious consequences than the lack of comprehension on the part of the general public and therefore must be countered all the more vigorously through explanation and enlightenment.

Considerations of this kind have caused me, pursuant to a request by a publisher, to attempt in this book to familiarize a broader public with a branch of science that has gone largely unnoticed or been insufficiently explored—a branch that holds a place of great importance among the natural sciences. The sections presented here treating the problems, methods, and findings of paleontology are therefore addressed primarily to the educated, interested layman and to those representing related branches of natural science and medicine as well, to aid them in forming an opinion. Furthermore, the positions taken on many current questions and controversies may be useful to many of my colleagues— geologists as well as paleontologists. Finally, this volume, not really a textbook, may serve as an introduction to the subject, especially for younger readers.

Of the many paleontological problems that might be considered, we have selected three groups of questions, which are listed as subtitles on the title page. The chapter on *geologic time* is rather brief, for this book was begun several years ago, and the author has since published a detailed discussion of this subject elsewhere [Schindewolf 1943]. The matter of geologic time is important in applied paleontology, which makes a significant contribution to the exploration of mineral deposits and, by extension, to the husbanding of our natural resources and safeguarding of our economic security. In addition, this chapter provides the basis for a complete understanding of fossils and their biological evaluation. Therefore, at least a cursory glance at the subject must be presented here; to that end, some of the material from the other work mentioned, now in its third edition, is repeated here.

On the other hand, basic questions of *organic evolution* are treated in considerable detail; here, paleontology has a critical contribution to make. For the first time, a relatively simplified presentation of and rationale for evolutionary concepts is given based on the author's paleontological observations and reflections; in recent years, these concepts have repeatedly been the subject of considerable difference of opinion. In this section, too—as well as in the last section treating *biological systematics*—much has been drawn from my earlier published papers.

There is, naturally, much material from other sources, which, however, cannot be identified more specifically, partly to keep the discussion from bogging down and partly owing to the impossibility of giving exact citations of literature to which my access, under the circumstances, was limited. At the end of the book, I give only a small selection of works that treat the subject in a comprehensive way; this bibliography will direct the interested reader to further, more specialized literature.

The manuscript of this book was written mostly in air-raid shelters in Berlin or under other unfavorable wartime conditions; many parts were destroyed and had to be written over again. By the end of 1944, it was finished. The complete collapse of our economy and our spirit prevented the book from being published immediately, as had been planned. As soon as the extremely difficult circumstances permitted, the *publisher* commendably did everything it could to bring the book into being and, as far as was possible, to present it in a form worthy of the endeavor. During the intervening period, the book was supplemented by information from more recent literature to the extent that I had access to it. I am deeply grateful for the kindness and understanding shown by the publisher!

Of particular importance are the numerous *illustrations.* When referring to them, please note that, for technical reasons, the halftone illustrations do not follow in sequence but are presented all together, printed on coated paper, at the end of the book. I owe many of the illustrations to the knowledgeable collaboration of A. Schulze, dec., F. Thiem, Mrs. E. Michels and Miss E. Brämer, dec., in Berlin, as well as to Mrs. G. Winter–v. Möllendorf, dec., in Frankfurt a. M. I am indebted to Professor W. Gothan, Berlin; Professor W. Janensch, Berlin; and Processor E. Voigt, Hamburg, for kindly providing me with other illustrations. Professors A. Heintz and L. Störmer, Oslo, generously granted permission to reproduce their magnificent wall chart of the evolution of the animal kingdom. Further, the following publishers generously made plates from their publications available: G. Fischer, Jena (figs. 2.20, 2.24, 2.36–39, 2.54, 3.7, 3.13, 3.57, 3.58, 3.95, 3.107; pls. 22B and 26); the Hohenlohe Publishers and Bookstore F. Rau, Öhringen (Württemburg) (figs. 2.50, 2.51, 3.109, 3.116–18, 3.123, and 3.150; pls. 24A, B, C, 25B, and 27C); E. Schweizerbart Publishing Company, Stuttgart (fig. 2.4; pls. 3B, 4A–H, and 5B); and B. G. Teubner, Leipzig and Berlin (figs. 2.46, 3.138; pls. 9A and 16C).

To those named above and to all others who had a share in the realization of this book, I offer my deepest gratitude!

Otto H. Schindewolf
Tübingen, August 1948

POSTSCRIPT

Most of the present book had been printed by September 1948; its appearance was planned for the end of that year. The currency reform caused renewed delays in the printing, which was moving along slowly in any event, and finally compelled the original publisher, which had commissioned the book, to curtail its publishing program.

After a long hiatus, the publisher E. Schweizerbart, at my instigation, declared that it was prepared to take the now homeless work under its wing and to assume the risks inherent in the publication of a book of this size at a time when conditions in the publishing business were difficult. In doing so, this publisher exhibited the same willingness to make sacrifices as it demonstrated immediately after the collapse of Germany by resuming, without delay, the almost century-and-a-half-old tradition of publishing geological/paleontological journals, thus performing a great service to our science. My sincere thanks, and, I hope, those of the readers as well, to the publishing company of E. Schweizerbart for finally bringing this book into being!

Otto H. Schindewolf
Tübingen, May 1950

1   The Nature, Task, and Place of Paleontology

DEFINITION OF TERMS, GOALS, AND METHODS

Paleontology is the *science of prehistoric life*—of the fauna and flora of the geologic past. It is concerned with the systematics, mode of life, spatial distribution, temporal succession, and phylogenetic evolution of life forms that have gone extinct and become fossilized—in short, of *fossils*.

There is no convenient German term for this science. The terms *Versteinerungskunde* (literally, the science of things that have turned to stone) or *Petrefaktenkunde* (literally, the science of things that have petrified) used to be employed. These are poor terms, for it is neither the state of being stone nor the process of having been turned to stone that is the critical mark of the objects (completely apart from the fact that the objects have not really "turned to stone" at all). Rather, the essential point is that the object was *once a living creature, an organism.*

More recently, an attempt was made to gain acceptance for the term *Vorwesenkunde* (literally, the science of prehistoric life), and it is sometimes found in older works, as, for example, in works by O. Volger and Ludwig Büchner.[1] The term is, indeed, descriptive of the core of our discipline but not totally comprehensible without further explanation. Moreover, there may really be hardly any need for a German term at all. Every layperson is familiar with the meaning of the terms *Chemie* [chemistry] and *Astronomie* [astronomy], even though the derivation of these terms is foreign. Thus, the term *Paläontologie* will present no obstacle once we are able to impart to the public the fundamental principles subsumed in the term.

Since the organisms studied by the paleontologist are sometimes vegetal and sometimes animal, the term "paleontology" can also be defined as the botany and zoology of the geologic past. In keeping with this, we distinguish the two subdisciplines of *paleobotany* and *paleozoology*. The question now is, what constitutes the meaning and informative value of paleobotanical and paleozoological research?

1. [Schindewolf's Bibliography is only a general list of references and omits the works of many of the authors he mentions in the text. See my Afterword for further discussion. The works I cite in parts 3 and 4 of the References should also be helpful to readers seeking more detail about the development of evolutionary thinking in German paleontology.—Ed.]

First, paleontology has the same goals as modern zoology and botany—to collect specimens, to arrange them in a way that yields an overview, and to describe them. The result of this work is an *enormous assemblage of the wealth of forms* exhibited today by the plant and animal world. It becomes apparent that many groups of the flora and fauna around us are impoverished remnants—shriveled twigs on once-flourishing shoots and sturdy branches of the tree of life. For example, contemporary brachiopods, cephalopods, crinoids, cartilaginous fishes, and reptiles are the meager remains of earlier, extremely rich and diverse groups of animals. Other groups of forms and even entire structural designs[2] have completely disappeared over geologic time and are no longer represented in our present fauna and flora.

The life forms of the present, then, constitute only a small portion of the enormous variety of forms that have inhabited the earth at one time or another. Paleontology, however, is in a position to fill out the contemporary spectrum of plants and animals in critical ways, to add countless new traits, and eventually, to set forth a complete understanding of many contemporary structures and phenomena based on past forms. *All living creatures are products of history, marked by their evolution.* The structures and functions, evolutionary history, and geographic distribution of modern organisms are therefore not determined just by the regimen governing their lives today, but much more by their evolution during the geologic past.

Therefore, we can say that paleontology finally enables us to see the natural, broader framework into which contemporary life forms can be placed. Thus, botany, zoology, and comparative anatomy are enriched and stimulated in important ways by the findings of paleontology. Furthermore, based on the peculiar nature of the objects studied, paleontology is able to make theoretical contributions to the problem of classifying organisms—to *systematics,* or better, to *taxonomy.*

When the paleontologist describes and classifies his material, he makes use of the basic research methods of zoology and botany: he studies the object from the point of view of morphology, comparative anatomy, and—as much as the object permits—developmental history; in other words, he does not limit himself to observation of the fully developed, finished form but also considers individual development. It is true that, in the application of these methods, the paleontologist is at an undisputed disadvantage compared with the scientist who is working with modern organisms. Whereas the botanist or zoologist has the entire plant

2. [Although the word *Bauplan* is in common use in current technical literature, I prefer to translate it for several reasons: Schindewolf did not coin the word; nor is it the only term he uses to refer to the concept (see the explanation of terms in Chap. 3; p. 203ff.); and most important, I would like to avoid the possibility of perpetuating any unintended connotations that the German term may have acquired for English readers. Throughout, I have translated *Bauplan* rather literally as "structural design," so that those interested in such things can track Schindewolf's usage of the term.—Trans.]

or animal, including its soft parts and tissues, available for study, this situation, with a few rare exceptions, is not true for paleontologists. The latter are confronted with remains that are always incomplete to some degree; it is usually only the resistant hard parts of the organism that come down to us in the rocks, whereas the ephemeral soft parts of animals and the delicate tissues of plants are not preserved.

Compared to the more complete possibilities for research available to biology, this is a significant limitation, but it is not as serious as it might seem at first glance. For example, although the paleontologist generally cannot carry out detailed histological investigations, although he cannot detect the nerves, muscle fibers, blood cells, and so on, of his fossil organism, he nevertheless has at his disposal enough good specimens to show that the extinct animal also had similar nerves, blood cells, and so on, which, of course, was never in dispute anyway. Moreover, the explication and systematic classification of organisms is not based on histological findings but on other characters that, in general, are accessible to the paleontologist in fossil material. Apart from that, conclusions based on analogy must to some extent replace the examination of the missing soft parts, although it is true that such analogies are often sources of all kinds of errors.

It goes without saying that experimental research is basically not a part of paleontology. Nonetheless, through detailed analysis of suitable specimens, noteworthy results can be obtained not only in the realm of physiology, but even—and one would hardly expect this—in the realm of psychology. In place of the artificial experimentation that is denied us, we have to try to discern from the fossil organism the effects of nature's large-scale experimentation, for example, the organism's reactions to changes in living conditions as recorded in the historical evolution. Unambiguous clues with regard to mode of life, spatial distribution, habitat, climatic conditions in the environment, cause of death, and so on, can be obtained partly from the fossil itself and partly from the enclosing rock; former marine bottoms or land surfaces with their ecological communities, traces of the lives of organisms, of their complex interactions and dependence on substrate conditions—all this can be read in the rock layers.

The disadvantages cited in the comparison with contemporary biology are abundantly compensated for by a special aspect inherent in fossil material that gives paleontological research its true direction and adds an entirely new dimension to the science of life: *the historical layering of fossil faunas and floras in their natural temporal succession.* The remains of these once-living creatures— the fossils—are embedded in various kinds of sedimentary rock, which succeed one another in the *same sequence in which they were originally deposited* and built up. By collecting the fossils from any one horizon, we obtain the extremely valuable possibility of *tracing historically the growth, transformation, and disappearance of former life forms* from the geologic past to their modern manifestations. The elucidation of the history of life constitutes the main field of research in paleontology. Paleontology is, then, a *historical science,* and this perspective lends it, and geology as well, a very distinctive character among the natural sciences.

RELATIONSHIP TO BIOLOGY: IMPLICATIONS FOR EVOLUTION

Nowadays, we base all biological observations on the *theory of evolutionary descent,* the concept that organisms are not the individual products of supernatural creation but rather diverged from one another developmentally in natural ways and are therefore linked by actual ties of descent. Paleontological findings have contributed more than a little to this logically compelling conclusion, for only fossils provide the hard evidence and proof of evolution as a historical process.

It is true that zoologists and botanists have also tried on occasion to construct evolutionary trees and engage in discussions of phylogeny based only on contemporary life forms, without taking the fossil material sufficiently into account; their results, however, necessarily remain in the realm of esoteric, abstract morphology and often faulty hypotheses that have nothing to do with actual history. When scientists attempt to arrange only living organisms in a phylogenetic sequence and claim to have traced true phylogenetic stages, the endeavor is, of course, basically on the wrong track.

Only paleontology, with its chronological sequences of fauna and flora, is able to present the grand scheme of evolution truly and, in keeping with the goal of any history, "to describe the way it was" (Leopold von Ranke). Whereas modern biology surveys only the (provisional) conclusion of the infinitely long evolutionary process, the paleontologist has countless examples of such states at his disposal, preserved in the various rock layers. Just as the anatomist puts his wax-plate models together piece by piece, using a great many individual sections and images to form a comprehensive, integrated whole, so the paleontologist connects the successive cross sections of organic communities to obtain a three-dimensional model of natural relationships. The static, individual frames become elements of a dynamic process: the evolution of organisms.

The goal of paleontology, however, is not just to *describe* the history, the development, and the transformation of individual phyla and morphological series, but also to go on to work out the *general laws* according to which phylogeny proceeds. Thus, paleontology supplies valuable building blocks for the *theory of organic evolution, or of general organic descent,* thereby claiming the right to assume a critical role in all discussions of the subject. As important as the results of genetic experimentation and developmental physiology may be for theories of descent, they remain only speculations on how it *might have been;* for they are only experimental models or analogies from which actual historic parameters are excluded. Since descent lies in the geologic past, only fossils can provide support for actual historical development and for the assessment of former evolutionary processes.

It may indeed be assumed with certainty that the laws discovered by genetics and developmental physiology were as valid in geologic time as they are in the present. Hence, the deductions arrived at by paleontology must not contradict the results obtained and facts established by those disciplines; rather, paleontology should incorporate those findings into its own thinking. Nonetheless, it appears, at least for the present, that the observed or deduced processes of evolution in its entirety are not fully explainable by the methods of genetics alone. Instead of drawing the only possible conclusion, that the factors identified experimentally are insufficient and need to be supplemented by additional concepts, some geneticists have thrown the baby out with the bath and concluded that there is no such thing as evolution, that this concept is perforce in error.

This conclusion, like all other objections to the findings of paleontology, must be decisively refuted, if only on logical grounds. Paleontology studies and explains a historical process of growth and development, a course of events that took place in the past. To do this, it uses the same methods as the science of history, namely, a deductive reconstruction of processes and events based on available resources and evidence. Insofar as deductions are logically incontestable and unambiguous, they must have validity as *historical facts,* which are not affected by genetic experimentation and cannot therefore be overturned.

Some geneticists have recently called the science of phylogenetics—the study of the evolutionary history of organisms—an experimental discipline. This, of course, is absurd. Since history refers to past unique events, it can never be experienced in actuality and tested experimentally. Thus, as far as their methods are concerned, paleontology and genetics operate on completely different planes; genetics is not competent to treat questions of the *history* of forms. It would be desirable if geneticists, while taking pride in their admirable results, would also recognize and respect the limitations to their application.

The logic of paleontology's position corresponds exactly to that obtaining in geology. As the paleontologist may not impinge on the laws of heredity or become entangled in contradictions of them, so is the geologist bound by general physical and chemical laws, which reason tells us were undoubtedly no different in geologic time.[3] What are different for the geologist are the special temporal and spatial circumstances of earthly events, which cannot be repeated experi-

3. Recently, in fact, geologists have in all seriousness raised the question of whether natural laws are indeed constant or whether they are not perhaps subject to change over geologic time. The problem is in itself not a new one and has already been answered definitively, it seems to me, by H. Poincaré in favor of an immutability of the laws governing natural events. If one defines the laws of nature as rules according to which processes always take place in the same way everywhere, there can naturally be no question of mutability and development over time. It is only our *formulation* of the laws of nature that is mutable. As soon as we learn from experience that the concept of a law is not universally applicable because it is somehow contingent upon time, then the law should be excluded from the formulation. This process of correction and reassessment results in true, universally valid, temporally independent, abstract laws of nature, with the prior, time-dependent, narrower concept representing just an exception occurring under special conditions.

mentally and which compel him to form conclusions based on other factors. Only when those particular factors are taken into account is the geologist able to pull together the abstractions of individual universal laws to compose a concrete, historical overview.

With regard to paleontology, it must also be considered that the extension of the biological perspective to include spatial and temporal dimensions too vast to be reproducible experimentally produces additional phenomena and effects that are not yet comprehensible inductively.

From the foregoing it emerges that, based on its methodology and the goals of its research, paleontology is a *biological* discipline. It forms an important branch of biology,[4] one that broadens considerably the biological view of the world. Now that biological thinking—the laws ruling life forms, questions of the origins of race and species—has assumed a central position in our contemporary world view, and rightly so, paleontology has a critical contribution to make. Only a comprehensive account of the growth and development of life will allow us to assess correctly our own position in nature within the framework of the organic world.

RELATIONSHIP TO GEOLOGY AND THE HUMANITIES

On the other hand, paleontology, through its historical mode of observation and the materials it studies, has the closest of ties with geology. There is an intimate reciprocity between the two sciences, a relationship of mutual enrichment. A thorough grounding in geology is the prerequisite for a correct, comprehensive interpretation of fossils, the peculiarities of their occurrences, and the means by which they were preserved; the reverse is also true: the interpretation of fossils is one of the most important fundamentals of geology.

Such interpretation provides the geologist with indispensable determinations of time and age, thereby establishing the basic premises for both pure and applied geology. There is no branch of theoretical geology and, furthermore, no systematic exploitation of mineral deposits, no rational provision of drinking water, no road or tunnel construction, no studies of dam or construction sites that are not ultimately based on paleontological findings. Prehistoric life affects even the formative processes of rocks and mineral deposits, giving paleontology a voice in these questions, too.

Similarly, the chronology so basic to the study of *prehistory* is also provided by paleontology.

The occurrence and distribution of fossils yields further implications for earth history: for the changing distribution of land and sea, the shifting of climatic zones, and so on. Marine fossils found on contemporary land masses show that

---

4. "Biology" is here understood in the broadest sense of the term, whereas in the title of this section, "Relationship to Biology," what is meant is the relationship of paleontology to the more focused discipline of the biology of living organisms [neontology].

those areas were once covered by seas. The remains of Tertiary groves of magnolias and laurel on Greenland or of araucarian forests beneath the ice of Antarctica tell us that in those areas the climate was once very different—considerably warmer than it is today. A statistical synopsis of such data from the animal and plant world allows us to establish the positions of the equator and the poles for individual segments of geologic time. The mode of life of earlier organisms provides clues to the special conditions under which the enclosing rock was formed; for geology, then, there are a great many further practical applications of fossils. Whereas we asserted above that within the system of the sciences, paleontology must be counted as a branch of *biology,* it should be emphasized, in contrast, that the earliest beginnings of paleontology lie with *geology* and that, even today, it is still closely tied to that science superficially, in the practical sense of teaching and research. Paleontology was not, as might be expected, founded by biologists attempting to expand the knowledge of contemporary plants and animals backward, in a manner of speaking, but arose, rather, as a close companion to geology, a necessary consequence of the commonality of the material studied.

The first fossils were collected and described by geologists and amateurs. Biological considerations were not an issue at all. The objects that had "turned to stone" were regarded as essentially geological in nature, as inclusions in rock, and were basically approached in the same ways as minerals and other inorganic structures. Very soon, however, people realized the overriding significance inherent in fossils for the determination of geologic time and age. Because of the close dependency of geology on paleontology, the original ties between these two sciences have remained until the present.

In conclusion, certain relationships remain to be mentioned, namely, those with the *humanities.* Early on, evidently since the Late Stone Age, humans have taken an interest in fossils, assigning them roles in religion and superstition, in ceremonies, customs, and folk art. There are many myths and fairy tales, pictorial representations and ancient monuments, connected with remarkable fossil specimens. Only the paleontologist can interpret them, tracing them back to the phenomena underlying their origins. Working with philologists and art historians, the paleontologist helps decipher the figurative meaning of such ancient representations, lifting the veil from many otherwise puzzling aspects of myths, customs, and folklore (see pls. 1 and 2A).[5] In a similar way, the paleontologist works hand in hand with the prehistorian to explain the cave paintings and depictions of animals made by early humans

5. Plates 1 and 2A, referred to here, as well as all halftones mentioned subsequently, are collected in one section at the back of the book.

2   Basic Questions in Geologic Time

THE BASIS FOR MEASURING TIME AND DETERMINING AGE

We have already said that a fundamental of geology, on which it stands or falls, is the possibility of determining the age of rocks and strata. Geology as a historical science with the goal of writing a history of the earth must have at its disposal a temporal system to date its events and arrange them within a time scale. Otherwise, geology would be a chaotic scrap heap of isolated facts and in no position to come up with systematic insights into the structure of the earth's crust and the mineral resources it holds. The same is true for paleontology, which must know the relative ages of the various faunas and floras if it wants to demonstrate their historical development.

THE ROCKS

The substance upon which the measurement of geologic time is based is the rocks we encounter at the earth's surface and to a certain depth below the surface through drilling[1] and the excavation of mines. Again and again, we observe in any given area different types of rocks, the origin and formative sequence of which must be determined.

If the rocks in question are layered sedimentary rocks, it is immediately evident that they were either deposited at the bottom of a body of water or, less commonly, on land, and consolidated, layer by layer, bed by bed, through the forces of gravity. With such a mode of origin, which involves a gradual succession of deposits, it is obvious that the lowest of the rock units is the oldest and that the layers lying on top of it must be progressively younger. This is the basic, elementary *law of superposition,* first formulated in 1669 by Nicolaus Steno,[2] but not applied in any practical sense until much later (Lehmann, Füchsel, etc.).

However, in any natural outcrop or in a quarry, there is always only a limited sequence of layers that can be assigned relative ages based on superposition. It

---

1. The deepest drilling to date, an oil well in Oklahoma, reached a final depth of 5,433 meters.

2. [Nicolas Stensen, 1638–86. Danish anatomist, geologist, and cleric, Steno was the author of *De solido intra solidum naturalites contento* . . . (On a solid body enclosed by natural processes within a solid [1669]): He was the first to propose the idea that fossils are the remains of ancient living organisms and that many rocks are the result of sedimentation.—Trans.]

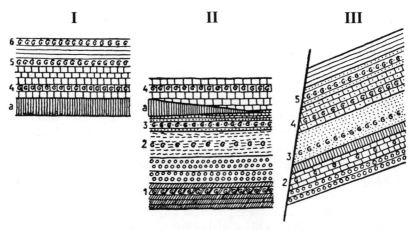

Fig. 2.1. Possible correlations of three geological profiles. Between exposures I and II, which were close together, a parallel can be established by means of the conspicuous index bank (*a*), which is at the same elevation in II as in I. However, within the exposure, the index bank wedges out, that is, gradually becomes thinner and disappears entirely, with the result that it can no longer serve as a comparative marker for the horizontal continuation of the stratigraphic sequence. Also, since farther away, in profile III, a bank of similar rock, although not as thick, appears at a low level of the series, it would be easy to establish an erroneous parallel based on it. Neither petrographic indicators nor the stratigraphic relationships offer any clue that profile III can be tied to the two others. The character of the rock and the thickness of the individual beds have changed markedly over the distance; furthermore, there is a dislocation, along which the sequence has dipped and tilted upward at an angle. A reliable parallel of the three profiles can really only be established based on fossil horizons 1–6, each of which contains characteristic index species regardless of the fluctuating character of the rock layers in which they are found.

has been estimated that from just the Cambrian—the first formation to contain abundant faunal remains—to the present, marine sediments have accumulated to an approximate overall thickness of one hundred kilometers. There are no natural outcrops or man-made exposures extensive enough to display this mighty stratigraphic sequence as it was laid down (to say nothing of the fact that in any single area, a complete series of all those layers never even formed). Therefore, it is necessary to combine a great number of shorter, separate profiles[3] to arrive at the composite stratigraphic sequence.

At times, the character of the rock offers help. For example, at the bottom of a quarry we observe a bed of rock of a particular thickness and some kind of characteristic petrographic structure (fig. 2.1, I*a*). In a deep ravine nearby, or in another quarry, we find the bed again, this time with another sequence of layers beneath it (fig. 2.1, II). We can then add this new rock series to the profile we found in the first quarry. This method, however, is only applicable to outcrops or

---

3. Geologic profile: strata section; for example, an outcrop that presents a series of layers in cross section.

exposures that are close together, and then, only sometimes. It fails over greater distances when immediate observations of the stratigraphic relationships of the individual profiles are not possible, when an index layer to use for correlation is lacking, or—and this is often the case—the rock changes in character or wedges out with distance.

Complicating the establishment of correlations and the linking of one series to another is the circumstance that strata are quite often no longer found in their original relationship and horizontal position but have been divided into small blocks by faulting and then been tilted, folded, or even turned upside down by mountain-building events. When the rock is in a perpendicular position, it cannot be determined without additional information which layer is the top one and therefore the youngest, and which the bottom and therefore the oldest. In over-turned strata, the tops and bottoms have been reversed and the spatial sequence of layers is just the opposite of its original temporal sequence.

In earlier times, it was believed that the age of a rock could be inferred just from its composition. It was assumed that certain *index rocks* such as graywacke or red sandstone were indicative of particular units of geologic time. At the end of the eighteenth century, guided by this principle, A. G. Werner proposed a classification of formations based on his observations in central Germany; he regarded his system as having universal validity. Subsequent research, which extended over a broader area, soon made it clear, however, that it is impossible to establish stratigraphic parallels over broad areas based on petrographic characteristics alone.

Today, we know that almost all sedimentary rocks (limestones, sandstones, shales, and so on) come in all colors and exhibit all kinds of petrographic peculiarities; that the same rocks have formed repeatedly in the different geological epochs; and that different rocks may form simultaneously and in immediate proximity. The character of the rock is not dependent on (or is only subordinate to) the temporal factor but is determined primarily by local conditions, namely, the particular erosional and depositional situation, which is uniform only within small, restricted areas; beyond that, it fluctuates widely.

THE FOSSIL RECORD

Enormous progress was made through the observations of William Smith, who, while doing field surveying in southern England around the year 1800, determined that each of the sequential layers contained fossils that were peculiar to it alone. Without concerning himself much with the nature of the fossils or their exact scientific significance, he attempted to determine the age of the strata purely empirically on the basis of his field observations, using the fossils as supporting evidence. Thus, Smith became the founder of precise *stratigraphy,* as the study of the sequence of strata is called. The old petrographically based

stratigraphy was now complemented by a stratigraphy based on organisms—
biostratigraphy—which is the only reliable method for the precise correlation
of strata. The broader consequences of this advancement of knowledge put it
quite on a par with the great discoveries in chemistry, physics, and medicine.

The theoretical ground for the superiority of biostratigraphy over lithostratig-
raphy is the fact that the former is based on a unique, irreversible process, independent of special local conditions, namely, the unidirectional transformation of
organisms. Only such a process, which, in addition, provides easily interpretable
time markers, could be of use in measuring time.

The processes of rock formation, on the other hand, are reversible, subject to
continuous reversal of direction. A limestone lying exposed on the surface
weathers and is gradually dissolved by water containing carbonic acid. The solution arrives at the sea, where calcium carbonate is precipitated once more to
form another limestone, and without further information one is unable to tell
when it was formed or the difference in age between it and whatever remains of
the parent limestone. As the cementing material of a sandstone is destroyed, the
rock itself disintegrates into loose grand grains, which can later be cemented
once more to form a new sandstone. The disintegration and rebuilding of rocks
is thus a continuous process, exhibiting no uniformity of direction, no particular
characteristic that might be related to time.

The situation with regard to the fossil flora and fauna embedded in the rock is
completely different. Fossils as elements of a continuous evolutionary process
are connected by the bonds of descent, and descent is irreversible. Thus, the
relative evolutionary levels of lineages of plants and animals provide us with a
useful means, and indeed the only means, of determining age. We compare the
fossil content of individual strata and, based on the likeness of particularly characteristic individual forms—the *index fossils*—or on the assemblage of fossil
organisms present, we deduce a likeness of age.

After countless individual observations have established the natural sequence
of index fossils, both faunal and floral, we have incontestable grounds for connecting isolated partial profiles with one another and for determining the formations that correspond to them temporally in other areas and even in other parts
of the world (see fig. 2.1). The basic temporal unit that can be understood and
delimited based on the presence of fossil organisms is called the *zone*. The joining together of individual zones results in the *chronology* of the geologic ages—
a frame of reference, the abstract temporal scaffold upon which individual events
and actual rock formations can be arranged. There are certain problems connected with this methodology, and we shall return to them later.

Obviously, the determination of age based on fossils can be only *relative;* it
provides information on the temporal relationships among the individual horizons. Our division into zones signifies, then, that the zone of ammonite y is
younger than the horizon of ammonite x and, on the other hand, older than the

strata lying above it, which contain ammonite z. It may appear at first that the relative nature of this temporal determination represents a significant limitation. However, this system of dividing time is completely adequate for all questions of theoretical and applied geology. The only thing that concerns us in geology is the determination of the *sequence* of events, which then enables us to distinguish between cause and effect based on process and consequence. Biostratigraphy accomplishes this task completely.

It is the same with regard to the exploitation of mineral deposits, of ores, coal, salts, petroleum, and so on: *absolute* age based on a certain number of years is of no importance at all. In this area, too, the only concern is to establish the relative age of the horizons in which these natural resources are found. For example, it is completely sufficient for us to know that Carboniferous coals appear in goniatite zones $w–z$, and to be able to recognize these zones beyond doubt and to locate them. This is the reason that in all geological exploration and drilling, fossils are of critical importance in the exploitation and managing of vast economic assets!

ATOMIC DECAY

There are, in addition, several other methods of dividing and measuring geologic time, which, in contrast to the method based on organisms described above, yield not a relative age but *absolute* numerical values. However, these methods, such as counting the annual layers in late glacial and postglacial banded clays or the attempts to incorporate the Diluvian [Pleistocene] Ice Age into the astronomical radiation curve, are relevant only for the most recent geologic past and are not applicable to older deposits.

For older deposits, other sensitive techniques have been worked out based on the regular, progressive decay of the atoms of certain radioactive elements and on the calculation of the amount of uranium-lead, thorium-lead, and helium in the respective uranium and thorium minerals. Here, too, the method is concerned with a directional process, which theoretically supplies the prerequisites for determining age.

For practical use, however, these dating methods are unfortunately not easy to apply, for they are too complicated and, because of the many possibilities for error, not precise enough. Often, the times arrived at contradict one another and, in any event, fall far short of what is needed for making fine distinctions. Nevertheless, they do yield ranges that are of considerable value.

For example, the application of those methods has yielded results showing that since the beginning of the Cambrian, when the first extensive and clearly interpretable fauna appeared, about 500 million years have passed; the results also provide welcome clues as to the approximate duration of individual systems and eras (Paleozoic: 300–340 million years; Mesozoic: 130–140 million years;

Cenozoic: 60–70 million years). We have, then, figures that provide a rough assessment of the speed of organic evolution (see fig. 2.2).

The much more extensive demands that geologists and paleontologists must place on age determination will never be met by these chemical and physical processes, and consequently, they will not be able to replace the considerably simpler, less ambiguous, and universally applicable methods of paleontology, even in the future. In general, fossils are common in rocks, and the expert often needs only a glance to tell the age of the rock under consideration.

The only limits to the utility of paleontological methods are in places where there are no fossils. This is the case primarily with the Precambrian series, and it is just there that the uncertainties and differences of opinion with regard to age begin, all the physical and chemical methods notwithstanding. If these ancient strata contained fossils, their age sequence would have long since been settled once and for all, as has been done for the more recent systems.

A SURVEY OF THE TEMPORAL DISTRIBUTION OF PLANT AND ANIMAL PHYLA

Before taking up some special questions concerning the measurement of time, I would like to present an overview of fossil material—its appearance, distribution, and aptness for determining age.

The primary basis for establishing divisions in time is the *fossils of animals*—and for good reason: they are more frequently found than plant remains, and they reflect a more rapid transformation, which means that individual vertical ranges are relatively shorter. On that basis, three great *ages, or eras,* of the animal world and its evolution have been distinguished: the *Paleozoic* (the oldest), the *Mesozoic* (the middle period), and the *Cenozoic* (the most recent).

In addition to these main divisions based on animal life, ages can also be characterized and delimited based on *plant fossils.* Thus, based on the main epochs of the evolution of plants, a *Paleophytic,* a *Mesophytic,* and a *Cenophytic* can be distinguished. Because the great turning points in the unfolding of the plant world do not coincide with those of animals, the boundaries within these two systems for dividing time differ, the main turning points for plant development usually preceding those for animals (see fig. 2.2). Whereas the upper limit of the Paleo*zoic* lies between the Permian and the Triassic, that of the Paleo*phytic* falls in the middle of the Permian; for in the New Red Sandstone (Lower Permian), most of the plant forms that characterize the Late Paleophytic die out and are replaced in the Zechstein (Thuringian–Upper Permian) by the enormous proliferation of conifers. Similarly, the sweeping transformation of the Mesophytic cycad and gingko flora to the equally form-rich angiosperm flora, which replaced it, took place during the middle of the Cretaceous, whereas radical changes in the fauna did not take place until the end of the Cretaceous.

Systems	Eras based on			Absolute time span in millions of years
	Animals	Plants	Mountain-building	
Quaternary	Cenozoic	Cenophyticum		0.8
Tertiary				60
Cretaceous	Mesozoic	Mesophyticum	Alpine (Cenogaicum)	80
Jurassic				35
Triassic				25
Permian	Upper (Paleozoic)	Paleophyticum	Variscan (Mesogaicum)	40
Carboniferous				70
Devonian				40
Gothlandian	Lower (Paleozoic)	Proterophyticum	Caledonian (Paleogaicum)	30
Ordovician				70
Cambrian				90
Algonkian (Proterozoic)	Proterophyticum			(540)

Fig. 2.2. Division of geologic history into systems and eras. The standard division is based on animal fossils, according to which the boundaries between the systems and the accepted eras are established. The chart displays the individual units in such a way that their correct relative time values are evident; there are two exceptions: the Algonkian, of which only the upper section is considered, and the Quaternary, which, because of its brief duration, must be represented in an exaggerated manner.

As we have said, because of their general applicability and considerably greater possibilities for making fine distinctions in the division of time, the zoological divisions are much preferred, the botanical divisions being mainly of interest only when paleoflora is being considered.

Finally, geologic time can also be divided based on significant *geological events,* for example, on the major mountain-building processes; when this system is used, the boundaries of the units are again different (fig. 2.2). Such a division, however, is not independent and autonomous. The temporal dating of all geological phenomena is based primarily on the course of animal evolution. This means that eras defined by geological events are only zoological divisions transposed into another system that is more suited to certain geohistorical considerations. In other words, *animal evolution forms the basis of any geological division of time.* Even the phytic system cannot dispense with animal evolution as a basis for comparison, for the botanically based system by itself is unable to determine age and establish stratigraphic correlations in marine strata, which, for the most part, are very poor in plants.

The Paleozoic, the oldest age to have abundant fossil remains, is preceded by yet another era, the Precambrian, which, according to its rock record, lasted about three times as long as the rest of the eras combined.

THE PRECAMBRIAN ERA

Only in the younger sections of the Precambrian era, the Proterozoic, or Algonkian, have sparse remains of lower organisms been found: more or less dubious lime-secreting *algae, radiolarians* (protozoa, or one-celled animals, with siliceous lattice-work tests, fig. 2.3), and *sponge spicules;* also burrows and tracks of *annelid worms, clams* (?), *snails,* and so on. *Crustaceans,* all kinds of other, as yet unidentified, forms, a few chitinous-shelled *brachiopods,* and if the dates of the finds reported from South Australia prove correct, large *annelids* and rather highly organized *arthropods* (pl. 2B) have also been found.

In contrast, the strata from the older Precambrian—the Azoic, or Archaic—are almost devoid of fossils. Regrettably, the outlook for the future with regard to making any substantial fossil finds in these old rocks is extremely slight. There are no more unaltered sedimentary rocks from this epoch. As a consequence of their great age and their occasional subsidence to greater depths, they have undergone all kinds of transformations brought about by the burden of layers deposited on top of them, high temperatures, and granitic intrusions. They have been completely recrystallized, and as a consequence of these structural changes any organic remains that they may once have contained have been destroyed without a trace.

Only the occasional appearance of carbonaceous or graphitic substances, the frequent occurrence of limestone intercalations in all members of the Precam-

Fig. 2.3. Radiolaria from the Proterozoic of Brittany. About 1,500 ×. (After L. Cayeux.)

brian, concentrations of phosphates, and similar indicators provide clues that organic life was once present there. Thus, the beginnings of the evolution of plants and animals are forever denied us. There is another reason for this: the earliest organisms were in all probability small, purely soft-bodied forms, which had yet to secrete a hard shell and, therefore, would scarcely have been preserved in any event.

THE PALEOZOIC ERA

Essentially, then, the range of paleontological research covers the post-Precambrian eras. Within this enormous segment of time, subdivisions have been distinguished: the formations, or *systems,* as they are usually called outside of Germany.

The oldest unit of the Paleozoic is called the Cambrian. (The names of the systems are often taken from the regions in which the strata in question exhibit typical development or were first studied. Thus, "Cambrian" comes from Cambria, the Roman name for what is today northern Wales.) In the Cambrian rocks, we encounter for the first time—in contrast to the extreme paucity of fossils in the Proterozoic—an abundance of well-preserved and clearly interpretable fossils. They are all marine fossils; continental forms are unknown.

Consequently, it is very probable that the *seas were the original home of life.* In any event, the oldest evidence of life we have consists only of *marine* organisms from shallow seas. The first signs of land dwellers, a supposed myriapod (millipede) and two genera of scorpionids, were found in the Gothlandian [= Silurian]; terrestrial animals—insects, pulmonate snails, reptiles—do not appear in large numbers until the Upper Carboniferous.

Further, it is noteworthy in this connection that in the oceans, every animal structural design was already present, whereas in fresh water, the echinoderms (starfishes, sea urchins, and allies), brachiopods, and tunicates (sea squirts) are lacking altogether, and the coelenterates (corals and relatives) and porifera (sponges) are only poorly represented; from the remaining phyla, other large groups such as the radiolarians, foraminiferans, graptolites, chitons, cephalopods, many types of "worms," and so on are all lacking. As for atmospheric dwellers, only arthropods, snails, and vertebrates are to be found. This circumstance, too, points in the same direction: the basic animal types had their origins in the sea, and only some of them, and they only very gradually, penetrated the other realms.

The majority of Cambrian fossil remains consists of *trilobites* (figs. 2.4, 2.5, 3.5, 3.6)—arthropods whose exoskeleton is divided into three parts both lengthwise (head shield, segmented thorax, tail shield) and crosswise; the name "trilobite" derives from this division. It used to be thought that these creatures were crustaceans; however, recent research has shown them to be more closely related to spiders, or, more precisely, to the larger group to which spiders belong—the Chelicerata. The wide distribution, great frequency of occurrence, and rapid rate of evolution of trilobites make them outstanding index and zone fossils. This is the reason that divisions of Cambrian time are based on these forms.

In addition to trilobites, there were primitive forms of *brachiopods* (fig. 2.6); these are remarkable for having chitinous shells, for most of the younger brachiopods of subsequent systems have calcareous shells (figs. 2.7, 3.3, 3.4, 3.68). In the Cambrian, there is almost no evidence of calcareous shell formation; hard parts are composed almost exclusively of horn or chitin.

It has often been assumed that this situation was related to a lack of calcium in Cambrian seas. It is much more likely, however, that the lower degree of organization and the special metabolism of Cambrian organisms were solely responsible; for the necessary amount of calcium would doubtless have been available to any animals capable of secreting calcareous shells. In seas of later

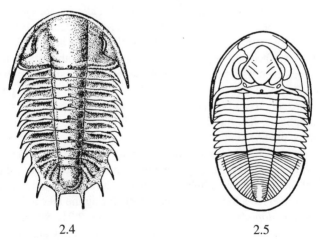

2.4   2.5

Fig. 2.4. Dorsal shield of the trilobite *Olenoides curticei* Walc. Middle Cambrian of North America. 1 ×. (After C. D. Walcott.)

Fig. 2.5. A more recent trilobite: *Dechenella* (*Dechenella*) *verneuili* (Kays.) from the Middle Devonian of the Eifel [region in the Rhenish Schiefergebirge, western Germany]. 1½ ×. Head and tail shields are heavily outlined to emphasize the differences between these two sections, each of which consists of several segments of almost fused plates, and the thorax, which consists of separate, movable segments. (After R. and E. Richter, modified.)

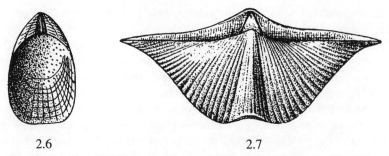

2.6   2.7

Fig. 2.6. *Lingulella davisi* (McCoy) as an example of a horn-shelled, hingeless brachiopod from the Upper Cambrian. 1⅖ ×. (From H. L. Hawkins.)
Fig. 2.7. For comparison: *Spirifer* (*Cyrtospirifer*) *verneuili* Murch., a younger type of brachiopod with a hinged, calcareous shell, from the Upper Devonian. 1½ ×.

date, and even still today, there were and are isolated chitinous-shelled brachio-pods (figs. 3.62, 3.63), which in spite of the undisputed abundance of calcium in their environment do not build calcareous shells! That this type of circular rea-soning leads to absurd conclusions is evident once one remembers that the shells of chitinous, unarticulated brachiopods are composed of alternating deposits of chitin, calcium carbonate, and calcium phosphate. Is that also supposed to be a result of changes in the composition of the sea water? Is it also based on corre-sponding rhythmic fluctuations in the amount of calcium available?

An extremely abundant, splendidly preserved Middle Cambrian fauna turned up in the systematic excavations of Charles Doolittle Walcott at Burgess Pass,[4] in the Rocky Mountains of Canada (British Columbia). There, with all the fine detail of body surface and appendages (extremities, mouth parts, hairs, and so on), sometimes even with preservation of the internal organs (vascular system, sensory organs), is evidence of *scyphomedusae* (jellyfish, pl. 3A), polychaetes (marine worms, pl. 3B and C), chelicerates (pl. 4A), crustaceans (pl. 4C, and fig. 3.7), onychophores, holothurians (sea cucumbers, pl. 4E–G), chaetognathans (arrow worms, pl. 4A), tunicates (sea squirts), and so on—all soft-bodied forms without skeletons. This rich fauna, the preservation of which was favored by the special conditions of fossilization present in the fine-grained shale, indicates that almost all of the invertebrate phyla were present during the Cambrian and must have already traversed a long evolutionary path during the Precambrian.

Even the *cephalopods* (see pl. 18A and B, and fig. 3.21), which were to be-come so important in later periods, are represented in the Cambrian by a few types, as were the primitive *sponges* (archaeocyathids, fig. 2.8), *pelmatozoans* (stemmed echinoderms, see figs. 2.18, 2.19), and, it is believed, the oldest corals.

4. [Charles Doolittle Walcott, 1850–1927, was a famous American paleontologist and director of the Smithsonian Institution from 1907 until his death.—Trans.]

Fig. 2.8. *Archaeocyathellus,* a representative of the archaeocyathids, simplified schematic drawing. About 2 × average size. (After H. Schmidt 1935.) The archaeocyathids, problematic fossils quite different from all Recent forms, have been reclassified again and again, sometimes called sponges, sometimes corals, foraminifera, or calcareous algae. More recent research shows that archaeocyathids are probably siliceous sponges.

On the other hand, vertebrates are completely lacking. It is true that a few years ago, it was reported that a supposed member of the agnathans, a primitive vertebrate related to the fishes, was found in the Middle Cambrian of North America, but the identification has been disputed by reliable sources.

*   *   *

In the two succeeding systems, the Ordovician and the Gothlandian (early name of the Silurian), a sweeping change in the faunal composition occurs, resulting in a clearly defined boundary with the Cambrian.

Trilobites continued to expand, producing an abundance of forms, but in new families and genera; here, too, they are an important means of classifying sedimentary strata. In addition, there is another group of forms that are of the greatest stratigraphic importance and are found extremely often, especially in the shaley facies of this system.[5] Flattened, sometimes looking rather like letters of the alphabet, they cover the surfaces of slabs of shale; their name—*graptolites* (figs. 2.9–14)—refers to their scriptlike appearance. Their place within the system was long in dispute. They were usually included among the hydrozoans (a group of coelenterates); however, the most recent anatomical studies (very carefully carried out by Kozlowski on excellent material that has been preserved in a three-dimensional state) show that graptolites are highly organized creatures related to the pterobranchs (to the class Enteropneusta, hemichordate wormlike animals that breathe through a pharynx).

In addition, there appear with great frequency and in a surprising abundance of types lime-secreting organisms: *corals* (already evident in some places unquestionably as reef builders, figs. 2.15, 2.16), *bivalves* and *snails, cephalopods*

5. By the term "facies" is meant the character of a rock as it reflects the sum total of the conditions under which it was formed and under which the life forms represented in it as fossils once lived.

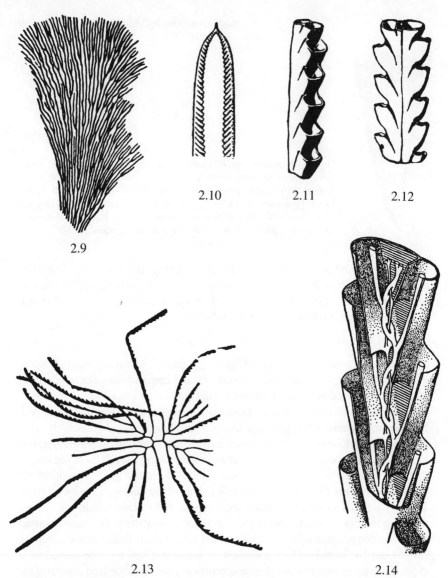

2.9

2.10   2.11   2.12

2.13   2.14

Figs. 2.9–14. Different types of graptolites from the orders Dendroidea (2.9) and Graptolidea (2.10–14).

2.9. *Callograptus salteri* Hall. Lower Ordovician from Canada. 1 ×.

2.10. *Didymograptus murchisoni* (Beck). Upper Ordovician from Wales. 1 ×.

2.11. *Monograptus dubius* (Sueß). Upper Gothlandian of the Baltic Provinces.

2.12. *Glyptograptus teretiusculus* (His.). Lower Ordovician of Sweden. 11 ×.

2.13. *Goniograptus thureaui* McCoy var. *postremus* Ruedem. Lower Ordovician of the United States. 1 ×.

2.14. Schematic drawing of *Orthograptus gracilis* (Rmr.), from the Upper Ordovician of the Baltic Provinces, showing a view of the complex inner structure. About 18 ×.

(From O. M. Bulman, H. L. Hawkins, and R. Ruedemann.)

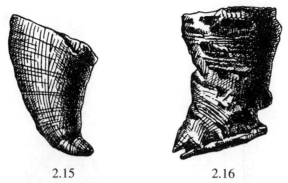

2.15   2.16

Figs. 2.15–16. Two corals (pterocorals) from the Gothlandian of Gothland. Both 1½ ×.
2.15. *Hallia mitrata* (Schl.) var.
2.16. *Kodonophyllum truncatum* (L.), lateral view, showing a broad surface grown onto a foreign body. (From Schindewolf.)

(at first only as nautiloids, early, more primitive precursors of the present-day *Nautilus,* see fig. 3.21), *bryozoans* (fig. 2.17), *brachiopods* (with calcareous shells, hinge formation, and brachidia, see figs. 3.3, 3.4), *pelmatozoans* (figs. 2.18–20), and less frequently, *echinozoans* (sea urchins), and plated *holothurians.*

The first representative of the *vertebrates* appears in the Ordovician: the oldest *agnathans* (animals without jaws.) Agnathans are closely associated with modern cyclostomes (hagfishes, lampreys) but differ from them considerably in outer appearance in that the agnathans have dermal armor and different body shape (figs. 2.25, 4.2, 4.3.)

A well-defined boundary separates the Gothlandian from the Devonian lying above it. *Graptolites,* so important up to this point, have become almost extinct,

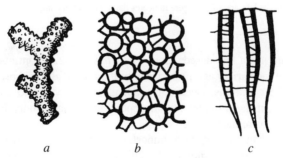

a   b   c

Fig. 2.17. Bryozoan of the suborder Trepostomata: *Hallopora ramosa* (d'Orb.). Ordovician of Cincinnati (USA). *a.* Colony. 1 ×. *b.* Tangential section with zooecia and mesopores. *c.* Axial section. *b* and *c* greatly enlarged. (After K. A. von Zittel 1924.)

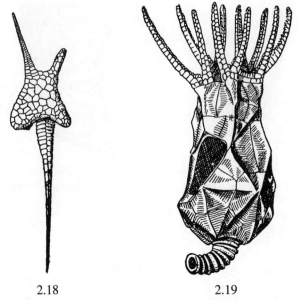

2.18   2.19

Fig. 2.18. *Dendrocystoides scoticus* (Bath.), a representative of the Carpoidea, an extinct class of pelmatozoans. Upper Ordovician from Scotland. ¾ ×. (After F. A. Bather.) The Carpoidea are pelmatozoans that are limited to the Lower Paleozoic. These sessile echinoderms were attached to a substrate by means of a stem; the sharply compressed theca is more or less bilaterally symmetrical. The stem is entire or partially bipartite, broader at the anterior end, and hollow. At the top left of the form illustrated there is a four-sided, plated, armlike process, a free extension of the ambulacral stems. The anus is at the bottom left of the theca.

Fig. 2.19. *Chirocrinus insignis* Jkl., a representative of the Cystoidea, also a class of Lower Paleozoic, primarily Ordovician, pelmatozoans. Lower Ordovician from northern Russia. (After O. Jaekel.) The Cystoidea are pelmatozoans with a radial structure only seldom clearly pentameral; their pouch-shaped or globular tests are perforated by a special pore system. The ambulacral radial grooves issue from the mouth and continue into the free, protruding, plated arms. A short stem is usually present and consists of thin, sometimes differently shaped, centrally perforated disks. The cross-hatching on the left side of the figure indicates the large anal opening.

leaving behind just a few descendants. *Trilobites* (fig. 2.5) continue on into the Devonian, but the variety and importance of the forms fall far short of what was true in the older systems. Declining gradually, they extend to the end of the Paleozoic. One lineage after another disappears until finally, in the Permian, only a few small forms remain.

On the other hand, a new, very important development is introduced during the Devonian, that of the ammonoids. This term refers to cephalopods that are related to the *Nautilus* of modern tropical seas but, because of various peculiarities, are considered as a divergent, independent entity. These ammonoids extend

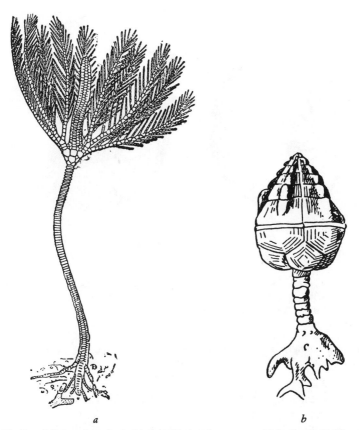

a   *b*

Fig. 2.20. Two different types of crinoids (sea lilies), pelmatozoans with bodies divided into calyx, stem, and free arms, and with an even, pentameral structure: *a. Botryocrinus,* from the Gothlandian of Northern Europe, a form from calm waters; it has a long stem, delicate arms, and an anal tube located high up, on the calyx. (In many crinoids of calm waters, the Liassic *pentacrines,* for example, the stem can be quite long, as much as 18 meters!) (After F. A. Bather.) *b. Cupressocrinus,* from the Middle Devonian of the Eifel, a distinct reef type of turbulent water with a rounded, compressed shape and short, broad arms that could be folded tightly together. ⅔ ×. (After O. Jaekel.)

throughout the later Paleozoic (Devonian, Carboniferous, Permian) and the entire Mesozoic (Triassic, Jurassic, Cretaceous).

The common occurrence and broad distribution of ammonoids and their rapid evolution, which results in the appearance of ever new and relatively easily identifiable forms, makes them almost ideal index fossils. Consequently, they have become the preferred faunal group for establishing stratigraphic divisions in the Late Paleozoic and the Mesozoic. In the Mesozoic, the animal appears in the form of the true ammonite, and in the Devonian, its more simply organized an-

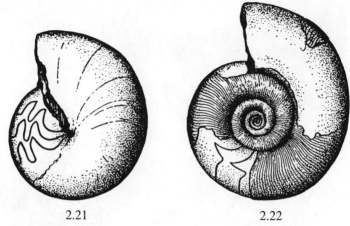

2.21   2.22

Fig. 2.21. *Sporadoceras münsteri* (v. B.), an index goniatite from the Upper Devonian. 1 ×.
Fig. 2.22. *Oxyclymenia undulata* (Mstr.), a widespread clymenian of the later Upper Devonian. 1 ×.

Goniatites are distinguished by a ventral siphon located on the outside of the spiral, whereas in the clymenians, the siphon is on the dorsal, or inner, side of the spiral. In outer appearance, the two groups of ammonoids differ (but not always) in that the shell of the goniatite is usually a thick disk to globular in shape with a narrow umbilicus, whereas the shell of the typical clymenian is a thin disk with a broad umbilicus, the inner whorls exposed. The surface of the shell shows arcuate or sickle-shaped growth rings. In places where the shell has been destroyed, the sutures (the folded edges of the septa dividing the whorls into individual chambers) are visible on the steinkern.

cestors (*goniatites,* figs. 2.21, 2.84–88; pl. 22A, pl. 32A and B); and *clymenians* (figs. 2.22, 2.89–91) are found.

The only limit to the utility of ammonoids in establishing stratigraphic boundaries occurs if the fossil is lacking, which would mean that the living conditions under which the layers were deposited were unsuitable for them. Ammonoids flourished only in calm, rather deep marine waters; consequently, they appear primarily in limey and clayey rock facies. In coastal areas, where the water was turbulent and dirty, living conditions would not have been suitable; therefore, in general, they are lacking in the clastic rock—sandstones, graywackes, and so on—formed under such conditions.

In such instances, other groups of animals must be used to classify sediments; in the Devonian, the richly represented *brachiopods* (figs. 2.7, 3.68) and *corals* (pl. 5C) are substituted. During that period and throughout the Paleozoic in general, brachiopods played the role later filled by bivalved mollusks and snails: they were typical elements of shallow-water and coastal faunas. In later times, they migrated to greater marine depths, just as did, for example, the sea lilies (fig. 2.20; pl. 6A), which are found throughout the Paleozoic and the Mesozoic in rocks formed in shallow areas. Today, however, sea lilies are deep-sea forms.

For the Devonian, there remain to be mentioned the *eurypterids,* or Giganto-straca (fig. 2.23), which are related to trilobites, and the still-living horseshoe crab *Limulus* (pl. 4B). Eurypterids begin with small forms back in the Cambrian and Ordovician but during the Devonian (and in places, as early as the Upper Gothlandian) became notable for their enormous size—to 1.80 meters in length. The *xiphosurans,* or horseshoe crabs, another order of merostomes (which includes the limulids) also appear in the Devonian (fig. 2.24). In addition, the Lower Devonian roof slates from Bundenbach (Hunsrück, in the Rhenish Slate

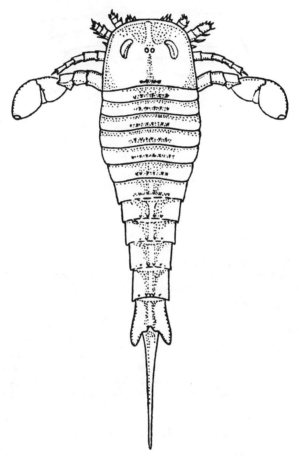

Fig. 2.23. *Eurypterus fischeri* Eichw., a representative of the order Eurypterida (Gigantostraca) from the Gothlandian of Ösel. Dorsal view. About 1 ×. The genus Eurypterus also appears in the Devonian, there attaining a considerably larger size than the one illustrated here. (After G. Holm.)

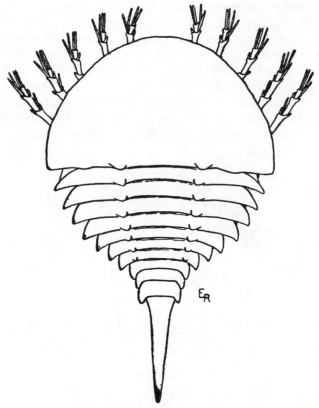

Fig. 2.24. *Weinbergina opitzi* R. and E. Richter, a xiphosuran from the Lower Devonian roof slate of Bundenbach, in Hunsrück. (After R. and E. Richter.) The xiphosurans, or horseshoe crabs, are members of an order of the subclass Merostomata; the only living relict is the genus *Limulus*. *Weinbergina* can be distinguished by the following: (1) the joint between the carapace of the cephalothorax and that of the tail spine is not fused to form a single shield, as in *Limulus* (see pl. 4B and fig. 3.62); (2) the cephalothorax has no extended posterior margin; and (3) its segmented appendages terminate in "brushes," whereas in the limulids, they each (except for the first and last appendages) bear a sickle-shaped terminal segment or a claw. Consequently, the Devonian genus is assigned to a special suborder (Synziphosura), which forms the ancestral stock of the limulids and is in turn descended from the lower Gigantostraca. The appendages are mostly covered by the cephalothorax; when specimens were x-rayed, however, their structure and the position of the appendages could be determined in every detail. The drawing, then, is not a reconstruction; the shape of the body has only been adjusted to compensate for the slight compression of the original, and the legs, partially displaced but beautifully preserved in the original, have been straightened out.

Mountains) have yielded splendidly preserved specimens of the curious subclass *Pantopoda* (or Pycnogonida, figs. 3.10, 3.12), which is also a spiderlike animal, and further, such remarkable groups of arthropods as the *Cheloniellida* (fig. 3.8) and *Nahecarida* (fig. 3.9), which will be mentioned further on.

Fishes are very much in ascendancy. In addition to the jawless *agnathans* (figs. 2.25, 4.2, 4.3) the first representatives of the true *fishes* appear in abundance; there are even several lineages each of placoderms and elasmobranchs

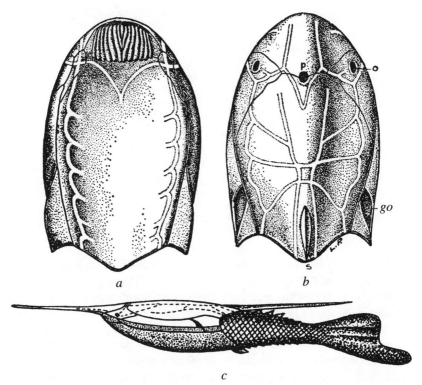

Fig. 2.25. *Pteraspis,* a member of the Agnatha, is found throughout the Lower Devonian. Agnathans are lower, fishlike vertebrates without jaws (the first ring of the pharynx has not yet developed into a mandibular arch as it has in fishes) and usually without paired appendages. The genus *Pteraspis* is the type of the order Heterostraci (subclass Pteraspidomorphi). *a, b.* Ventral (*a*) and dorsal (*b*) views of the anterior part of the body, which is covered with large bony plates. The double dotted lines indicate the well-developed sensory canals, present in all these lower forms. The mouth opening is located to the front of the row of narrow, elongate plates at the forward edge of the large ventral shield. *o* = eye socket (orbit); *p* = pineal foramen; *go* = external gill opening; *s* = dorsal spine. About 2 ×. (After J. Kiaer from A. S. Romer 1933.) *c.* Reconstruction of another species of *Pteraspis,* with a long rostrum in front of the mouth opening. Unlike the forward section of the torso, the rear section is covered with small scales. About ⅓ ×. (After W. Gross and E. I. White.)

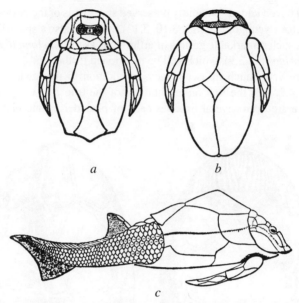

a   *b*

c

Fig. 2.26. *Pterichthys,* a Middle Devonian armored fish from the placoderm group (subclass Antiarchi) with closed cranium, capsular body armor, scaled tail, and a very peculiar bony "pectoral fin" (arthropterygium). *a, b.* Dorsal (*a*) and ventral (*b*) views of the head and pectoral shields. *c.* Lateral view of the whole animal. ½ ×. (After R. H. Traquair.)

(cartilaginous fishes), the first ones appearing at the upper limit of the Gothlandian (figs. 2.26, 2.27), as well as of teleost fishes. Teleostomes, the group that includes, in particular, the true bony fishes with apical mouth openings, first appeared in the Lower Devonian (pl. 4A; figs. 2.28, 3.143), flourished during the Carboniferous, and continue to expand today. For fishes, then, the Devonian was a time of radiation, of enormous diversity. There is another very important newcomer in the Upper Devonian: the first primitive *amphibians* (stegocephalians, figs. 2.29, 2.30), the oldest evidence of tetrapod lineages, are found there.

With regard to *plants,* the wide distribution of *psilophytes* (figs. 2.31, 3.94, 3.145), primitive terrestrial plants, should be mentioned. They begin in the uppermost Gothlandian but do not become prominent until the Devonian. Furthermore, in the Upper Devonian we also find the first representatives of the higher vascular cryptogams, the *pteridophytes* (fig. 2.32).

As has already been stated, the stratigraphic classification of the subsequent Carboniferous and Permian systems is based on ammonoids, which continued to evolve rapidly, developing various new organizational features. Here too, just as in the Devonian, the ammonoid facies are matched by facies characterized by an abundant *brachiopod* and *coral* fauna. These two groups of animals are an im-

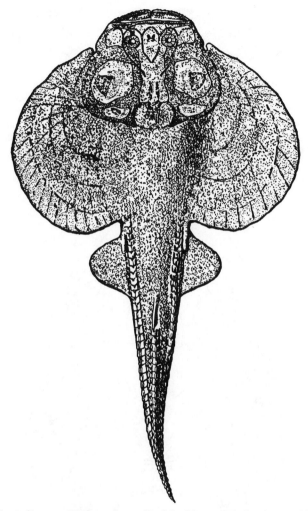

Fig. 2.27. *Gemündina stürtzi* Traqu., a flat, raylike fish with extremely flared pectoral fins, from the Lower Devonian roof slate at Bundenbach, in Hunsrück. Dorsal view. ½ ×. (From F. Broili.) In spite of the surprising similarity of form, there is no close relationship between *Gemündina* and the ray, which did not appear until much later, in the Jurassic; the similarities of body shape are only superficial and are related to the similar, bottom-dwelling mode of life. *Gemündina,* a placoderm, occupies a subclass all its own (Rhenanida). The surface of the body is covered with small denticular plates (tesserae); the tail portion is covered with scales arranged in rows. In the region of the skull and pectoral arch, large dermal bones have developed beneath the epidermal denticles.

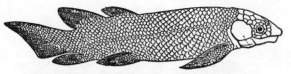

Fig. 2.28. *Dipterus,* a Devonian bony fish from the subclass Dipnoi. About ¼ ×. (After R. H. Traquair, from A. S. Romer 1933.) Both the dipnoans (lungfishes) and the crossopterygians (lobe-finned fishes) are distinguished by having a nasopharyngeal passage and are therefore often contrasted with the rest of the bony fishes, the actinopterygians, as the Choanata. With regard to this character, they are close to the tetrapods. Nevertheless, because of their highly specialized jaw structure, there is no question of dipnoans being the ancestors of amphibians and other tetrapods.

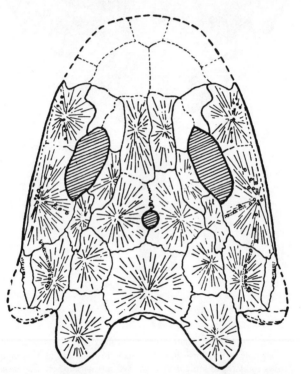

Fig. 2.29. Skull of the stegocephalian *Ichthyostegopsis* Säve-Söderb., one of the oldest of tetrapod genera, from the later Upper Devonian of Greenland. Dorsal view. It is characteristic of the group of lower amphibians known as stegocephalians that, in contrast to the higher amphibians, the cranium consists of solid membrane bone, the only openings being the orbits, the nostrils, and the parietal foramen (opening of the epiphysis, "the pineal eye"); and that there are large, open ear slits on the posterior rim. In the Upper Devonian family Ichthyostegidae, to which the genus shown here belongs, the exterior nostril openings are located on the edge of the palate on the ventral side of the skull, whereas in the tetrapods they are dorsal. (After G. Säve-Söderbergh, from F. von Huene.)

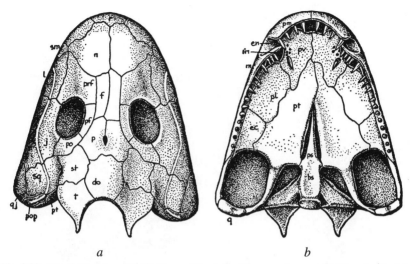

a   *b*

Fig. 2.30. Dorsal (*a*) and ventral (*b*) views of the skull of the second genus of ichthyostegids, *Ichthyostega* Säve-Söderb., from the Upper Devonian of Greenland. About ⅓ ×. (See the explanation with fig. 2.29.) *in* = internal nasal opening; *en* = external nasal opening, also located on the underside of the skull. The terms for the other bones of the skull are not of concern here. The light, lengthwise lines at the edges of the dorsal side represent mucous canals (corresponding to the sensory lines of fishes), which are very characteristic of all lower amphibians. (After G. Säve-Söderbergh, from A. S. Romer 1933.)

portant aid in the classification of strata in places where ammonoids are lacking.

Another group useful in stratigraphic classification is the *foraminiferans* (figs. 2.33, 2.34), one-celled organisms with chambered tests from the phylum Protozoa. They appear here for the first time in geological history in truly gigantic forms, which exceed by far the average size of their microscopically small relatives.

Plants, too, came to be of stratigraphic importance during the Carboniferous. Because of the favorable climatic conditions prevailing, there was a profuse development of *pteridophytes* (ferns, club mosses, and horsetails, see pl. 24A, B, C) and *pteridosperms* (seed ferns, fig. 2.35); remains of these plants make up the thick deposits of coal for which the system is named. Fern fronds and other plant remains are also found, usually very well preserved, in the rocks above and below the coal and are an important tool for recognizing individual coal seams and determining their ages. *Freshwater clams* have also been used successfully for the same purpose, for they appear frequently in the shales intercalated between the deposits of coal.

Other invertebrates deserving mention are the *insects* (figs. 2.36, 3.139); the *spiders,* which are conspicuous for their large size; and the *sea urchins,* which

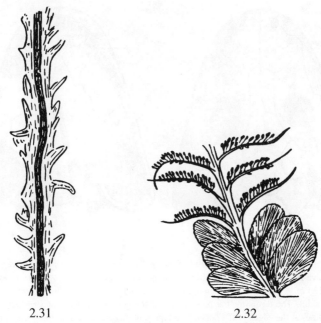

2.31   2.32

Fig. 2.31. Stem fragment of a *Psilophyton* sp., from the Bohemian Devonian. The genus illustrated is a typical representative of the psilophytes, the lowest vascular cryptogam (spore-bearing plant) and oldest terrestrial plant. These are irregularly branched plants that resemble algae (seaweed) superficially and were formerly thought to be algae. They are either completely naked—leafless— or have spinelike, unveined appendages. The stems, therefore, served both as organs of assimilation and as supports; at the tips, they bore the relatively large sporangia, or spore containers (see fig. 3.94). Within the stem, there is a small vascular bundle (protostele), which stands out clearly in the illustration here. (From W. Gothan.)

Fig. 2.32. *Archaeopteris hibernica* (Forb.) Daws., from the Upper Devonian of Ireland, an ancient fern with relatively large fan-shaped, veined leaves (Archaeopterides). Fertile portion of the frond with sporophylls at the tip. (From H. Potonié and W. Gothan 1921.)

In addition to these large-leaved ferns, other groups of pteridophytes already evident in the Upper Devonian were the club-mosses *Cyclostigma* and the horsetail-like forms of *Asterocalamites*. The late Devonian plant assemblage is consequently more closely tied to the Carboniferous flora than to the older Devonian flora.

appear here for the first time in a rich diversity of forms (figs. 2.37, 2.38). Toward the end of the Carboniferous, however, the various types of these Paleozoic sea urchins, some highly specialized, are already extinct, with the few exceptions of the cidaroids and a few lines of lepidocentroids, which carried on the line.

The break between the Devonian and the Carboniferous stands out very sharply in the evolution of the *fishes*. The ancient armored fishes of the placoderm group do not go beyond this boundary; from then on, their role is assumed by the proliferating *selachians* (sharks and their kin) and the *actinopterygians*—

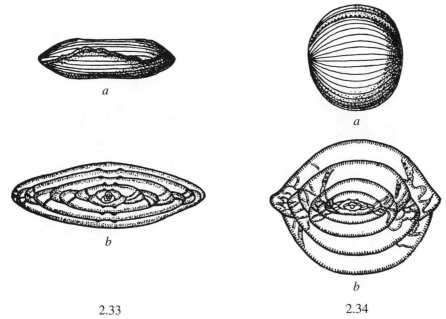

a

a

b

b

2.33   2.34

Figs. 2.33–34. Slightly schematic representation of the two main types of late Paleozoic large Foraminifera.
2.33. *Fusulina.* Upper Carboniferous. *a.* Lateral view. About 3 ×. *b.* Axial longitudinal section.
2.34. *Schwagerina.* Upper Carboniferous to Upper Permian. *a.* Lateral view. About 2 ×. *b.* Axial longitudinal section.
(After R. Wedekind 1937 and J. J. Galloway.)

Fig. 2.35. Frond fragment from *Neuropteris heterophylla* Brongn., a form with fernlike foliage but forming true seeds, making it related to the lower gymnosperm class Pteridospermae (seed ferns). Characteristic of the genus illustrated are the tongue-shaped leaflets with a fan-shaped system of radiating veins. (After H. Potonié and W. Gothan 1921.)

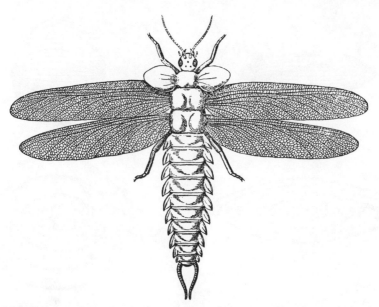

Fig. 2.36. *Stenodyctia lobata* Brongn. from the Upper Carboniferous of France, a form of ancient flying insect (Palaeodictyoptera) with undifferentiated fore- and hind wings, almost identical in size and structure, that could not be folded back; three identical, unspecialized pairs of legs; a pair of short, blunt wings on the first segment of the thorax; and lateral appendages on the segments of the abdomen. 1 ×. (After A. Handlirsch.)

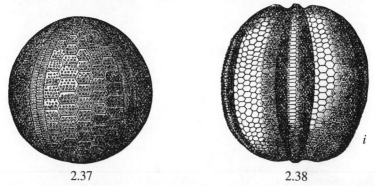

2.37   2.38

Figs. 2.37–38. Two Carboniferous sea urchins from the group "Palechinoidea" (family Palaeechinidae), whose "regular" pentameral test consists of many rows of varying numbers of plates; the test has not yet attained the form found in more recent echinoids ("Euechinoidea"), where there are two even rows of plates each in the ambulacra and the interambulacra.

2.37. *Palaeechinus elegans* McCoy. Lower Carboniferous of Ireland. 1 ×. Test globular, ambulacra narrow, with 2–6 rows of small, depressed plates, interambulacra with 4–16 rows of plates. (After Bailey, from J. Wanner.)

2.38. *Melonechinus multiporus* (Norw. and Ow.), Lower Carboniferous of North America. ⅔ ×. Test melon-shaped, with broad, depressed interambulacral areas (*i*). Ambulacra composed of 6–12 rows of plates and interambulacra of 3–11, for a maximum of more than 100 rows of plates, compared with the constant 20 rows present in more recent sea urchins. (After G. Steinmann, from J. Wanner.)

the group of the "true" modern fishes. Furthermore, an important event occurred in the realm of the vertebrates and should be noted—the first appearance of *reptiles,* which evolved from lower stegocephalian stock.

*   *   *

The Permian system constitutes the end of the Paleozoic, the age of ancient animals, and in fact, there is at that point a break of major importance in faunal evolution.

In the Permian we find the last of the *trilobites,* which were so thoroughly characteristic of the Paleozoic. Large groups of *hydrozoans, brachiopods, crinoids,* and *bryozoans* of the old stamp die out and are replaced, some toward the end of the Permian and some at the beginning of the subsequent Triassic system, by other, more advanced types. The last representatives of the *blastoids* (fig. 2.39), a Paleozoic group of pelmatozoans, or stalked echinoderms, are found here.

The ancient *corals* of the bilaterally symmetrical, four-radiate "*Tetracorallia*" type extend as far as the close of the Permian; at the upper limits of that system, they are transformed to the type of the form still living today—the radial, hexamerous stony corals (fig. 3.45). Within the *sponges* and the *sea urchins,* too, sweeping transformations were at hand as the Permian gave way to the Triassic.

With regard to vertebrates, the Permian and the Carboniferous combined are the high point for *stegocephalians* (fig. 2.40). The groups *Embolomeri, Rhachitomi, Lepospondyli,* and *Phyllospondyli* are completely limited to these two systems; only the *Stereospondyli* manage to hold on into the Triassic (fig. 3.70). *Reptiles* (figs. 2.41, 3.122, 3.123), with a number of different taxa, attained broad distribution in the Permian continental deposits of Europe, North America, Asia, and South Africa. Several of these ancient groups died out in the Permian and were replaced in the Triassic system by numerous new forms. In short, we encounter almost everywhere a radical contrast between the old and the new.

Fig. 2.39. *Schizoblastus permicus* Wann. Upper Permian of Timor (Indonesia). 2 ×. (From J. Wanner.) The species is one of the most recent forms of the class Blastoidea, which appeared in the Gothlandian as descendants of the cystoids, developed a great many forms in the Lower Carboniferous and the Permian, and vanished toward the end of the Paleozoic, leaving no descendants. The small, unstalked theca consists of three cycles of overlapping plates: the basals (not visible in the illustration), the radials (*r*), and the deltoids (*d*); between them lie the interradial areas. The mouth opening is located at the center of the apex and is covered with a few small plates, which are fused to form an oral cap (*oc*).

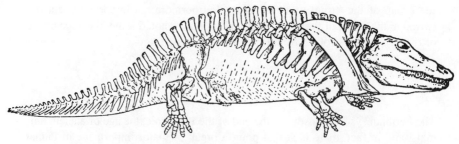

Fig. 2.40. *Eryops,* a stegocephalian (order Labyrinthodonti, suborder Rhachitomi) with rhachito-mous vertebrae (see fig. 3.154), from the Lower Permian of Texas. (From W. K. Gregory.) The form shown here is a typical representative of these large amphibians, which began as early as the Upper Carboniferous but are primarily characteristic of the Permian, where they are widely distributed. The body reached a length of up to 2½ meters; the skull, to 60 centimeters. The cranium is closed; the number of bones in the skull is still large, as in all primitive amphibians (93). The eyes are small and situated far back on the head. The extremities are short but very sturdy; the pectoral girdle, in particular, is large and solid. This animal was primarily terrestrial.

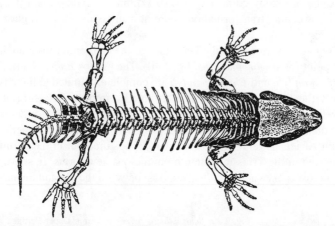

Fig. 2.41. *Seymouria,* one of the most primitive, most ancient of the reptiles known thus far. Lower Permian of Texas. Dorsal view. About ⅙ ×. (After S. W. Williston.) The genus belongs to the order Cotylosauria, a group of primitive reptiles that appeared for the first time in the Upper Carbonifer-ous; it is still very close to the stegocephalians by virtue of the structure of the skull (see figs. 2.29, 2.30, 2.40, 2.45). As in the stegocephalians, the membrane bones form a cranium that is completely closed except for the eye and nasal openings, the parietal foramen, and the large, open ear slits. Thus, temporal openings, characteristic of all higher reptiles, are still lacking. Nevertheless, based on the structure of the vertebrae and certain peculiarities of the pectoral and pelvic girdles, *Sey-mouria* is classified as a reptile.

In the *plant world*, where evolution is always a step ahead of that in the animal world, the far-reaching renewal turns up as early as the middle of the Permian system, when the age of cryptogams is cut off by the new era of the *gymnosperms*.

THE MESOZOIC ERA

If the entire Paleozoic can be characterized as being the age of trilobites, then the Mesozoic is the age of *ammonites*. They are represented in the three systems—the Triassic, the Jurassic, and the Cretaceous—by an enormous abundance of forms, and their importance for determining age exceeds that of all other groups of animals. From the Triassic on, their organization shows unmistakable progress in comparison with corresponding Paleozoic precursors (pl. 6; figs. 3.14, 3.32, 3.151).

In addition to the ammonites, another branch of the cephalopods—the *belemnoids* ("thunderbolts," see figs. 2.42–44)—gain in importance; these cephalopods belong to the still-living group Dibranchiata (squid). They evolved during the Triassic from orthocerans (members of the nautiloids) with external shells by transferring shell formation to the inside of the body. In addition, as was mentioned earlier, a sweeping transformation of many other animalian lines takes place during the Triassic system.

In a striking contrast to the Paleozoic, *clams* come sharply to the fore, and from this point on, they and *snails* inhabit the shallow marine littoral previously occupied by *brachiopods*. The high point for brachiopods has come and gone, their former diversity appearing much thinned out in the Mesozoic. Already in the Triassic, *crustaceans* are represented by modern-looking forms (*decapods*, pl. 8A). Furthermore, the "living fossil" *Limulus* (pl. 4B; fig. 3.62) was already present, as was the lungfish *Ceratodus,* which had a worldwide distribution during the Mesozoic and survives today only as a relict in a closely related genus (*Neoceratodus,* fig. 3.141) limited to a tiny habitat consisting of two rivers in Queensland, Australia.

Of the primitive *amphibian* branch of the *stegocephalians,* only a single group remains (the *stereospondyls,* figs. 2.45, 3.1, 3.71), now flourishing with gigantic forms; it ultimately dies out, however, at the Triassic-Jurassic boundary. The modern order *Anura* (frogs) shows up for the first time in the Lower Triassic with an interesting primitive representative. Reptiles enjoy a mighty ascendency in the number of new types, in body size, and in the expansion of realms inhabited (land, lakes, oceans). Of the *reptilian* orders still in existence, the *rhynchocephalians* (with the single Recent genus *Sphenodon*), the *squamates* (lizards and snakes), and the *turtles* all appeared for the first time during the Triassic (see

Figs. 2.42–44. Structure and reconstruction of the belemnites.

2.42. *Belemnitella mucronata* (Schl.), index fossil of a particular section of the Upper Cretaceous (Mucronate beds). The illustration shows a rostrum—that part of the belemnite shell most frequently found—in its natural size; often this is all that is found. The rostrum is a solid, conoidal or cigar-shaped, radially fibrous, calcite sheath, pointed at the posterior end. At the blunt end, there is a conical cavity (alveolus), which holds the more fragile phragmocone, a structure that has usually disintegrated or fallen out. The belemnite in the illustration (ventral view) shows the deep hyponomic sinus in the region of the alveolus and the distinct impressions of vessels, proof that the rostrum was internal, completely enclosed by the mantle.

2.43. A slightly stylized longitudinal section showing the inner structure of the hard parts of the belemnite. The phragmocone, the conical, chambered element of the shell, which is carried inside the rostrum, is shown intact. Only the phragmocone is homologous with the chambered parts of nautiloid and ammonoid shells, whereas the rostrum as an outer envelope is a neomorph. The back of the phragmocone is dorso-ventrally cut away to show, in addition to the narrow chambers (faintly indicated), the ventral, marginate siphon. At the top front the phragmocone is shown three-dimensionally once more, with the wall (conotheca) partially intact; the growth lines on the dorsal side describe a steep convex curve rising toward the front. The extension of that portion where the curve rises constitutes a third element of the shell, the pro-ostracum—a bladelike, very fragile dorsal extension of the wall of the phragmocone. The pro-ostracum is seldom preserved, and if found, is always fragmentary. It corresponds to the dorsal living chamber, or conch, of the other cephalopod shells, whereas in the belemnites, the ventral part is replaced by the muscular mantle, that is, by soft parts. Moreover, the pro-ostracum is homologous with the uncalcified gladius of Recent Teuthoidea (squid).

2.44. The complete shell and its relationship to the soft parts of the animal are shown in a reconstruction depicting the animal as transparent. (After A. Naef.) Here, ⅙ natural size, is a juvenile *Megateuthis gigantea* (Schl.) from the Dogger. (Many belemnite animals grew to the very considerable length of 4–5 meters!) The position of belemnites in life was horizontal; they were nektonic forms that swam with the posterior tip of the body pointed forward.

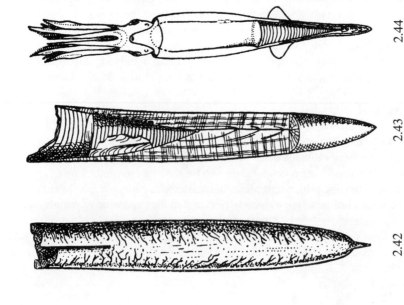

2.44

2.43

2.42

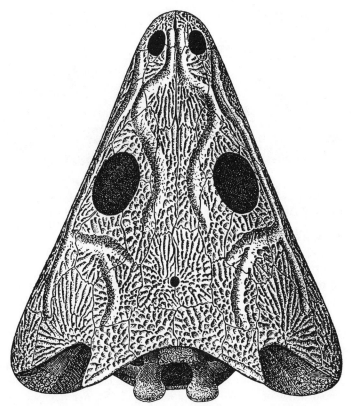

Fig. 2.45. Skull of *Lyrocephalus euri* Wim., a stegocephalian (order Stereospondyli) from the Triassic of Spitzbergen. ⅗ ×. (After G. Säve-Söderbergh.) As in the other ichthyostegids shown in figs. 2.29 and 2.30, the cranium is still completely closed; the only apertures, seen in this illustration of the dorsal side of the skull, are the nasal openings, which issue on the ventral side at the edges of the palate. The mucous channels in the forward part of the skull form a typical "lyre," as is often the case with stegocephalians. For the brain and the nerves of the head, see fig. 3.1.

fig. 3.71). In addition, the first representatives of the mammals are found in the Upper Triassic—small *aplacental* animals of no importance at all for the time being.

*   *   *

The Jurassic system is sharply set off from the Triassic as a result of the mass extinction among *ammonites* at that point. All the rich abundance of Triassic lineages, some of which were highly specialized, died out—with a single exception, which then served as the starting point for a new radiation of enormously diverse and rapidly transforming lines of ammonites (see figs. 3.32, 3.152). They and, in second place, the *belemnites* (figs. 2.42–44) once again form the pre-

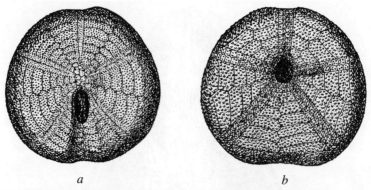

a   *b*

Fig. 2.46. *Hyboclypeus gibberulus* Ag., an "irregular," or bilaterally symmetrical, sea urchin of the Middle Jurassic of France. *a.* Upper side, with the anal opening (periproct) located posterior to the apex in a groove. *b.* Underside, with the mouth (peristome) displaced slightly forward from the center. 1 ×. (After G. Cotteau, from E. Stromer von Reichenbach 1909.)

ferred group of animals for establishing divisions of time. A turning point similar to the one observed for ammonites at the Triassic boundary can also be established in the evolution of the *nautiloids* (see fig. 3.32). In addition, the oldest *Teuthoidea,* squidlike animals, appear.

Also making their first appearance in the Lower Jurassic are the irregular *sea urchins* (figs. 2.46, 3.13), several of the modern orders of *insects,* and the true, modern bony fishes, the *teleosts* (fig. 2.47), even though the fish fauna of the Jurassic and of the Mesozoic in general is still dominated primarily by the older, lower ganoid fishes of the *Holostei* group (fig. 2.48). *Reptiles* (see fig. 3.71) continue to flourish, bringing forth at the end of the Jurassic and the beginning of the Lower Cretaceous the mightiest terrestrial creatures of all time (pls. 28 and 29). The *Crocodilia,* the last of the four surviving orders of reptiles, are

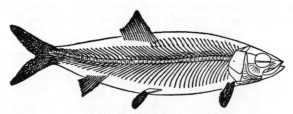

Fig. 2.47. *Leptolepis dubius,* a primitive teleost first from the [Upper] Jurassic. ⅓ ×. (From A. Smith Woodward.) This genus is one of the oldest forms of the bony fishes, which are dominant today. Teleost fishes are descended from the ganoid fishes (Holostei). The scales are thin, elastic, and rounded (cycloid scales; in the pictured form, however, there is still a thin fused coating on the scales); the vertebral column is completely ossified, the skull is lighter, with a more open structure than is found in the lower actinopterygians.

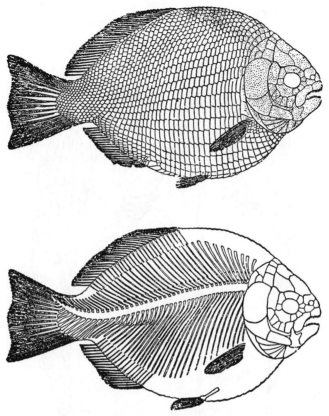

Fig. 2.48. *Dapedius,* a high-backed Jurassic representative of the Actinopterygii, order Holostei (family Semionotidae). ¼ ×. (From A. Smith Woodward.) The Holostei, ganoid fishes, are common throughout the Mesozoic; their numbers decrease sharply from the Tertiary on, and today, the order includes only the two North American genera *Lepidosteus* and *Amia.* The upper picture shows the body encased in a covering of large, thick, ganoid (fused) scales with the typical rhomboid shape. In the lower picture, the scaly outer covering has been removed to show that in this primitive actinopterygian, the centra of the vertebrae are not yet ossified.

represented from the Lower Jurassic on; beginning at that time, reptiles also conquered the air, expanding their range of habitats (*Pterosauria,* figs. 2.49, 2.50, 3.140; pl. 27B). Other flying animals appeared in the Upper Jurassic—the first *birds* (fig. 2.51).

*   *   *

There were two noteworthy events in the Cretaceous: the heavily armored ganoid fishes (*Holostei*), previously dominant, were superseded by the more mobile,

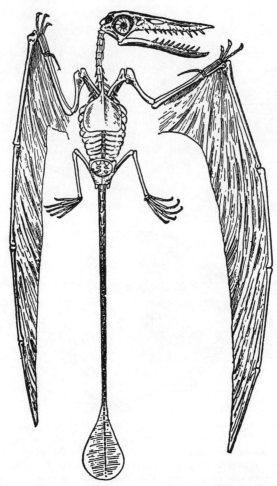

Fig. 2.49. *Rhamphorhynchus gemmingi* v. Mey., a pterosaur from the Upper Jurassic of Franconia [southern Germany], distinguished by its long tail with terminal paddle, short metacarpals on the flight fingers, and complete separation of the eye, preorbital, and nasal openings. About ¼ ×. (After S. W. Williston.)

agile bony fishes (*Teleostei*), which have dominated marine and freshwater environments since the Upper Cretaceous; and the *angiosperms* (pl. 9B) flourished, bringing a thoroughly modern character to the flora.

The undisputed masters of land and sea were still the *reptiles*. However, their remains are not distributed widely enough or found often enough in individual outcrops to be useful in the classification of strata. Stratigraphy is based almost

Fig. 2.50. *Pterodactylus*, another type of pterosaur, from the Upper Jurassic of southern Germany; with short tail, long metacarpals on the flight fingers, a differently shaped wing, and without complete separation of the eye, preorbital, and nasal openings. Reconstruction of the creature as it hunted, flying low over the sea. (After O. Abel, from R. Reinöhl 1940.)

Fig. 2.51. *Archaeornis siemensi* (Dames), the famous "first true bird" from the Upper Jurassic litho-graphic limestone of Eichstätt (Bavaria), the original of which is in the Berliner Museum für Natur-kunde. (The second specimen of an early Jurassic bird, with toothed jaws and long, lizardlike tail, now in the British Museum in London, belongs to another genus: *Archaeopteryx lithographica* v. Mey.) About ¼ ×. (After W. Dames, from F. Reinöhl 1940.)

exclusively on *mollusks,* primarily *ammonites* and *belemnites,* supplemented by *sea urchins* (pl. 9A) and small forms of *foraminiferans.*

Among the *bivalved mollusks,* the rudists (fig. 2.52), an aberrant group di-vided into several families, should be emphasized; they begin as early as the Jurassic but do not attain broad distribution and stratigraphic significance until the Cretaceous. These forms became sessile and formed reefs; they are often

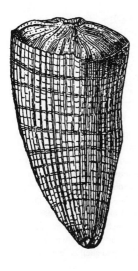

Fig. 2.52. *Hippurites gosaviensis* Douv. Upper Cretaceous of Gosau, Upper Austria. ½ ×. (After K. A. von Zittel 1924.) The valves of the shell are extremely unequal: the right valve is an elongate cone growing from its tip; the left valve has been transformed into a shallow, porous cover.

found in association with reef corals, but there are also reefs composed entirely of rudistids, for with regard to environment, they were less particular than corals.

THE CENOZOIC ERA

Like the Mesozoic era, the Cenozoic begins—or the Cretaceous ends—with another profound change in the animal world, one that justifies speaking of the end of an era. The ammonite line, once so important, dies out without descendants at the boundary between the Cretaceous and the Tertiary. The belemnites also disappear or, rather, are fundamentally transformed through the degeneration of their internal hard parts. Other invertebrate phyla lose many of their groups, some more extensive and characteristic, some less so.

The *reptiles* experience a widespread, radical decline. Mass extinction occurs in many lineages from various terrestrial and marine habitats; the giant dinosaurs become extinct, and all that remains of the once flourishing, diverse lines are scanty remnants (see fig. 3.71). Whereas we have fossil evidence of approximately thirty extinct orders of Paleozoic and Mesozoic reptiles, only four orders remain in the Tertiary, and they are still represented in modern fauna. The dominance of the reptiles is now superseded by that of *placental mammals,* which begin in the Upper Cretaceous and expand rapidly from the Tertiary on, bringing forth in a frenzy of evolution a sudden proliferation of all of the structural designs in existence today (fig. 3.70).

Although it is mammals that give the Tertiary its special character, they are of only limited usefulness in establishing divisions of time and correlations among strata. Much more important in these respects are *bivalves* and *snails,* which

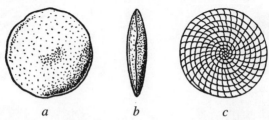

a   *b*   *c*

Fig. 2.53. Slightly schematic drawing of an Eocene *Nummulites* in side view (*a*), frontal view (*b*), and medial longitudinal section (*c*). These large foraminiferans grow in some cases to almost five times as large in diameter as the specimen illustrated. Because of their spiral tests and the transverse chambering of the whorls, these forms used to be included with the ammonites. However, there is one difference that is immediately apparent—in ammonites, as in all cephalopods, the chambers are run through by the siphonal tube (see figs. 3.20, 3.25, 3.32, and 3.34–36). Apart from that, the organization of the animal that builds the shell is not at all similar: Foraminiferans are representatives of the Protozoa, primitive, unicellular animals; their tests are secreted from the protoplasm of a single cell. Cephalopods, on the other hand, are highly organized multicellular animals with extensive division of labor among individual cells and groups of cells. The cephalopod shell is secreted by the epithelium of the mantle, a complex composed of many cells.

populated the oceans and bodies of fresh water in enormous numbers, leaving their shells behind in their respective deposits. Even the unicellular *foraminiferans* can be used successfully to determine age. Once again, as in the Carboniferous and Permian systems, they bring forth easily recognizable large forms (fig. 2.53). For the rest, the overall character of the animal and plant worlds began to look almost modern.

During the Tertiary, and particularly toward the end of that system, temperatures began to decline; the Ice Age was approaching, inaugurating the last, comparatively very short division of time in the history of the earth and its life—the Quaternary.

The sweeping climatic changes brought about a transformation in the composition of faunas and floras, primarily affecting, however, only geographic ranges and species composition. The decisive event of this period is the origination of the *human race,* the beginning of its triumphant advance on the world.

A CALENDAR OF GEOLOGIC TIME

The appearance and temporal distribution of the individual groups of plants and animals can be seen graphically when all of geologic time is placed within a framework familiar in daily life. Recently, L. Rüger took on this interesting task and came up with a "calendar of prehistory."

In his diagram, that portion of the earth's history for which there is a rock record—about two billion years—is given the value of one year. The individual systems and the major groups of organisms are then arrayed along the months of this symbolic year according to the available absolute dating (fig. 2.54).

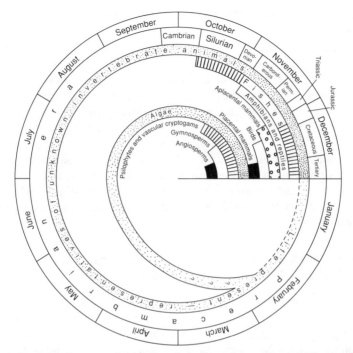

Fig. 2.54. Proportionally correct division of geologic time and the temporal distribution of animal and plant phyla, using the twelve months of the year as a frame of reference. (From L. Rüger.)

In the diagram, the Precambrian extends to September 23; only from that time on do we have abundant fossil documentation of the animal world. This late date makes it clear that by the early Cambrian, the vast majority of the major invertebrate lines had already taken definitive shape and assumed separate status. Whether life was already present in "January" or not we do not know; but by the end of "summer," enough time had passed for the evolutionary levels attained during the Cambrian to be reached.

During the first third of October, we encounter the oldest forms of vertebrates, the agnathans, which, in the diagram, are included under "fishes"; toward the end of October, the first terrestrial plants appear. At the beginning of November, the amphibians show up, and at the end of that month, the aplacental mammals; placental mammals did not make their conspicuous appearance until much later, toward December 20. The first human remains date from about 600,000 years ago, which corresponds to the final two and one-half hours of the diagram; the modern human, species *Homo sapiens,* has only been around for a scant half hour. The "world history" of humans, traceable back for little more than 6,000 years, is compressed into the last one and one-half minutes, and a human life lived to the ripe old age of 80 years has the relative time value of one and one-fourth seconds.

This abbreviated time scale and the numerical values we perceive in it demonstrate impressively the enormous extent of time encompassed by the geological past, at the same time graphically showing the relative brevity of the period of time that serves as the source of our observations of organic evolutionary processes. Nevertheless, during that time, remodeling and innovation on a vast scale took place, providing the basis for an understanding of evolutionary events—how they happened and the paths they followed. However, the chart also admonishes us to caution as we evaluate the observable processes by which races form today; the unit of time over which these processes can be observed is extremely brief, and the events themselves are quite insignificant when compared with the course of phylogenetic evolution, which extends over an enormous span of time.

SOME PARTICULAR QUESTIONS CONCERNING THE PALEONTOLOGICAL CLASSIFICATION OF TIME

After this quick overview of the occurrence of the principal groups of plants and animals over geologic time, it may appear that dating with the aid of fossils is a very straightforward, easy thing to do. Whenever a trilobite is found in a quarry, it can only mean that those rocks date to the Paleozoic, and if this trilobite is, for example, a member of the genus *Paradoxides,* we must be in the Middle Cambrian. A clymenian from the large ammonoid group is an unequivocal indication of the Upper Devonian, and after the specimen has been identified down to genus and species, our pronouncement on dating becomes even more accurate. Appearances notwithstanding, this paleontological method of classifying time and establishing relative ages also has its problems and theoretical points of controversy.

In fact, the method is indeed simple to apply and incontestable as long as only relatively crude approximations are at issue. There are a number of very characteristic, short-lived genera of ammonites, trilobites, brachiopods, and so on, that are very easy to memorize and provide a fairly precise chronology.

But paleontology, in the temporal evaluation of fossils, shares with all science the fact that its research is never complete, that new problems and objectives continually arise, requiring renewed, expanded, and more thorough examination of the material. The demands of stratigraphy are much greater than they used to be; rough, approximate dates based on a few index fossil types are no longer adequate. Questions of modern geology as well as the practical requirements of the mining industry demand dating that is as sophisticated as it can possibly be.

The example given above referred to the Middle Cambrian—a term applied to an enormous span of time; the era lasted so long that to identify something as being Middle Cambrian is, for most purposes, not saying much at all. In order to yield a date of practical application, the species of our *Paradoxides* must be determined as carefully as possible and put in the proper place within the evolutionary line to which it belongs. When this is done, and the life span of this species has been ascertained with precision, an inference can be drawn as to a particular *zone* of the Middle Cambrian.

But efforts are made to go beyond even that. Extremely meticulous investigations, particularly those involving variation statistics, show that the stamp of a species at its first appearance is not exactly the same as it will be at the end of its range in time. The gradual, albeit always slight, morphological transformation of the organism within the framework of the species as we conceptualize it may yield yet other time markers.

Further, it is possible to connect the duration of two zone fossils, *a* and *b,* that occur one after the other with that of a third, *x,* whose range partially overlaps the ranges of the first two; thus, a unit of time can be isolated in which only *a* appears at first, another in which *a* and *x* are found together, a third in which *b* and *x* are associated, and finally a fourth, in which only *b* prevails. In this way, zones can be divided into even smaller units of time, called *subzones.*

To accomplish this, the first task, an essential one, is to collect the fossils from a suitable sequence of strata very painstakingly, layer by layer, in order to ascertain the appearance and disappearance, as well as the succession, of individual small transformational steps. Then, similar collections are made in other areas to see whether the same succession is present elsewhere; even so, the question remains open whether, without further evidence, a corresponding sequence may be used to substantiate simultaneity in the strictest sense.

Most important, however, is the meticulous morphological investigation of the material collected in order to arrive at an incontestable determination of species, subspecies, and incidental mutants and a correct assessment of their phylogenetic position. It is not enough just to examine the superficial appearance of the fossil; it is much more important to investigate the internal structure and the course of the individual developmental history, to exclude the possibility of error based on convergence or homomorphs—superficial similarities of form among organisms that are not closely related. Questions of mode of life, of ecological and climatic conditions, and so on are taken into consideration; in short, even stratigraphy requires that all the tools of modern paleontology be put to use.

Therefore, the distinction that used to be made between paleontology as a purely biologically based science and the practice of applying the index-fossil concept in the manner of a craft is no longer valid. There is really only one way to approach fossils: rigorously scientific biological principles must be the foundation. The result of all these investigations is a name that describes clearly the position of the fossil in question with regard to genus and subgenus, species and subspecies. Only after these purely paleontological preliminaries have been carried out can the names be put to use as incontrovertible elements available for stratigraphical application.

THE CONCEPT OF ZONES

The zone is the basic unit for the classification of time; it is defined as the absolute life span or, put more precisely, the period of time covered by the observed vertical range, of a species as determined by the comparative investigation of as many profiles as possible. It includes a span of time short enough that the index species would not have undergone transformations taking it beyond the limits that define it as a species, but long enough for it to have dispersed horizontally over a broad area. When, in a profile, the species in question is replaced by another, the zone of the next species begins.

Nor is it a matter of choice *which* species are used to establish zones. There are groups of animals that change extremely rapidly and others that evolve very slowly. Since the objective of classifying time is to achieve the finest distinctions possible, it follows as a matter of course that the animal groups evolving most rapidly will be chosen. These are the phyla and orders that have been particularly emphasized in the survey presented in the previous section.

It should be noted further that the intensity of transformation fluctuates over time and, therefore, so does the stratigraphical value of the individual groups and lineages. A phylum that provides extremely valuable index fossils during one particular unit of time may be stratigraphically insignificant during the next; it must then be replaced by another group of animals, one that offers better possibilities for subdividing the unit of time in question.

From the definition given above, it follows that we consider the zone to be a *purely temporal concept.* Therefore, zonation has an abstract, temporally qualitative character, distinct from the geological processes that occurred during the unit of time in question and from the actual composition of the rock strata that were laid down during any given zone. Zonation provides a temporal structure comparable to the reigns of the Roman caesars or the dynasties of Egypt, which are used to date events of world history.

For example, in the Middle Jurassic, we speak of the *Parkinsonia parkinsoni* zone, meaning by that the time span covered by the existence of this ammonite, which we know from experience to be a suitable zone fossil. In designating the zone as such, we completely ignore the fact that in one area, this zone is represented by marl, and in another by clays, sandstones, or iron oolites. These are circumstances that are significant not for time, but for space, and have more to do with geology in the broad sense, or regional geology. We also ignore the fact that locally, in many strata of this zonal unit, the characteristic index ammonite may be lacking for reasons having to do with the facies—with adverse living conditions at the site of deposition. Nevertheless, we speak of the *Parkinsonia parkinsoni* zone, in the sense of its being a unit of time within which these strata fall.

In such instances, dating is possible if the typical representatives of lower and higher zones occur below and above the rock series in which the index ammonite is missing; with the help of these different fossils, the stratum in question can be positioned correctly within the zonal system. In other instances—if, for example, the standard zone fossils are lacking over a considerable vertical distance—we must take a detour. We first establish a classification based on other organisms that may be present in the facies in question, on brachiopods, for example. Then we look for areas where the brachiopod facies interfingers with the ammonite facies and merges with it. At that point, as a result of the intermingling of the two faunal elements, it is possible to associate the brachiopod segment with the ammonite series and fit it into the zonation system.

Paleontological zonation is thus *chronology,* a purely temporal system of classification, and not actually stratigraphy, for it disregards the actual formation of the individual strata or beds. Not until the actual stratigraphic sequence is inserted and comparisons have been made does zonation become stratigraphy and by extension, geology—the history of the earth—which fills out the individual units of time with their corresponding historical content of sedimentary conditions, transgressions and regressions of oceans, mountain-building processes, glaciation, and so on.

Today, many authors conceive of the zone differently, ascribing a *spatial* significance to it. They see it in terms of the rocks themselves—rocks with the same complement of fossils. The way a term is understood, however, is a matter of definition, and consensus should follow the first usage if at all possible. The creators of the concept of zonation are Alcide d'Orbigny (1852)[6] and A. Opel (1856), both of whom assigned a temporal, abstract meaning to the term *zone.*

In particular, the outstanding German scientist Oppel (1856, 3) gave a detailed rationale for the methods he used to correlate strata. He described the goal as being "to investigate the vertical range of each individual species at different localities, disregarding the mineralogical composition of the layers," and termed the result "an idealized profile, in which elements of the same age, although in different areas, are always characterized by the same species." This should make clear what Oppel meant by the concept of zonation that he had introduced: the focus of his approach is not on the rock but on the life span of the index species, which yields an idealized, or abstract, temporal system of zonation, one that does not take local and spatial factors into account. It is advisable to adhere to this original, temporal reading of the term.

6. [Alcide Charles D'Orbigny, 1802–57, a follower of Georges Cuvier, was a French naturalist who established the value of invertebrate fossils for stratigraphy. He traveled in South America (1826–34) to collect natural history specimens and was professor of paleontology at the Musée d'Histoire Naturelle in Paris, a post created especially for him.—Trans.]

*   *   *

Another question frequently discussed is whether it is more advisable to base zonation on individual, selected species or on an entire faunal assemblage; in other words, should we use *species zones* or *faunal zones?* These should not, however, be regarded in any way as alternatives. We have already said that in order to refine our system, we sometimes use, in addition to the actual index species of the zones, a second or even a third species. Just this, then, results in a certain portion of the total fauna being taken into consideration and is a transition to the faunistic method, which, moreover, never makes use of the entire organic world but involves only a selection, leaving out primarily the long-lived, persistent forms.

Both ways are feasible, and the individual situation determines which procedure will achieve the objective and should therefore have preference. With ammonites, for example, the very brief life span of the different species allows the successful delimitation of well-defined, distinct species zones. With other groups of animals—foraminiferans, clams, and snails, for example—whose forms change more slowly, various transformational series are combined. Thus, when the individual species of a single evolving lineage are too long-lived to yield a temporal distinction of sufficient resolution, the weight of evidence must lie with the changes in the faunal assemblage.

The absolute time value of a faunal zone may be different from that of a species zone based on a single, short-lived species. But since the rate of evolution fluctuates—for members of different phyla and even within the same phylum at different phases of evolution—even the absolute temporal duration of individual species zones may be completely different. Therefore, the duration of the faunal zone, perhaps also different, is of no consequence at all. The task consists of subdividing time as finely as possible, and all available possibilities must be called upon to serve this goal. However, paleontological chronology can never be more than *relative*.

SYNCHRONY OR MERELY EQUIVALENCE?

When we follow zones across broad areas—when we equate rocks containing the same index forms but located in various parts of the world—what proof do we have that we are really seeing simultaneity of origin? A preliminary question that arises in this regard is how matching sequences of species and faunas, the basis of zonation, occur at all. There are two conceivable possibilities, and indeed, both are found in nature: (1) Either the evolution and appearance of the index forms occur within a single, restricted area, for example, in a particular ocean basin (since the index fossils we deal with are mostly marine organisms), and from that center of origin, the organism spreads throughout the rest of the ocean; or (2) we are dealing, not with a unique evolutionary event tied to a single locality, but with a polytopic evolution, one occurring in two or more disjunct areas. In this instance, the parent stock of an evolutionary line would have achieved broad distribution through migration and then, at various localities, would have undergone transformation of the same kind, or at least in similar directions, through the potence of its genotype.

We shall see later in another connection just how extensive parallel evolution is among organisms, which means that such a process of independent, equidirectional development at different localities is, in itself, thoroughly possible. However, parallel evolution cannot be a significant factor in the origin of identical sequences, at least in the long term, because increasing divergence of the evolutionary paths could be expected. Therefore, in the later units of the stratigraphic sequence we would not find the same forms or even very similar ones.

Much more frequently, therefore, we encounter *successive waves of migration* of the individual faunas and species from a particular original stock. Such dispersal takes time, of course, and as the distance from the evolutionary center increases, the appearance of the index form in the corresponding rocks is progressively later.

This concern was raised for the first time eighty years ago by Thomas Henry Huxley. He maintained that there was not the slightest possibility of determining the duration of migration. Therefore, there could be no reliable means of assessing the age relationship between two strata located far apart; the correlation of the fossil content and of the sequence of the fossils in question would still not be proof that the two strata were the same age. According to Huxley, it would be theoretically quite conceivable that a Devonian fauna and flora of the British Isles lived contemporaneously with a Silurian group of organisms in North America and a Carboniferous fauna or flora in Africa. He would maintain, therefore, that we should not speak of *isochronic* strata (strata of the same age) but only of *homotaxial* strata (horizons that are of *equal value* with regard to the relative order of arrangement and their fossil content).

Since that time, objections of this kind have been repeated again and again, although not as strongly as Huxley raised them in his probably intentionally exaggerated argument. The justification for these objections must be considered. Let us therefore take some living organisms and compare the times required for their populations to spread, the animals sometimes moving on their own, sometimes being passively transported.

It is known that the small marine snail *Littorina littorea* (pl. 10A) spread along the Atlantic coast of North America from Halifax to Cape May, a distance of a good thousand kilometers, in less than fifty years. Based on the observed contemporary rate of migration, this species would have covered about 160,000 kilometers since the postglacial *Littorina* age began, about six thousand years ago, and consequently, could have circled the equator four times had geographical conditions permitted.

Other investigations have been carried out with starfish, which are certainly not the most mobile of animals. Marked individuals were released and found some time later as much as one thousand kilometers from the place of release. That is an enormous distance to cover in a fraction of an individual's life span.

The most significant spread of marine invertebrates is the passive planktonic dispersal of larval stages by ocean currents. The speed of the different contemporary ocean currents has been determined by means of drift bottles to be an average of fifteen to sixteen kilometers per day. These figures were then used to calculate how long it would take a species whose members were sexually mature only once a year and, like certain echinoderms, had a larval period of thirty-three days to disperse from a hypothetical original habitat on Cape Horn throughout the Antarctic realm by traveling with the modern circumpolar current. It turns out that under favorable conditions, they could cover the coast of South Africa in about fourteen years and reach the shores of New Zealand—about twenty thousand kilometers away—in forty years.

Other marine creatures have a shorter larval stage. Brachiopods, for example, swim and drift about in the ocean for only about eight days. But if they were taken up by a current of the force of our modern Gulf Stream, those eight days would see them carried about one thousand kilometers away. If, when the migration is over, living conditions are favorable, the larvae may settle down, mature, and produce new larvae, which will be transported in turn.

Diffusion over enormous distances, then, requires such a surprisingly brief period of time that even when considerable allowance is made for less favorable migratory and settling conditions, this temporal element is of absolutely no consequence at all when compared with the interminably longer duration of zones, reckoned, on the average, at several hundred thousand years. Even on land, in bodies of fresh water, for example, where possibilities for dispersal are more restricted and regions are distinctly isolated, certain creatures manage to disperse with astonishing speed. An example is the well-known mitten crab (fig. 2.55), a

Fig. 2.55. Chinese, or mitten, crab (*Eriocheir sinensis* Milne-Edw.), male, with thick pelage on the penultimate segments of the chelae, which gives the crab its name. Reduced. (After P. Kuckuck and A. Hagmeier.)

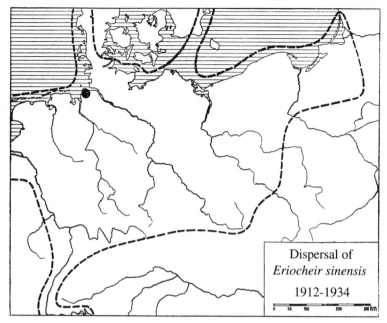

Fig. 2.56. Dispersal of the mitten crab in central Europe during the period between 1912 and 1934. The starting point of its migration is indicated by a black dot between the mouths of the Elbe and the Weser rivers. (From Schindewolf 1945.)

freshwater animal introduced [into Europe] from China; within a time period of about two decades, it came to occupy a large, consistent range in western Europe (fig. 2.56).

For all practical purposes, then, the dispersal times for index fossils can be disregarded; *we may absolutely count on simultaneity* and not just on equal value *of disjunct strata with matching fossil content,* not down to the year, perhaps, or to the century, but still only with temporal differences that, in view of the scale in question, are completely meaningless.

Furthermore, the repeated appearance of the fossil sequence, always identical and sharply delimited, at widely disparate points on the globe is certain proof of the rapid, cosmopolitan spread of the index organisms. If the migratory periods had been very long and of irregular duration, inconsistencies and overlaps would appear; the individual index forms would be heavily intermingled and would not appear equally well sorted everywhere in the same sequence. To go back to Huxley's example, it is completely out of the question that distinctly separate waves of a completely uniform faunal series, one indicative of a particular period of time, would roam the seas as a unit for vast periods of time and arrive at some point considerably later, geologically speaking, in exactly the same sequence it was in when it started out.

MICROPALEONTOLOGY—ITS PLACE AND IMPORTANCE IN STRATIGRAPHY

In recent years, a supposedly new and separate branch of paleontological re-search—"micropaleontology"—has been the subject of much comment, mak-ing it seem appropriate to devote a few words to the subject. It is concerned with the investigation and stratigraphic interpretation of the most minute of fossilized creatures, which necessarily involves the use of microscopes.

The organisms studied are primarily the unicellular foraminiferans (figs. 2.78–83) and the little bivalved crustaceans known as ostracodes (figs. 2.57, 2.58). In addition, minute parts of larger organisms—for example, the calcare-ous skeletal elements of the holothurians (figs. 2.59–65), the pedicellariae of sea urchins (figs. 2.66–69), conodonts (toothlike structures of unknown systematic position (figs. 2.70–75), the spores and pollen of plants (fig. 2.76), and so on—are picked out of rock samples, and again, considerable enlargement is necessary for observations to be made.

Applied "micropaleontology," which is called "microbiostratigraphy," is of great practical value; in particular, the study of foraminiferan faunas has proved invaluable to the petroleum industry. The advantage of these forms lies in the fact that they occur in such great numbers of younger sedimentary rocks. When a deep, small-diameter borehole is drilled, the chance that the resultant core will yield an ammonite or a snail or clam shell large enough to be evaluated is rela-tively slight, whereas the small rock sample usually contains large quantities of foraminiferan shells, which can then be cleaned and used to establish a relative date for the strata.

This has been going on during the last twenty-five years or so on a large scale, as drilling for oil has increased. According to American estimates, there are al-

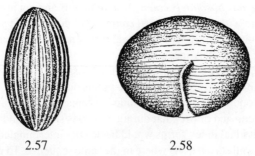

2.57   2.58

Figs. 2.57–58. Two ostracodes (bivalved crustaceans) from the central European Upper Devonian. 2.57. *Richterina* (*Richterina*) *striatula* (Rh. Richt.). Upper Upper Devonian from Kadzielnia, near Kielce (Holy Cross Mountains, Poland). 45 ×. 2.58. *Entomis* (*Richteria*) *serratostriata* (Sdbg.). Lower Upper Devonian from Weilburg (Rhenish Mountains). 22.5 ×. (After H. Matern.)

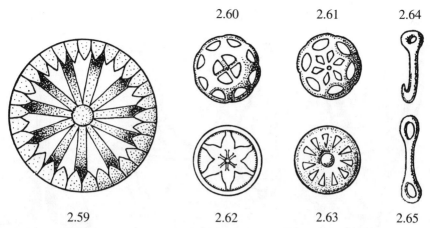

2.60   2.61   2.64

2.59   2.62   2.63   2.65

Figs. 2.59–65. Calcareous skeletal elements of holothurians (sea cucumbers). Apart from isolated and, in some instances, still doubtful impressions of entire animals from the Cambrian (see pl. 4E and G) and the Upper Jurassic, what we usually find are the almost always very regular and beautifully formed calcareous wheels and spicules, which are, or were, embedded in the tough, thick body wall. They are truly "Natural Works of Art" [*Kunstformen der Natur,* title of a book by Ernst Haeckel]. The designs are extremely diverse and are very diagnostic for the individual genera and species, making these skeletal elements apt for use in systematics and, to a certain extent, for stratigraphy.

2.59. Wheel from *Myriotrochus.* Early Upper Jurassic of Württemberg. 200 ×.

2.60. Wheel from *Protocaudina traquairii* (Ether.). Lower Carboniferous of Illinois. 50 ×.

2.61. Wheel from *Palaeochiridota plummerae* Chron. Upper Carboniferous of Texas. 50 ×.

2.62. Wheel from *Chiridota curriculum* Schlumb. Eocene of France. 250 ×.

2.63. Wheel from *Protocaudina geinitziana* (Spand.). Zechstein [Upper Permian] of Germany. 250 ×.

2.64. Spicule from a synaptid holothurian (*Ancistrum*). Upper Dogger of Württemberg. 45 ×.

2.65. Spicule from a dendrochirotic holothurian. Lower Liassic of Württemberg. 80 ×.

(After Th. Mortensen, and after C. Croneis and J. McCormack.)

most one hundred paleontologists and their assistants working in just the California petroleum industry. This number is graphic testimony to the great economic importance of paleontology, and we point with pride to the relevancy for our daily lives of the work performed by paleontologists.

* * *

However, we must dispute the enthusiastic claims made by many professional petroleum paleontologists that here we have a new and revolutionary methodology, an independent branch of research and *the* paleontology of the future.

1. First, it should be pointed out that in the past, before micropaleontology as such existed, foraminiferans were, of course, quite definitely not ignored. The large forms already mentioned in another context (figs. 2.33, 2.34, 2.53) from

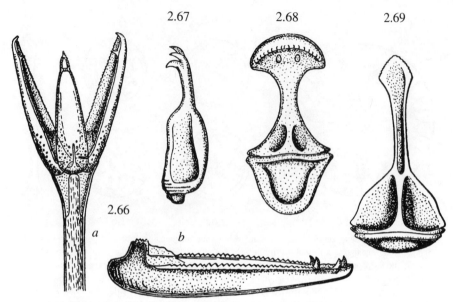

2.67   2.68   2.69

2.66

a   *b*

Figs. 2.66–69. Pedicellariae (pincerlike structures) of various sea urchins.
2.66. Globiferous pedicellariae of the Recent regular sea urchin *Eucidaris clavata* Mrtsn., from St. Helena. *a.* Head of an open, entire pedicellaria to show the arrangement of the individual valves, which, as fossils, are usually found isolated. 90 ×. *b.* Individual valve. A drawing made by G. C. Wallich, in 1862, reproduced here as a curiosity; Wallich described this minute object, not even half a millimeter long, as the lower jaw of a new, extremely tiny vertebrate. 150 ×.
2.67. Valve of a globiferous pedicellaria of the regular sea urchin *Hemipedina.* Upper Upper Jurassic of Württemberg. Side view. 60 ×.
2.68. Valve of an ophicephalous pedicellaria, probably from an irregular sea urchin. Lower Upper Jurassic of Württemberg. Internal view. 175 ×.
2.69. Valve of a rostrate pedicellaria, probably from the irregular sea urchin *Holectypus.* Lower Upper Jurassic of Württemberg. Internal view. 175 ×.
(After Th. Mortensen.)

the Carboniferous, the Permian, and the Tertiary have always been important as index fossils. But there has long been an interest in small foraminiferans, too. Alcide d'Orbigny and A. E. Reuss, to name only two, published papers of fundamental importance of the foraminiferan faunas of both the Cretaceous and the Tertiary one hundred years ago, in 1840, and quite deliberately used the fossils to date strata. Since then, numerous other scientists have steadily and systematically continued work in the same vein. Ever since 1874, foraminiferans have been used to date drilling cores from exploitable mineral deposits. Micropaleontological investigation, therefore, is not exactly a *new achievement of the petroleum industry,* even though the industry undoubtedly gave it considerable impetus.

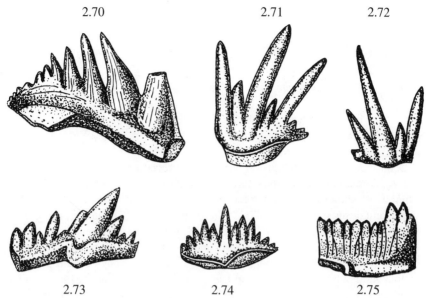

2.70   2.71   2.72

2.73   2.74   2.75

Figs. 2.70–75. Different forms of conodonts. All 25 ×.
2.70. *Neocoleodus spicatus* Brans. and Mehl. Internal side view. Middle Ordovician of Colorado.
2.71. *Chirognathus alternatus* Brans. and Mehl. External view. Middle Ordovician of Colorado.
2.72. *Chirognathus varians* Brans. and Mehl. External side view. Middle Ordovician of Colorado.
2.73. *Polygnathellus emarginatus* Brans. and Mehl. Side view. Middle Gothlandian of Missouri.
2.74. *Chirognathus multidens* Brans. and Mehl. Exterior view. Middle Ordovician of Colorado.
2.75. *Spathodus primus* Brans. and Mehl. Side view. Middle Gothlandian of Missouri.
(From E. B. Branson and M. G. Mehl.)

Ever since C. H. Pander (1856) gave conodonts their name and described them in large numbers for the first time, these microscopically small tooth- or platelike structures have usually been interpreted as being the teeth of primitive fishes. Zittel, Rohon, H. W. Scott, and other authors have regarded them as being annelid remains; F. B. Loomis thought they were the teeth of gastropods. Against the fish origin, it has been argued repeatedly, and recently also by E. A. Stensiö, that the structures and the pedestals on which they rest do not exhibit normal bony structure but rather differ greatly in their internal microstructure from vertebrate teeth and corresponding structures. A few years ago, H. Schmidt found a group of variously shaped conodonts, which until then had had different names, in their original association and saw them as the visceral arches (mandibular plates, hyoid teeth, frill teeth of the gill arch) of an unarmored placoderm.

Very recently, W. H. Hass has studied the internal structure of conodonts in detail using thin sections. It turns out that the older (inner) layers of the laminated skeletal elements show multiple evidence of fractures and other kinds of disturbances that have been repaired by subsequent deposits of calcareous layers. Therefore, the conodonts must have been enveloped in secreting soft tissue and, in any event, could not yet have served as teeth at the time the disturbances occurred. If they had functioned as teeth later, after breaking through the soft tissue, they would have to show superficial traces of wear that has not been overgrown, and this is not the case. Hass concludes that conodonts were internal support structures for tissues occurring either on the surface of or within the bodies of some kind of marine animal—tissues that were exposed to stress or pressure.

The stratigraphic utility of conodonts is independent of their position, as yet unclarified, in animal systematics. They are found throughout almost the entire Paleozoic, from the Lower Ordovician to the Permian, and in some horizons, are represented frequently in extremely typical assemblages, making them useful for establishing correlations among strata. (As long as we are not sure of the individual affinities of the different types and of their location within the organism, their taxonomy and naming remains, of course, purely artificial and utilitarian; this in no way interferes with the use of the formal names in geology and stratigraphy.)

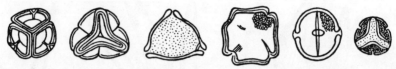

Fig. 2.76. A few types of pollen from the German Tertiary (Eocene, Miocene), which have been used successfully to date brown coal, or lignite. Much enlarged (300–850 ×). (After H. Potonié.)

2. Further, the claim that these investigatory procedures incorporate a distinct, independent discipline of paleontology—one that deserves to be singled out as "micropaleontology" or "microbiostratigraphy"—is to be refuted from the standpoint of scientific systematics. "Micropaleontology" has neither a uniform, self-contained field of research nor particular, specialized investigatory methods. Its objectives are the same as those of "macropaleontology," namely, morpho-logical-anatomical investigation, taxonomic designation, phylogenetic classifi-cation, and finally, stratigraphic application of fossils. To be able to stand on its own as a separate field of science, "micropaleontology" would have to produce certain general results that exceed and are not derived from the results of "mac-ropaleontology." This, however, is not the case.

The only difference lies in the fact that the animal groups or the plant remains treated by "micropaleontology" happen to be small, their study requiring the technological aid of a microscope, which, however, is also indispensable in the study of large fossils. Apart from their sizes, the objects of study have little else in common; they are not objectively linked in an intrinsic, natural way but come from the most heterogeneous plant and animal phyla.

No independent scientific discipline, however, can be based on a tool or on a mechanical procedure such as the preparation and cleaning of samples; the only adequate rational basis lies in the objective uniformity of the material studied, a particular and appropriate research objective, and specific results that are all its own. Pictorial renditions of the objects of study for use in scientific description are at least as important to paleontology as microscopy is, yet we do not say that the technology of illustration constitutes a special paleontological discipline!

Just as in an earlier context we could not sanction a distinction between sci-entific paleontology and the practical analysis of index fossils, we *cannot rec-ognize micropaleontology as a separate discipline.* Instead, we strongly emphasize the *consistency and indivisibility of paleontology* as the comprehen-sive science that uses the methods of biology to investigate past life, regardless of whether the research is pure or applied and regardless of the size of the objects being studied.

3. With regard to the future significance of the study of microfauna and -flora, it should be said that the endeavor will undoubtedly continue to make an ex-tremely valuable contribution to the practical correlation of strata, especially once the often purely stratigraphical-statistical methodology has been expanded

to include a corresponding biological dimension. However, this methodology will probably only be important in instances in which other measures fail, for the small foraminiferans are by no means ideal as index forms. The life spans of the individual species are much longer than one wants for a true zone fossil. Here, as with the other animalian groups that change slowly, one can only obtain useful results by taking the entire faunal assemblage as a basis and evaluating the frequency and association of a large number of species (figs. 2.77–83).

Furthermore, the paleontology of foraminiferans is by no means a universally applicable method. In the Paleozoic and the oldest Mesozoic, foraminiferans are much less frequently found—probably mostly because of the subsequent destruction of their fragile tests—than in younger formations. At the least, they are

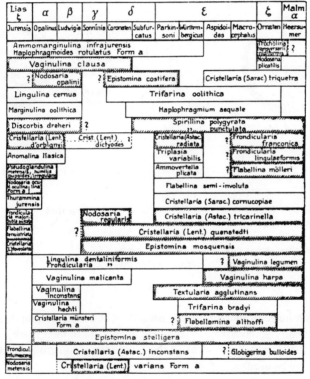

Fig. 2.77. Index Foraminifera in the Brown [Middle] Jurassic of northwest Germany. (After H. Bartenstein and E. Brand.) The criterium for determining the age and classification of individual layers in the ammonite zones shown above is not a particular species, for each one usually occupies a considerable range within a whole series of horizons of different ages; rather, it is the *communities* of species, which become evident in the vertical rows on the table. The names of the forms that are always found are outlined in cross-hatching to make them stand out. Question marks indicate provisional uncertainties with regard to the range of the species in question.

2.78   2.79   2.80   2.81

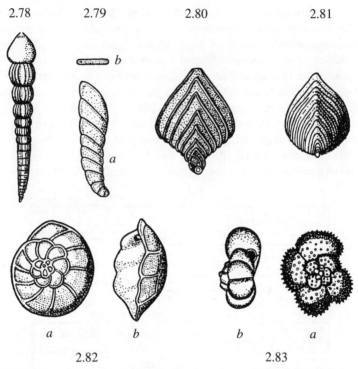

a   b   b   a

2.82   2.83

Figs. 2.78–83. Some of the types of foraminiferans shown in the table in figure 2.77. Enlarged (5–15 ×).

2.78. *Nodosaria.*

2.79. *Vaginulina. a.* Overall view. *b.* Frontal view from above, with aperture.

2.80. *Flabellina.*

2.81. *Frondicularia.*

2.82. *Epistomina. a.* Dorsal view. *b.* Side view with aperture.

2.83. *Globigerina. a.* Dorsal view. *b.* Side view with aperture.

(After J. J. Galloway, in R. Wedekind 1937.)

not easy to remove from extremely consolidated and metamorphosed rocks and are consequently not available in sufficient quantities. In their place, ostracodes (fig. 2.57, 2.58), which do occur in the older strata, sometimes in enormous numbers, may be used.

But again, as with the young Mesozoic and Tertiary foraminiferans, such forms are only a substitute to be used in instances in which other, more suitable index forms are lacking. A single goniatite or trilobite yields much more precise temporal information than the richest ostracode fauna; likewise, an identifiable ammonite fragment permits a more exact and, more important, quicker, more effortless dating than does the interpretation of a foraminiferan fauna.

Finally, in a certain sense, dating based on microorganisms is possible only by means of a detour. The standard zonation of the relevant divisions of the geological systems is based on goniatites, trilobites, ammonites, and so on. The recognized foraminifera or ostracode units usually have only a limited, regional validity because the organisms themselves are tied to a narrow range of facies; before the units can provide binding, absolute evidence for dating, they must first be intercalated into the standard system of classification, which is generally valid over a broad geographic area, by determining what parallels exist.

This criticism is not intended to disparage the paleontological use of small organisms, for where they are applicable, they have proven to be very successful. Rather, this objective evaluation of micropaleontology serves only to point out its correct place within the overall framework of paleontology, at the same time putting its claims of importance and exaggerated hopes for the future back into proper perspective.[7]

There is no doubt that the elegance of the investigatory procedures deserves admiration. One example is the clarification of the history of the forests and moors of the most recent geologic past through the interpretation of the spores and pollen of plants; these results are very important and would hardly have been obtained with this degree of statistical certainty by any other means. Even in objects that have already been prepared—a skull, prehistoric specimens, and the like—pollen analysis can still be used successfully in determining age. The tiny quantity of peat, loam, or clay that remains attached to the piece in any depression or hole is usually enough to use for dating.

Pollen and spores have also been used with remarkable success in classifying brown coal and identifying seams of coal. *Nonetheless, there is no doubt that the future of our discipline lies only in small part in the field of "micropaleontology."*

DELIMITING THE LARGER UNITS OF TIME

The task of classifying time does not end with demarcating zones and arranging them in sequence; as with any system of classification, the further objective is a comprehensive, hierarchical system of dominant and subordinate categories, which combine units that belong together and contrast those that are further apart. Thus, zones—the lowest-ranking unit of time, which we have already discussed—are grouped into *stages,* the next highest-ranking, and more inclusive, unit.

In the case of zones based on a single species, we concur with Rudolf Wede-

7. An addition made during corrections to the manuscript: It has been said recently (E. Triebel 1947), that micropaleontology is to the overall field of paleontology as microchemistry is to chemistry. I agree with this completely, although from my own point of view and in a somewhat different sense; for microchemistry, too, is just a certain technological expansion and special way of working within the framework of its overall discipline, which does not mean that it represents an independent methodology, such as, for example, physical or physiological chemistry!

kind in defining the stage as the life span of the genus from which the zone fossil comes. For example, in the Upper Devonian, we speak of a *Manticoceras* stage, which is based on the goniatite genus *Manticoceras* and includes a series of zones that are characterized by species of this genus that are short-lived but have wide horizontal distribution. The boundaries of stages based on genera are, in general, very distinct; we shall see an example of this further on. If we are dealing with faunal zones, the stage boundaries are determined by major faunistic discontinuities that contrast a sequence of closely associated individual faunas with a previous and a subsequent consistent faunal series.

Groups of stages become *divisions within systems,* which in turn combine to become *systems,* several of which constitute an era. (The practical and historical course of classification, however, has mostly proceeded in the other direction, beginning not at the bottom but with increasing subdivision of the more comprehensive units to ever smaller and finer ones.)

How, then, are the boundaries of systems and eras established? Theoretically, this question is not difficult to answer: To establish the boundary between two systems, it is appropriate to use those groups of organisms that play the dominant stratigraphic role within the systems and provide the standard by which the various units are classified. Moreover, it is best to take the first appearance of a new form as the basis, for new forms usually appear on the scene suddenly and in large numbers. A boundary defined in this way is therefore more sharply defined than one based on the extinction of older types, since dying-out is generally a slower, more gradual process, and stragglers can sometimes extend on into the new time period.

In practice, however, there are sometimes difficulties and differences of opinion. One may occasionally argue over whether this or that turning point in organic evolution offers the most appropriate boundary. But when all is said and done, it is not so important that a system begin with one zone or another as long as one strives for a solution that is, of course, as rational as possible, one reflecting the realities of the natural situation. Such matters can be decided by consensus; it is far more important that the careful age determinations and correlations provided by zonation remain fixed.

Marine Invertebrates

The organisms upon which subdivisions *within* the systems are based are, with very few exceptions, taken from marine fauna, and of those used, most are shallow-water invertebrates. One reason for this is that the vast majority of sedimentary rocks are deposits of shallow seas (scarcely any deep-sea formations from the geologic past are still in existence) and contain fossils belonging to that kind of habitat. In contrast, continental rocks play only a very minor role. Under these circumstances, it naturally would make no sense to set up a basic system of subdivisions based on continental forms, for it would not be applicable to the majority of rocks.

The main reason, however, is that fossils of marine invertebrates are extremely common and have a broad horizontal distribution, which means that their ranges were not narrowly restricted by climate and geography. This is especially true for the freely moving forms. Upper Triassic strata of the Alps and California, for example, have numerous species and almost all of their genera in common: areas that are about 10,000 kilometers apart as the crow flies are almost 20,000 kilometers apart in terms of animal dispersal. The same faunal sequences and the same species of Jurassic ammonites found in the Jura Mountains of western Europe have also been found in Syria; at the mouths of the Indus River; in the archipelagos of Sunda and Molucca; and in New Guinea, Australia, Africa, and North and South America.

As a whole, the marine environment is essentially more uniform, more homogeneous, than the terrestrial environment, which falls into many tightly circumscribed climatic, faunal, and floral provinces, and where geographic isolation plays an important role. Just think of Australia, where the continental fauna is decidedly eccentric, whereas the organisms inhabiting the coastal waters differ very little from those found in other parts of the Indian Ocean.

Furthermore, the effects of climatic differences were clearly less pronounced in ancient seas, at least during the Paleozoic and the Mesozoic, than is the case today. Instead of the regional distribution of marine fauna, which we see in our own time, a more worldwide distribution was the rule in the past. This distributional pattern is largely related to the existence of the Tethys Sea, which persisted on into the late Tertiary; this was an extensive, probably globe-girdling "mediterranean" sea, which formed an east-west link between the oceans, providing an unobstructed westward equatorial current and establishing equable temperatures.

Furthermore, preservational conditions are considerably worse for terrestrial organisms; they are usually not buried immediately, as in oceans, but remain exposed to subatmospheric agents for long periods of time, subject to bacterial decay and the depredations of other animals, which may devour them, gnaw on them, or carry them off. The remains of freshwater animals are swept away by rivers; vertebrate skeletons become disarticulated and the elements are widely dispersed. All these reasons underlie the *superiority of marine invertebrates for establishing zones,* with the logical consequence that they also form the basis for the *delimitation of systems and eras.*

Plants and Vertebrates

On the other hand, it has been proposed recently that the boundaries of the larger units of time be based on evolution in the world of plants, with further time markers to be drawn from the evolutionary course of vertebrates. According to this system, the commonly used temporal units of Paleo*zoic,* Meso*zoic,* and Ceno*zoic,* which are based on the animal world, would basically be replaced by the eras of the Paleo*phytic,* Meso*phytic,* and Ceno*phytic,* which are conceived differently. The rationale is as follows: The development of the plant world is the first-ranking, central organic event, and that of the animal world, especially of the vertebrates, was intimately dependent upon it.

One may or may not believe that this is correct (it is undoubtedly a biased overstatement; for marine fauna, at least, there is no such dependency)—the necessity remains that we have to work with the classification of time in the most practical way, and plants and vertebrates do not offer this possibility. Where plants have accumulated as a consequence of special conditions and appear in a more or less uninterrupted sequence, as in the Upper Carboniferous, they are valuable in stratigraphy and in the identification of seams of coal. But these are only exceptions. In marine strata, which are the most important, terrestrial plants that were occasionally swept out to sea occur much too infrequently for a useful system of zonation to be based on them or—and this is more important—for any given stratum to be fitted into it.

The same argument applies to vertebrates. In particular depositional situations, they may also be used for chronological-stratigraphical purposes; the reptiles in the Permian and Triassic continental deposits of South Africa and the Tertiary mammals are examples. In general, however, the rarity and limited geographic distribution of such fossils precludes their usefulness in this regard. But to use the beginning of the individual groups of vertebrates as time markers would mean establishing the subdivisions of time on a basis that is all too uncertain and fluctuating. The dates for the first appearance of the various vertebrate classes and orders are usually based on scattered, individual specimens, and each new day could bring forth amplification and rearrangement.

For example, the physical remains of the oldest amphibians (figs. 2.29 and 2.30) have only recently been found in the Upper Devonian; they had previously been known only from the Carboniferous on. The order of the frogs had long been known only since the Upper Jurassic; recent finds have put their appearance back as far as the Lower Triassic. The coelacanthid group of fishes (from the larger unit of the crossopterygians) was thought to have died out at the end of the Cretaceous (cf. fig. 3.66), and it was one of the biggest surprises of recent years when a *living* (!) representative of this group was caught off the coast of South Africa (fig. 3.65).

Revolutionary finds of this sort are possible only with vertebrates, for only in exceptional instances do we have uninterrupted sequences of vertebrate faunas; most fossils occur sporadically and isolated, widely disparate in time and place. For vertebrates, the cliché about the incompleteness of the fossil record is quite justified, but when applied in the same sense to marine invertebrates, it must be emphatically rejected. The possibility that something like a living ammonite might be dredged up today—this would be exactly analogous to the recent catch of a coelacanth—can be completely ruled out. We can say with absolute certainty that ammonites became extinct at the end of the Cretaceous and more recent fossils are not anticipated. We are also certain that there will be no such surprises with regard to the other marine invertebrates used for time classification. That is why only they are able to provide a sound basis for the establishment of zones and stages. And since that is so, it is impossible to see why a different standard should be introduced for delimiting systems and eras, one that would replace the language now used to describe the components of systems.

Abiologic Events

On the other hand, there are scientists promoting the use of abiologic, geological events to delimit the larger units of time. Mountain building, the periodic heaving and subsiding of the earth's crust, the vast marine transgressions and regressions and the concomitant displacement of shorelines, or cycles of sedimentation, that is, the rhythmically recurring deposition of sequences of rock, should satisfy the objective because these events supposedly offer more precise, distinct time markers than organic evolution does. This trend is most widespread in North America and, in a certain sense, represents a return to the old petrostratigraphic way of doing things, which tried to exploit the observed effects of inorganic events in the very same way.

But how can dating by means of orogenies and transgressions be accomplished without the help of fossils? A fold is a fold, an unconformity an unconformity[8]—they have no inherent temporal character of their own. They look the same whether they formed during the Carboniferous or the Tertiary. Only when I have found a Carboniferous goniatite or a typical Tertiary mollusk fauna above the unconformity can I determine its age.

Likewise, there is nothing at all in the character of the rock in transgression conglomerates that hints at their age; their appearance is determined by local factors—the respective erosional and depositional planes. Not until something like a hoplite (a member of a typical ammonite group) is found do we learn that the transgression conglomerate belongs to a particular horizon of the Lower Cretaceous, whether it be in Europe, South America, North Africa, India, or anywhere else.

The dating of crustal movements, of the effects they have had on the rock, is, in every instance, then, derivative; the observed effects, in themselves, say nothing about the age of the rock and therefore cannot be directly of use as time

---

8. Unconformity: the topping of a tilted series of strata by another dissimilar one, in contrast to normal, parallel depositional layering. The feature arises as a result of orogeny and is an indicator of such.

markers. In light of this obvious circumstance, it is incomprehensible why the original paleontological-chronological classification, logically the only possible one, should be supplanted by another or translated into a different kind of formulation. This only makes sense, as we have already pointed out, for particular geological problems; as a generally authoritative, basic system for classification, however, divisions established with the help of geological events are unacceptable.

It is not in the least evident, furthermore, how this indirect approach could increase the exactitude of determining age and boundaries. If organic evolution is not supposed to yield unimpeachable time markers, then tectonic events, which are dated based on organic evolution, provide even less!

Most important, however, are the grave concerns that argue against establishing the universally valid system boundaries based on geologic events, inasmuch as such events never have worldwide scope and are not strictly contemporaneous. There is no such thing as a sudden, abrupt orogeny—a primary event happening all over the earth at once. The upthrusting of mountains always takes place along a relatively narrow zone and can be broken down into a series of individual phases, each of which may have a stronger effect in one area than in another; one area may reflect an earlier phase and another area a later one. Cycles of sedimentation also exhibit significant differences in individual depositional basins and nothing at all in the way of temporal concurrence.

By their very nature, extensive transgressions cannot be global in scope; they are countered by simultaneous marine regressions in other areas. In addition, they generally advance very slowly and migrate gradually from place to place, with the result that a transgression plane exhibits no temporal uniformity but rather cuts at an angle through various temporal units. Strictly speaking, then, a boundary based on such a feature is valid only for the initial profile.

It may be claimed that index fossils must also extend their range by migration and, therefore, that even matching fossil content cannot prove the strict correlation of the strata in question. We have already dealt with this argument, having established that migrations of the index forms can, for all practical purposes, be ignored because of their relatively very brief duration. In any event, this much is known, that these organisms spread much more rapidly within their unobstructed marine environment than a large-scale transgression would; and further, during this span of time they themselves evolve and are transformed. For it is the fossil sequence that allows us to trace the heterochronicity of transgression planes and the different times of their onset at any given place.

As justification for the use of inorganic time markers, it is usually claimed that organic evolution and the paleontological classification of time are ultimately determined and influenced by physical environmental factors, by crustal movements, displacement and changing patterns of habitats, and conditions of facies and sedimentation. Physical changes in the earth's crust would therefore be the primary events, with organic transformations contingent upon those factors.

There is obviously an intimate reciprocity between the organic world and its inanimate environment. Nevertheless, apart from local encroachments in a given region, which are completely irrelevant for the main flow of organic evolution, immediate effects on organic evolution of the processes we are talking about here are nowhere in evidence. We shall come back to these circumstances in our later observations on phylogeny.

Aside from the impracticability of adopting tectonic events as the means of drawing boundaries, there is no theoretical basis for doing so, either. Furthermore, the procedure should be rejected as illogical in that it mixes two different principles: that of a classification based purely on organic evolution and that of the dating of geologic events. It is an impermissible amalgamation of the standard of measurement and the object to be measured, of objectively independent entities, which, to avoid circular reasoning, must be kept well separated.

*   *   *

It has often been maintained, of course, that even when the divisions into zones and stages are arrived at paleontologically, the depositional situation and other geological factors play a decisive role. Particularly in those instances in which very discrete faunistic or floristic breaks occur, it should be inferred that there are gaps in the stratigraphic sequence—interruptions in sedimentation or destruction of strata already in existence.

There is no way, the argument continues, that we could infer from our profiles, which are to a large extent incomplete, the natural life spans of genera and species, for we have only segments of preserved floras and faunas to work with. Between them are periods of time during which the organisms must have persisted, but intermediate forms are not found in rocks where we might expect them, leaving the false impression that the floras and faunas in question are sharply discrete. My comment is this: The existence of such eventualities must indeed be reckoned with; nevertheless, it seems that their importance has been decidedly overstated. At the least, the existing sources of error can be eliminated by comparing a sufficiently large number of profiles.

We would like to consider, as an example of a series of stages composed of units that are clearly distinguishable by their faunas, the *cephalopod classification of the Upper Devonian* of the Rhenish Mountains as proposed by Rudolf Wedekind. There, atop the Middle Devonian *Anarcestes* stage (see fig. 2.84) lies the *Maenioceras* stage, characterized by the sudden massive appearance of the goniatite genus *Maenioceras* (fig. 2.85). (We recall that goniatites are representatives of the lower ammonoids, that is, the cephalopod stock. Since the names are not important here, it is unnecessary to describe the genera in question in more detail; the illustrations in the chart are sufficient to support the point being made.)

Atop the *Maenioceras* stage, again separated by an extremely sharp boundary, lies the *Manticoceras* stage, which has been designated as the beginning of the

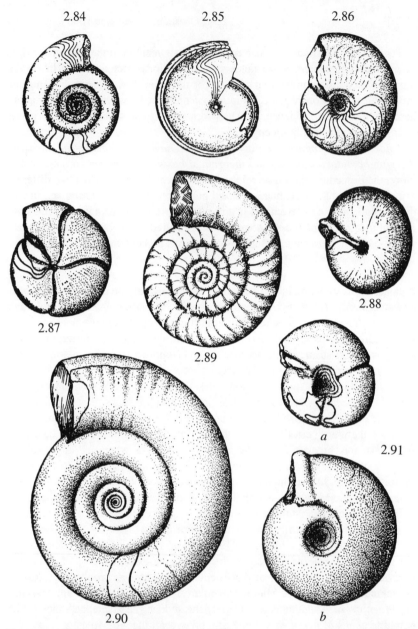

2.84   2.85   2.86

2.87

2.89

2.88

a

2.91

2.90   *b*

Figs. 2.84–92. Index ammonoids from the Middle Devonian, Upper Devonian, and earliest Lower Carboniferous; each species is limited to a particular horizon and abruptly replaces the species preceding it. All of the forms pictured came from the Rhenish Mountains but are found worldwide, exhibiting the same stratigraphic pattern everywhere.

2.84. *Anarcestes.* Lower Middle Devonian. 1 ×.

2.85. *Maenioceras.* Upper Middle Devonian. 1 ×.

2.86. *Manticoceras.* Upper Devonian, Stage I. 1½ ×.

Upper Devonian. The eponymous genus *Manticoceras* (fig. 2.86) appears abruptly and develops an abundance of species, which aid in establishing a finer degree of zonation. In the Rhenish profiles, this wealth of forms disappears at a clearly defined boundary, above which there appears on the scene suddenly, without any transition, a new goniatite fauna, one characterized by the genus *Cheiloceras* (fig. 2.87) and its allies. The *Cheiloceras* stage based on it is in turn superseded by the *Prolobites-Platyclymenia* stage, in which not a single species of *Cheiloceras* is to be found. This stage is characterized by the massive appearance—without any transition—of the peculiar genus *Prolobites* (fig. 2.88), with which the first representatives of the clymenids (fig. 2.89) are associated.

Although until that point the *goniatites* have been outstandingly useful as index fossils, serving as the foundation for the subdivision into stages and zones, they lose their utility in the later Upper Devonian. Their intensity of transformation diminishes sharply, not to resume until the Lower Carboniferous; in the later Upper Devonian, they are quiescent. During that period, only a few new forms are brought forth, and these are so long-lived that not much can be done with them stratigraphically. They must be replaced by other, more suitable forms.

A new group indeed turns up—the *clymenids,* also a cephalopod lineage—which quickly proliferates in a multiplicity of forms and dies out again soon after, at the boundary between the Devonian and the Carboniferous. It provides the index forms for two other late Upper Devonian units (figs. 2.90 and 2.91),

2.92

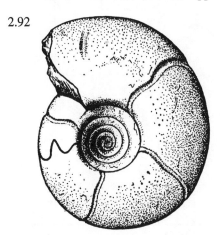

2.87. *Cheiloceras.* Upper Devonian, Stage II. 1½ ×.
2.88. *Prolobites.* Upper Devonian, Stage III. 1½ ×.
2.89. *Platyclymenia.* Upper Devonian, Stage III. 1½ ×.
2.90. *Orthoclymenia.* Upper Devonian, Stage IV. ¾ ×.
2.91. *Wocklumeria.* Upper Devonian, Stage V. *a.* Juvenile with triangular shell silhouette. 1½ ×.
b. Adult specimen with normal rounded spiral. 1 ×.
2.92. *Gattendorfia.* Lower Carboniferous, Stage I. 1 ×.

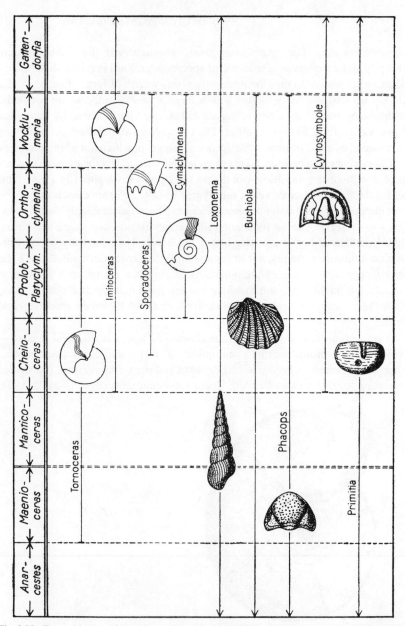

Fig. 2.93. Temporal range of some invertebrates in the Middle and Upper Devonian and the earliest Lower Carboniferous. In the column on the left, the index genera of the individual stages are indicated. Typical representatives of all these genera are shown in figures 2.84–92. The next columns show the vertical ranges of the goniatite genera *Tornoceras, Imitoceras,* and *Sporadoceras,* and the clymenian *Cymaclymenia.* The genus *Prionoceras* has almost the same stratigraphic range as *Imitoceras.* Other clymenians of similar longevity to *Cymaclymenia* are *Cyrtoclymenia* and *Oxyclymenia.* The difference between the short life span of the stage genera and the relatively long one of

Continued

the *Orthoclymenia* and the *Wocklumeria* stages, which are sharply distinct from each other and from the previous *Prolobites* stage and the subsequent *Gattendorfia* stage of the Lower Carboniferous (fig. 2.92).

It should be particularly emphasized that the abrupt faunistic transformation just described takes place within one invariable facies and that, further, the individual faunas did not occupy different habitats. Nevertheless, because of the large jumps—unmediated by any connecting links—between the sequentially appearing index fossils, it has been suggested that the type sections for the aforementioned system of divisions may be incomplete.

Later experience, however, has shown that this Devonian sequence is valid not only for the Rhenish Mountains, but has been corroborated wherever the corresponding strata appear or have been more closely studied, for example, in the Harz Mountains, in the Upper Franconian–East Thuringian area, in the Sudeten, the Carnic Alps, the Pyrenees, in Brittany, in southern France, the Polish Holy Cross Mountains, in the Urals, in Kazakhstan, Novaya Zemlya, North Africa (French Morocco), western Australia, and further, with regard to some of the stages, also in North America and China (where the rest of the stratigraphic units are developed as a different facies).

In all the areas named, we find the same sequence of cephalopods and the same distinct boundaries as described above. Nowhere, however, have horizons with transitional fauna, which might be intercalated between the stages named, been discovered. Since it cannot be seriously assumed that the postulated gaps appear everywhere in the world where the sequence is found at exactly the same levels in the profile—this eventuality is completely excluded by the very different local geological situations in which the individual occurrences are found—

---

the rest of the ammonoids named is extremely large and noteworthy. This temporal pattern becomeseven more striking if the individual species are taken into consideration, which we are unfortunately unable to do in the chart, owing to lack of space. The species of the index genera are limited to portions of the individual stages and form the zone fossils for the finer divisions. Among the other ammonoids, however, are certain species (for example, *Tornoceras simplex, Sporadoceras posthumum, Cymaclymenia striata, Oxyclymenia sedgwicki, Cyrtoclymenia angustiseptata,* and so on) that not only extend over different zones but even over several stages, demonstrating the continuity of layers that are otherwise sharply delimited based on their index forms. The observable steady, gradual, step-by-step transformations of these genera and species prove that there can be no appreciable gaps in the horizons that would account for the abrupt modifications of the index forms and for the distinctness of the divisions between the zones and stages.

The other invertebrates are even more long-lived. The snail *Loxonema* extends from the Ordovician on into the Upper Triassic; the bivalve *Buchiola,* from the Gothlandian on through the entire Devonian; the trilobite *Phacops* (indicated in the chart by just the head, as is the other trilobite genus *Cyrtosymbole*) also has a life span extending from the Gothlandian to the end of the Devonian; the ostrocode genus *Primitia* runs from the Ordovician to the Permian. Among these, there are species (for example, *Buchiola palmata* and *B. retrostriata*) that are distributed throughout the entire Upper Devonian and many species of ostrocodes that are found throughout a large number of stages of the Upper Devonian.

we are forced to conclude that the profile of our Rhenish Mountain model, the subdivision of which is based on cephalopods, is basically complete. *The sharp, abrupt modification of forms at the boundaries of the individual stages is determined purely biologically by the ammonoids' high rate of evolution and is not an artifact produced by gaps in the stratigraphic record.*

The abruptness and lack of transition, however, applies only to the index genera and species selected for classification. There are other genera of goniatites and clymenians that transform less rapidly and, correspondingly, extend throughout several stages. This is particularly true of the fauna associated with the ammonoids—for the long-lived clams and snails, the corals, and the late Devonian trilobites, which changed less rapidly than their predecessors (fig. 2.93). These faunal elements are used to establish the continuity of the individual stages and zones and, moreover, to prove it. The observations described are of great importance for assessing the so-called gaps in the fossil record and for understanding the sudden, unmediated appearance of new types of forms; these problems will be taken up further along in the book.

It seems, at least with the cephalopod-based stratigraphy, that the situation is basically the same everywhere. Even the ammonite stages of the Jurassic and the Triassic show these distinct, abrupt boundaries, and they, too, are in evidence worldwide.

We come to the conclusion, then, that the various objections to the theoretical basis for classifying time according to paleontology and to the practical implementation of this method are not sound. The irrefutable fact remains that *only fossils as elements of an autonomous, directed, and irreversible course of evolution can provide a serviceable time scale for ranking geological events* and that, with regard to chronology, they are basically superior to periodically recurring inorganic processes.

This assertion, mind you, is related to the working out of the general *chronology* of prehistory. The fact that applied *stratigraphy,* used in field mapping by geologists, for example, is predominantly based on local petrographic criteria, is a thoroughly different matter. This type of application, tailored to local conditions, should not be confused (although it often is) with the basic methodology of dividing time purely on paleontological grounds. In principle, this kind of applied stratigraphy is also based on fossils, since the rock units shown on the maps are not just petrographic entities but rather rocks of a *particular age,* and the determination of the age is based on fossil content observed at the locality or elsewhere.

3   Basic Questions in Organic Evolution

As we explained briefly in the first, introductory, chapter, paleontology is of great, even critical, importance for problems of phylogeny, and by extension, for the larger issue of the theory of organic evolution. Only paleontology has command of dated prehistoric documentation, which alone can substantiate evolution over time and explain the courses it has taken. Therefore, the zoologist J. Schaxel (1922, p. 49), who has contributed much to the clarification of biological theories, was correct when he said that one would expect careful evolutionary research to be grounded in paleontology. "The course of evolutionary history will be read directly from [paleontological] data, and this is the only way in which the evidence of past life becomes relevant to the present."

But in reality, this is not the way things developed historically. Because paleontology was still a relatively undeveloped field when the theory of evolutionary descent first appeared, leadership was at first completely in the hands of the biology of living organisms. However, according to Schaxel's certainly candid point of view, biology [in the sense of neontology], has repeatedly violated paleontology "by not simply allowing the historical record to speak for itself but rather requiring it to confirm circumstantial morphological evidence and offer proof of the phylogenetic stages" postulated by biology.

The concept of evolution as such is an ancient one. The Greek philosophers Anaximander and Empedocles probably saw some such principle operating in the organic world, although hardly in the sense in which we understand it today. Furthermore, at the beginning of the nineteenth century, long before Charles Darwin (1859), to whom we usually ascribe authorship, the concept was given scientific expression by Jean-Baptiste de Lamarck (1802, 1809, 1815);[1] and in Germany, too, all kinds of promising beginnings are found. And yet, especially at first, the concept had no effect at all on paleontology or biology.

The decisive breakthrough did not come until Darwin made his appearance. The triumphal march of the theory of evolutionary descent was under way, sustained at the outset almost exclusively by zoologists, anatomists, and botanists. Only later did it find its way into paleontology, and at first, efforts were limited

1. C. F. Kielmeyer (1793); G. R. Treviranus (1805, 1831); F. Tiedemann (1808–14); J. F. Meckel (1815, 1821); F. S. Voigt (1817); A. M. Tauscher (1818); J. G. J. Ballenstedt (1818); and others.

to applying the new insights of modern biology indiscriminately to the fossil material.

As time went on, paleontology was successful in contributing countless important pieces of evidence to support the theory of evolution in general. On the other hand, as the evidence accumulated, it led to many observations that did not fit the concepts derived from living plants and animals. Only recently have paleontologists reflected on the special, unique quality of their material and tried to interpret it in an unbiased way, uninfluenced by other disciplines. In so doing, they have come up with certain ideas that contradict the interpretations of the now-flourishing science of genetics,[2] and as a consequence, paleontology has been subjected to all kinds of chilly rejection.

Two objections in particular are raised again and again that are said to restrict the evidential value of paleontological data:

1. the assertion that paleontologists have available to them only the mineralized hard parts of extinct organisms, which are not adequate for making phylogenetic assertions and are not sufficiently unambiguous; and
2. the cliché about the gaps in the fossil record has been used over and over to discredit fossil material and has become ingrained.

We first want to take up these two questions of the *qualitative and quantitative preservation of fossils,* since the ability of paleontological evidence to hold up and to be regarded as trustworthy rests to a large extent on clearing up these issues in a satisfactory way.

THE AUTHENTICITY OF THE FOSSIL RECORD

THE PRESERVATIONAL STATE OF FOSSILS

It is undoubtedly a distinct disadvantage for the paleontologist that the material he investigates seldom includes a whole animal with its soft parts—that he must be satisfied with whatever remnants the conditions of fossilization (which may be more or less favorable) have left behind. But does this also mean, despite the indisputable advantage inherent in the historical character of fossils and in their clear-cut chronological arrangement, that they are to be considered of little value in reconstructing the course of evolution?

Now, upon closer, unbiased inspection, it turns out that the truth of the matter is not nearly as unfavorable as it might appear at first glance. In many instances, paleontological material has been *preserved in an astonishingly complete and*

---

2. "The more evidence brought to light by paleontology, the sharper will be the conflict between the view of the history of organisms based on evidence and the one asserted by biology." This sentence was written by a *biologist*—J. Schaxel (1922, p. 50), whom I have already cited several times—which lends it particular weight!

detailed state, offering extensive views of the structure of soft parts, too, even though these were not or are no longer preserved in their original condition.

Vertebrates

This is true primarily for vertebrates. We mention as an example the brilliant investigations of *cephalaspids*—ancient, lower agnathans (jawless fishes)— carried out by E. Stensiö (see fig. 4.3). He was able to reconstruct the brain and the circulatory and nervous system of the head in every detail, down to the most delicate branchings, and to draw conclusions as to the presence of electrical fields and other extremely interesting details (pl. 10B; pl. 11). A penetrating anatomical analysis using modern procedures (serial sections and wax models) made it possible to reconstruct the soft parts of these Paleozoic organisms in almost as much detail as one would find in a living animal.

As a result, morphological-phylogenetic comparisons could be made on an absolutely reliable basis between cephalaspids and their closest living relatives, the petromyzonts (lampreys, fig. 4.4). Other Paleozoic agnathans and fishes, Me-

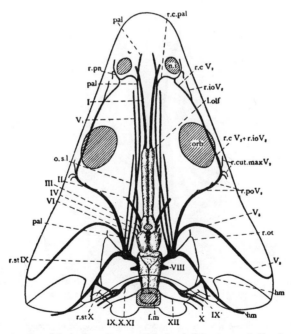

Fig. 3.1. Reconstruction of the brain and nerves of the head of the stegocephalian *Lyrocephalus euri* Wim., from the Triassic of Spitzbergen. About ½ ×. The brain case of this species is shown in figure 2.45. The Roman numerals and other symbols indicate nerves I–XII with their various branches. (After G. Säve-Söderbergh.)

sozoic stegocephalians (fig. 3.1), and so on, have yielded similarly outstanding results; and the extent to which the anatomy of former soft parts can be reconstructed is often more a question of methodology and hard work than of preservation.

In the *mammalian* orders, dentition and skeleton form the main basis of systematics and phylogeny. For fossil mammals, these elements are present in more or less complete states of preservation, and where individual parts are missing, sufficiently reliable conclusions can be drawn based on the principle of correlation, by referring to the corresponding parts of living relatives.

Furthermore, very important ideas about the structure and form of soft parts can be gained from the skeleton. Steinkerns, or casts, of the cranial cavity provide clues as to shape, size, and organization of the brain (fig. 3.2). Muscles, ligaments, and cartilage have left indications of their points of attachment on the individual bones; from this evidence the structure and extent of musculature can be deduced with a great degree of probability. This kind of methodologically unexceptionable reconstruction has been carried out in an exemplary manner by, among others, W. K. Gregory and H. F. Osborn with titanotheres, early Tertiary

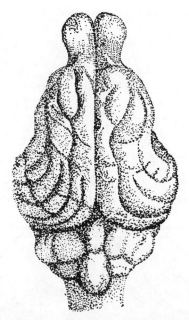

Fig. 3.2. Brain of *Amphicyon geoffroyi,* a Tertiary canid (Oligocene or Miocene) of France. ¾ ×. Like all other Tertiary canids, the species illustrated shows a smaller, distinctly narrower forebrain than is found in Recent forms. Nonetheless, the grooves of the cerebrum (four convolutions on each of the two hemispheres) correspond completely to those in the brains of modern dogs. (After P. Gervais, from Tilla Edinger.)

ungulates of North America. These two scientists produced reconstructions that provide the basis for broad conclusions even about the physiognomy of these extinct animals (pl. 12A and B).

That in these forms, as in almost all other instances, the actual muscles, skin, and blood vessels with their contents have not been preserved can hardly be considered a drawback, for no one doubts that extinct mammals were covered with skin just like their modern counterparts, that blood flowed in their veins, and that their musculature was structured in a similar way and had the same functions. We know this well enough from a few lucky vertebrate fossils that are exceptionally well preserved, if such proof were really needed. But for taxonomy, for the incorporation of the form into the system, these characters mean almost nothing.

Moreover, when circumstances are favorable, the soft parts of fossil mammals and other vertebrates are present in as complete a form as one could wish. Most people are familiar with the Pleistocene mammoths from the Siberian ice, which have been preserved as if in a natural icehouse, with skin and hair intact. Their soft parts have maintained a freshness such that even blood tests could be performed, which gave evidence that mammoths were close blood relatives of Indian elephants. Further, we would like to refer to the valuable find of woolly rhinoceroses, also from the Ice Age, at Starunia, in the eastern Carpathians, south of Lemberg; the hide of this animal could be prepared and mounted just like the hide of a Recent mammal (pl. 13A). In this instance, the complete preservation of the soft parts was due to the conservational properties of salt water and petroleum—in other words, the animal had been pickled naturally. These two finds are of particular importance in that they teach us that the animals, related to living inhabitants of the tropics, were admirably adapted to the Ice Age climate by virtue of their thick coats and heavy fat deposits.

However, there are also vertebrates of *much greater geological age* that sometimes show excellent preservation of the soft parts: O. Reis was the first to carry out histological examination of the well-preserved muscle tissue of fish from the Jurassic Solnhofen Limestone. Similarly propitious fish fossils occur in the Upper Cretaceous of Lebanon, and B. Dean was even able to prove the existence of transversely striated muscle in a shark from the Devonian of North America.

The mummified corpses of trachodons, members of a genus (*Trachodon*) of large reptiles from the Upper Cretaceous of North America, are famous for the preservation of their skin and sometimes even of the stomach contents, which allows the last meal to be reconstructed. Fragments of the hides of the giant sloth of Patagonia (*Glossotherium*) are available for examination. For ichthyosaurs from the Liassic Posidonia Shale of Holzmaden, in Württemberg, soft parts and skin are preserved at the least as impressions or outlines; and for the Jurassic pterosaurs of southern Germany, important information about bodily form, the shape of the fins or wings, the tail, and so on is yielded.

Further, the *fish* in the Lower Tertiary lignite of the Geiseltal, near Halle am Saale [Germany] are so excellently preserved that E. Voigt, in an exemplary study, was able to elucidate not only the osteology but also the soft parts and their histology. The musculature showed outstanding preservation of the transverse striations and the fibrils. Periosteum (connective tissue) could be detected and studied; in numerous instances, pigmentation of various kinds (pl. 13B) has been preserved, the actual pigments still present, apparently in their original, unaltered form (lipochrome and melanin). Voigt could say with justified satisfaction that the fossil freshwater fishes he had described "were in many respects better understood than many living species."

These forms also provided valuable clues as to the mode of life and environment: the age of individuals could be determined based on the rhythmic growth of the otoliths ("ear stones"); stomach contents allowed inferences to be drawn as to diet; the presence of the remains of those fish in crocodile coprolites (fossilized excrement) told what the enemies were, and so on. In short, for these forms, a well-rounded picture of their morphology, anatomy, and histology, and of the environmental conditions under which they lived was obtained.

Further, Voigt (1935, 1937) was able to report on preserved soft parts of amphibians, reptiles, and mammals from the rich Geiseltal locality: on epithelial cells with nuclei (pl. 14A), chromatophores (pigment cells, pl. 14B), transversely striated muscle tissue (pl. 15A), cartilage (pl. 15B), fat cells, blood vessels with erythrocytes (red blood cells) preserved, the pelts of various mammals, and on and on. With regard to taxonomy, as already mentioned, these interesting discoveries are of no overriding importance, yet they are not simply "curiosities," an opinion recently voiced. Through evidence of this kind, we have finally come to full knowledge of just how much of an organism can be passed down under favorable conditions of preservation. *The basic feasibility of carrying out histological investigation of fossil material is thereby established;* this supposed barrier, said to have separated the research capabilities of paleontology from those of zoology, has fallen.

Invertebrates

For invertebrates, too, there is much evidence of soft parts in an excellent state of preservation; for example, E. Voigt (1938) has also described Tertiary insects (pl. 16A) and worms. As in the case of Tertiary vertebrates, the discoveries made based on this invertebrate evidence correspond in general quite closely to what one might expect from Recent relatives and provide, further, interesting specific information (see the captions to pl. 14B and pl. 16A).

Moreover, they offer the prospect—using methods and special technological procedures devised for the investigation of soft parts—of successful determination of the zoological classification of extinct groups, heretofore impossible owing to the lack of living representatives to use for comparison. The prerequisites are all there, for as we have already seen for vertebrates, the preservation of soft parts is not limited just to the most recent formations from the Tertiary and the Quaternary; we also have evidence from considerably older eras.

As examples from the invertebrate realm, there are the squids from the Lithographic Limestone and the Liassic Posidonia Shale of southern Germany and the starfish and arthropods from the Lower Devonian Hunsrück Shale, at Bundenbach (figs. 2.24, 3.8–10, 3.12). We refer further to the fossils from the ancient Cambrian Burgess Shale of British Columbia (pl. 3A, B, C; pl. 4A, C, E, G; pl. 5A; fig. 3.7), so beautifully preserved in all their detail; there, it is precisely the outstanding preservation of soft parts that allows us to realize that we are in the presence of an extremely rich invertebrate fauna.

In one particular instance, it was the consideration and careful preparation of soft parts that resulted in an important success: a group of animals that no one had been able to clarify was finally placed in its proper taxonomic position. We are speaking of the *graptolites* (figs. 2.9–14), which had been moved around incessantly within the system, classified at one time or another with the hydrozoans, the alcyonarians, and the bryozoans. They never seemed to fit in properly anywhere until the detailed investigations of Kozlowski provided incontrovertible evidence that graptolites are to be placed with the pterobranchs, a group of enteropneustas ("acorn worms"). It is evident that the examination of soft parts is not basically off limits for paleontologists. And so, we can well entertain the hope that some well-preserved specimen will eventually clear up other faunal groups such as the conularians, the hyoliths, and so on, which thus far have eluded an indisputable interpretation.

There is still a general tendency to grant vertebrates a special importance when it comes to questions of phylogeny and to supporting other kinds of biological assertions. Invertebrates, in contrast, are widely mistrusted for these purposes. Those holding this view stress that what is usually preserved of them, the hard parts, only represents the external skeleton, and they believe that such remains provide insufficient clues for the understanding of internal organization and offer too few diagnostically useful characters.

There are even some vertebrate paleontologists who stick to the presumptuous view that paleontological theory can only be based on vertebrates and that invertebrates provide only unreliable, insufficiently convincing conclusions. This attitude must be decisively rebutted, for it endeavors to exclude unfairly a large field of research, which, because of its superiority to vertebrate research in other respects, is at least its equal when it comes to purely biological matters.

Brachiopods

It is certainly true that a fossil brachiopod shell (see figs. 2.6 and 2.7 for some examples) offers us only the external skeleton; but what an abundance of clues to internal structure it provides! First, we get the external form, the shape of the umbo, the interarea, and the anterior margin, the folding or corrugation of the two valves into furrows and ridges, the structure of the shell, its sculpture, and the growth lines, which indicate possible changes in form during ontogeny.

From the presence or lack of a pedicle opening, we can infer the size and shape of the pedicle, which has no hard parts. The closure of the pedicle opening by plates of the most various kinds provides valuable clues for assessing kinship. The same is true for the hinge and articulating structure, with its teeth, dental sockets, and complex brachial supports—the hinge plates, medial septa, the spondylium, septalium, and all sorts of other special structures.

The internal surface of the valves (fig. 3.3), or the surface of the cast formed from them, shows the points of attachment and the impressions of muscles: the position, size, and shape of the pedicle, adductor (closing), and divaricator (opening) muscles can be seen, as can blood vessels from the mantle, the genital chords, and under certain circumstances, even the brachia. The brachidia (brachial supports), which support the fleshy lophophores and are so important for taxonomy, have been preserved and are either immediately accessible naturally or can be made visible through suitable methods of preparation (fig. 3.4). The great complexity of these structures provides reliable characters for the clarification of kinship and phylogenetic relationships.

In short, from the hard parts available to us, from the indications just listed and from many others, we are able to obtain a very complete picture of the

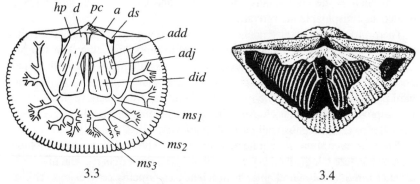

3.3   3.4

Figs. 3.3–4. Internal structural elements of the brachiopod shell.
3.3. Inner side of the ventral valve of *Dinorthis* (*Plaesiomys*) *subquadrata* (Hall). Upper Ordovician of North America. *hp* = hinge plate; *d* = delthyrium; *pc* = pedicle callosity; *a* = area; *ds* = dental socket; *add* = adductor (closing muscle) scar; *adj* = adjustor (pedicle muscle); *did* = diductor (opening muscle); ms_1, ms_2, ms_3 = primary, secondary, and tertiary mantle sinus, folds in the mantle that contain the blood vessels and some of the sex organs. (After C. Schuchert and G. A. Cooper.)
3.4. *Spirifer striatus* Sow. Lower Carboniferous of Ireland. Dorsal view of the shell. The dorsal valve is broken away to show the calcareous brachial supports (helicopegmate or spiriferoid type) inside the hollow shell. (After Th. Davidson.)

structure of these forms. Not that every single individual would have all of these characters in a good state of preservation—but what one specimen or one state of preservation does not reveal, another will. With these forms, as with almost all invertebrates, there is the great advantage of having an abundance of material available for study.

Furthermore, in this case we are fortunate in that there are living representatives of the stock to which we can refer for comparison. By beginning with the forms that are most closely related to Recent forms and proceeding step by step to the types that have undergone pronounced transformation, we also arrive at an incontestable interpretation of forms that are structurally completely different from modern ones.

Thus, when the paleontologist carries out evolutionary investigation of brachiopods and deduces general patterns from the results, he is on thoroughly solid ground. Nevertheless, he need not fear comparisons with Recent brachiopods. The systematics of Recent brachiopods is still, at least for the present, based primarily on hard parts, which is all the evidence we have for most forms. According to Helmcke, of the approximately 260 Recent species, only 5 have thus far been studied anatomically!

On the other hand, what an extraordinary enrichment of our knowledge of the total multiplicity of forms and of the various basic types brought forth by this class, only a few of which have survived to the present, is offered by the consideration of fossil brachiopods! Today, of the over 1,200 genera and subgenera known for brachiopods, only 65 are still in existence. And in the future, the number of genera of fossil brachiopods will continue to rise, making the ratio of fossil to Recent forms even more one-sided.

This increase through fossil material is, moreover, not just *quantitative,* but also, and primarily, *qualitative.* Entire orders and suborders, such as the Palaeotremata, Orthacea, Clitambonacea, Pentameracea, Atrypacea, Spiriferacea, and Rostrospiracea, each with many families and genera—and very remarkable structural peculiarities—have become extinct. In contrast, in today's oceans, only the Rhychonellacea and the Terebratulacea are present with any abundance of forms. We really must ask ourselves, then, with regard to evolutionary deductions, where are the greater gaps in knowledge of the subject matter—in paleontology or in zoology?

Arthropods

Let us look further, at the arthropods. The systematics of Recent forms rests primarily on characters of the exoskeleton and the organs closely connected with it; the characteristics of these structures are basically accessible to paleontologists, too. Here again, a comparison with the research methods of zoologists shows no unbridged gulf of any sort.

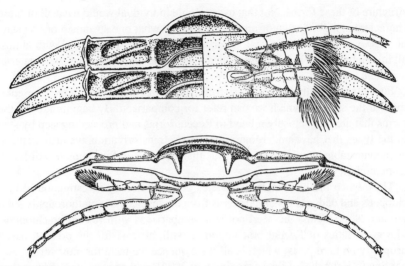

Fig. 3.5. Enlarged reconstruction of two body segments of the trilobite *Ceraurus pleurexanthemus* Green. Middle Ordovician of Trenton Falls, New York. Top: Ventral view. On the left side of the drawing, the ventral portions of the tergites have been cut away to show the inside of the upper surface. On the right, the biramous limbs can be seen, consisting of walking legs (telopodites; left off in the lower segment) and filamentous gill branches (praepipodite). Bottom: Frontal view. (After L. Störmer.)

The reconstruction is based on serial sections, glass and wax plate models, with the help of which L. Störmer, laboriously but successfully, was able to clarify all the structural elements. The results show that the limbs of a trilobite are not the extremities of a crustacean but rather the ancestral type of chelicerate limbs.

We might refer, for example, to the careful investigations of Leif Störmer, who was able to clarify down to the finest details the complex structure of the head and appendages, the gill branches, and so on, of *trilobites* (fig. 3.5). These observations and the results of the zoologist P. Schulze, who studied Recent comparative material, have furnished proof that the trilobite phylum, which became extinct way back in the Paleozoic and has no immediate Recent relatives, is connected to the chelicerates, to spiders. Furthermore, in several instances, the developmental history of the trilobite from the egg through the various larval stages to the adult carapace can be followed (fig. 3.6).

In *insects,* too, the exoskeleton is of primary importance, and paleontological finds fit in completely with zoological principles of classification. Some fossil forms, often magnificently preserved with unparalleled fineness of detail, show the veination of the wings (figs. 2.36, 3.139); the extremities and oral appendages can be studied, the tracheae are often recognizable, the phylogeny of the stridulatory organs (used for chirping) in crickets can be followed, and on and on.

Purely histological characteristics of the soft parts have no part at all in the morphology of insects or of arthropods as a whole. But even in this area, there are exceptional circumstances in which paleontologists, too, can make discoveries. Thus, Voigt (1938) was able to detect many details of the tracheal system, musculature (pl. 16A), glandular tissue, gut, sex organs, and so on in Tertiary insects from the Geiseltal lignite and the Samland [Baltic] amber.

In those instances in which characters of the internal structure of Recent insects had occasionally been used for taxonomic purposes, there was, remarkably, broad agreement with the systematics that had been based purely on morphology. For example, H. E. Hinton recently investigated for the first time the internal anatomy of the Mexican water beetle, from the family Elmidae. His findings are almost completely in agreement with the current classification based on morphology. In only two instances did new genera have to be proposed based on particular anatomical peculiarities; yet even there, external differences could

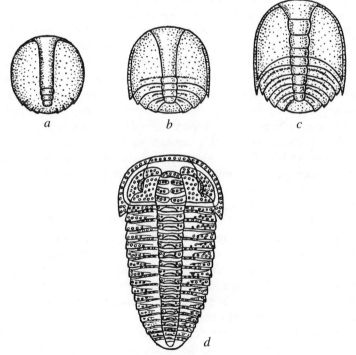

Fig. 3.6. Some stages in the ontogeny of the trilobite *Sao hirsuta* Barr. Middle Cambrian of Bohemia. *a.* Anaprotaspis stage. 30 ×. *b.* Metaprotaspis stage. 30 ×. *c.* Paraprotaspis stage. 30 ×. *d.* Mature stage. 1 ×. (After J. Barrande and C. E. Beecher.) For many trilobites, up to thirty molting stages are known, which exhibit the main stages of ontogeny—protaspis, meraspis, and holaspis, in uninterrupted sequence.

be ascertained that paralleled the internal ones, ensuring a morphological identification of the respective taxonomic units. Such experiences offer impressive proof that morphological clues provide a reliable basis for setting up the natural system and the phylogeny.

However, the organs of copulation and the mouth parts, of such particular importance for the details of zoological insect systematics, have not been preserved in the fossil record. And yet, this lack does not weigh too heavily when it comes to working out the larger taxonomic and phylogenetic units. Here and in similar situations, supplementary investigations of characters that are also available in fossils must be carried out on Recent material; this will aid the paleontologist and benefit zoology as well, by contributing to an understanding of living forms.

Moreover, even in old Paleozoic *eurypterids,* genital appendages have been preserved in excellent condition and can be investigated, as can jaw structures, remains of integument with fine sensory hairs, bristles, pores, the attachment points of muscles, and so on, with the result that many of these forms can be described in the same, complete way as Recent arthropods.

In the case of *crustaceans,* we have already discussed the genus *Burgessia,* from the Middle Cambrian Burgess Shale, in British Columbia, in another context (pl. 4C). Superbly preserved specimens show the investigator not only the general shape and articulation of the body but also delicate appendages and numerous internal structural features; consequently, the animal can be assigned an incontrovertible place in the system. This famous locality has yielded many other crustaceans, members of both the lower and higher groups (entomostracans and malacostracans), with equally outstanding preservation of every detail. There is evidence here that many of the taxa living today were already present in that far distant past, and conversely, sharply divergent branches were present that have since become extinct.

Let us look at the genus *Marella,* for which there is abundant fossil material. The species shown in figure 3.7 is a swimming form of most delicate structure, yet details can be seen as clearly as one could wish. The head shows an extended shield with long processes directed backward and four pairs of appendages. Each of the next twenty-four segments bears two appendages; in our drawing, the inner, walking legs (endopodites) are covered by the outer, swimming branches (exopodites) with their fringe of gills. These appendages are segmented in a very primitive way, having not yet transformed to branchiopods, as is the case with more recent representatives of their closest relatives, the phyllopods. Also primitive is the lack of differentiation in the thorax and the limbless abdomen; these changes had already taken place in another crustacean of the same age, *Burgessia* (order Phyllopoda, suborder Notostraca). Consequently, *Marella* is contrasted with the Notostraca by being given a suborder of its own, Marrellida; alternatively, many authors have included it with a few related forms and place

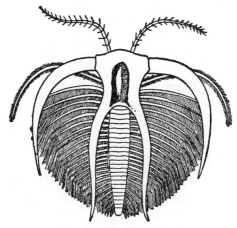

Fig. 3.7. *Marrella splendens* Walc. Middle Cambrian of British Columbia [Burgess Shale]. Dorsal view. 3 ×. (From R. Richter.)

them all in an order called Marrellomorpha, which is on the same level as the phyllopods and malacostracans. This group is probably part of the stock leading to the higher crustaceans, or at least is close to it,

In addition to the notostracans with a cephalothorax, there is also evidence as early as the Middle Cambrian of the suborder Anostraca, which has no cephalothorax. These forms resemble modern *Branchipus* in shape and also already have the same branchiopoda. Furthermore, it is noteworthy that all these ancient types were marine, whereas their modern descendants or relatives live in fresh or brackish water.

Similarly, the Lower Devonian roof slates of the Hunsrück area (Bundenbach and elsewhere), already mentioned several times, have yielded many arthropods with body appendages and details of structure in an unsurpassed state of preservation. The structural types of crustaceans and spiders we know of from later systems and modern times have been considerably augmented by these earlier forms. *Cheloniellon* (fig. 3.8), for example, is a form of such peculiarity that a separate subclass of the Cheloniellida had to be set up for it.

It is further noteworthy that the genus *Nahecaris* (fig. 3.9), of which there are many specimens, included the largest representatives of the order Phyllocarida. Not only the more general organization—segmentation, the structure of the limbs, the stalked eyes, and so on—but also even such particulars as the sensory bristles on the antennae, the fringing on the legs and furca, and other structures can be seen clearly in every detail. Within the crustaceans, the genus stands alone in that *both* antennae, the antennules as well as the antennae, bear two ringed flagella each. *Nahecaris* occupies a group by itself (Nahecarida) within the Phyllocarida, the order that represents the roots of the higher crustaceans.

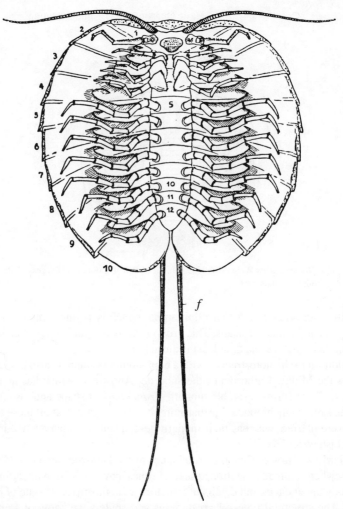

Fig. 3.8. *Cheloniellon calmani* Broili. Lower Devonian roof slate of Bundenbach, in Hunsrück (Rhenish Mountains). Ventral view. ⅗ ×. (From F. Broili.) In this illustration, almost nothing has been reconstructed; the details shown have *actually been preserved* and observed! The only changes are that the antennules, legs, and the furca (*f*), consisting of two long whips, all somewhat displaced before burial through the action of bottom water currents, have been depicted in their original symmetry. Numbers 1–3 refer to the "segments" that form the head: 1 = the first "segment," with uniramous antennules, antennae (with antennal glands at the bases), and labrum (upper lip, epistom), formed from the fusion of three original metameres; 2 = second "segment," with mandibula and maxillula, originally developed from two separate metameres; 3 = third "segment," with maxilla, formed from a single metamere. Numbers 4–10 indicate thoracic segments: 4 = body segment with the maxillipeds, the proximal segments of which are each enlarged to form a leaf-shaped coxa (gnathobase) with a serrate edge; 5–10 indicate body segments with six identical pairs of biramous appendages, each of which consists of a walking leg and a gill-bearing swimming leg. Numbers 11 and 12 are two abdominal segments with the two most posterior limbs.

In all other crustaceans, the head shield is unsegmented, having originated through the fusing of

Continued

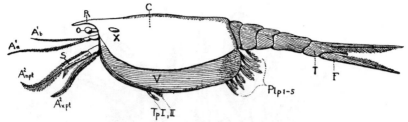

Fig. 3.9. *Nahecaris stürtzi* Jkl. Lower Devonian roof slate of Bundenbach, in Hunsrück. Lateral view. ⅔ ×. (After F. Broili.) The univalved carapace has a spinelike process (rostrum) at the forward end. Beneath it and to each side lies a stalked eye. Then come the two antennae (antennula and antenna), each with two flagella. Beneath the velum of the carapace, two of the eight thoracopodes (thoracic legs) protrude; at the posterior end of the velum, there are five biramous pleopodes (abdominal legs). The eight abdominal segments terminate in a spine (telson) and a sturdy forked tail (furca), which probably served as a steering mechanism during swimming.

Further, the curious group of sea spiders (pantopodes or pycnogonids) is found in the Bundenbach Shale represented by *Palaeoisopus* (fig. 3.10) and *Palaeopantopus* (fig. 3.12); these forms combine many of the characteristics of crustaceans with those of arachnoids (spiders). Apart from all external body characters, there is evidence of such peculiarities of internal structure as the penetration of the gut diverticulae into the legs, which is analogous to the situation in Recent forms. For the rest, these forms, too, augment considerably the morphological and phylogenetic information that can be drawn from their living descendants.

Echinoderms

The situation with echinoderms is as favorable as that with arthropods. For echinoderms, too, the skeleton offers an abundance of characterizing features, most of which bear a close physiological relation to the organization of the soft parts. The echinoderm test provides information about the structure of the fundamentally important ambulacral water-vascular system, the course of the circulatory and nervous systems, the position and shape of the gut, the oral and anal openings, the genital organs, and many other particular traits (fig. 3.13).

The diagnostic hydrospires of the blastoids and the taxonomically important

six simple metameres. In *Cheloniellon,* however, the oral appendages are divided among *three* movable "segments," which, as stated above, formed through the fusion of six individual metameres. Because of this segmentation of the head into three unequal segments, *Cheloniellon* has a very primitive character and occupies a completely isolated place within the lower crustaceans; this caused Broili to establish a subclass for it alone called Cheloniellida, which is in addition to the entomostracans and the malacostracans. More recent research, however, suggests that the cheloniellids should be classified among groups of the spiders (chelicerates).

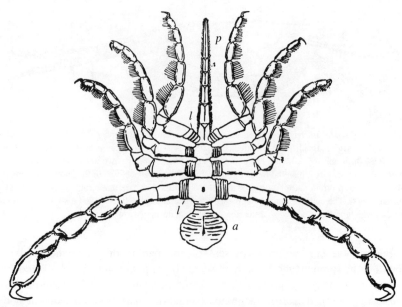

Fig. 3.10. *Palaeoisopus problematicus* Broili. Lower Devonian roof slate of Bundenbach. Ventral view. ½ ×. (From F. Broili.) *p* = proboscis; in the resting position, it was directed down and backward at an angle, extending somewhat beyond the first two segments. Numbers 1 and 8 indicate respectively the most anterior and posterior of the eight body segments. Of these, the first four are thickly edged with bristles but have no appendages. Each of the last four bears a pair of claw-bearing legs; the first three pairs are almost the same size, but the fourth is considerably more robust and differs in shape, position, and number of segments. *l* = lateral continuations of the body segments, reinforced by cross-ribs. *a* = abdomen, relatively very large, leaf-shaped, composed of at least five segments. The question mark refers to a distinctly independent segment of uncertain determination that appears in each of the three anterior pairs of appendages.

In contrast to Recent representatives of the pantopodes (fig. 3.11), which have four or five body segments and always only a single tiny abdominal segment, *Palaeoisopus* is distinguished by a considerably larger number of metameres (8 + 5), which definitely characterizes it as primitive. This feature provides a basis for the view that the ancestors of pantopodes are to be found among the annelids. Further, living forms of *Palaeoisopus* lack the characteristic sculpture of parallel ridges on the lateral extensions of the body segments seen in *Palaeoisopus* and *Palaeopantopus,* shown in figure 3.12. Because of this difference, F. Broili placed the two ancient genera in a group called Palaeopantopoda, as opposed to the Recent group Pantopoda.

pedicellariae of the sea urchins (figs. 2.66–69) are capable of being fossilized. The external oral gills of the sea urchins and their masticatory apparatus, when not themselves preserved in the fossil record, have at least left recognizable indications of their presence on the test. Other important clues are provided by the usually magnificently preserved histological structures of the skeletal elements.

Even the *ontogeny* of fossil sea urchins can be investigated (figs. 2.37, 2.38, 2.46, 3.13; pl. 9A), for all of the plates of the test ever produced by the animal,

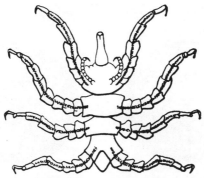

Fig. 3.11. For comparison: a Recent sea spider, *Pycnogonum littorale* Ström., a small form 13 millimeters long that lives along the coasts of Germany and in northern seas among the algae. About 3 ×. In addition, there are considerably larger marine forms, which grow to a half meter in length.

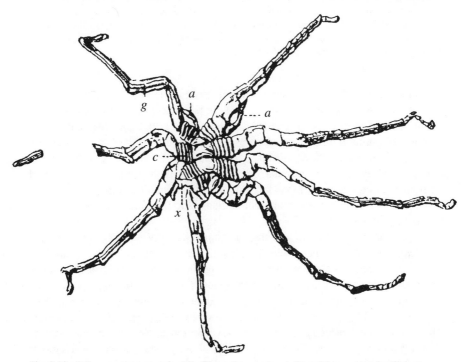

Fig. 3.12. *Palaeopantopus maucheri* Broili, another representative of the pantopodes (Palaeopantopoda) from the Lower Devonian Bundenbach Shale. 1 ×. (From F. Broili.) *g* = gut running through the legs. *a* = small appendages at the anterior end of the body disk. *c* = cross-ribs on the lateral extensions of the body segments. *x* = a portion of the body not yet explained when the specimen illustrated was first published; based on new evidence, it has turned out to be an abdomen composed of *three* segments. Thus, there are certain differences between this animal and *Palaeoisopus* (fig. 3.10) with regard to the number of segments, which indicates that pantopodes were showing different differentiation and developmental levels as early as the Lower Devonian.

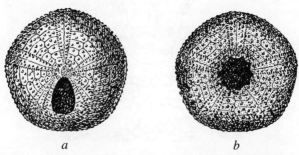

a   *b*

Fig. 3.13. *Pygaster umbrella* Ag., an irregular, or bilaterally symmetrical, sea urchin from the Upper Jurassic of France. 1 ×. (After G. Cotteau, from E. Stromer von Reichenbach 1909.) *a.* Dorsal view with genital plates and large, sievelike madreporal plate in the center, below which is the large, anal field (periproct), shaded dark. Further, the increasing size and differentiation of the ambulacral and interambulacral plates, from the crown to the periphery, can be seen. *b.* Ventral view with central oral field (peristome), the notched outer edge of which indicates the presence of exterior oral gills. (Compare this with fig. 2.46, *Hyboclypeus,* with a differently shaped central plate and a smooth-edged peristome, without notches to accommodate gills.)

from its juvenile stage on, remain for its entire life in the same sequence in which they were deposited. Nevertheless, a striking peculiarity of individual development can be seen in the fact that the plates of the radial and interradial rows begin at the summit of the test and move from that point to the opposite side. As the rows grow longer, they grow larger and undergo increasing secondary differentiation. Growth of the globular or ellipsoid test is greatest in the equatorial region, and consistent with this, the plates at that point are the most well developed; the degree of this development is diagnostic for each species and for its phylogenetic position.

Therefore, in a sea urchin test, there is no parallel between the degree of plate specialization and the age at which the plates were formed, as is true in normal ontogenetic development, where the most recently developed structure shows the highest degree of differentiation. In sea urchins, rather, the degree of differentiation in each row decreases from the equatorial plates toward the apex— toward the point where the subsequently formed plates originate; in each row, the most recently formed plates are the least elaborate and persist in their original lower rudimentary state. These relationships should be taken into consideration in the application of the biogenetic law—when evaluating ontogeny from a phylogenetic standpoint.

Madreporarians

The systematics of stony corals is based almost exclusively on the hard parts— the calcareous skeleton—and, in particular, on the structure of the septal apparatus, as well as the arrangement of the mesenteries, longitudinal fleshy partitions separating the radial chambers of the enteron and enveloping the calcareous septa. The structure and developmental sequence of the mesenteries can be deduced with sufficient reliability in fossil forms, too, through detailed analysis of the septal apparatuses (figs. 3.55 and 3.74). Thus, paleontologists do not find themselves at any significant disadvantage compared to zoologists, who, for purposes of taxonomy and phylogeny, make relatively little use of the soft parts at their disposal.

It is entirely possible, then, to determine without question the phylogeny of the extinct groups of Paleozoic pterocorallia ("tetracorals") and to assign them a place within the system of Recent forms; we shall return to this subject in a later chapter. Compared with vertebrate paleontology, this kind of phylogenetic investigation of corals has an inestimable advantage in that it is supported by the complete fossil record of the ontogenetic development, from the first secretion of the calcareous skeleton to senescence.

Mollusks

A favorite subject of phylogenetic research is the *cephalopods,* a class of mollusks that begins in the Upper Cambrian and extends to the Recent. We have already seen that these forms provide the most important index fossils for the Late Paleozoic and the Mesozoic, partly because they are so common and partly because they evolve extremely rapidly. These are properties that make them appear to be especially suitable for addressing phylogenetic questions, particularly since their shells also contain an excellent record of the complete individual course of development, from earliest shell formation on. Here again, the objection has repeatedly been raised that the external shell by itself provides insufficient and unreliable clues.

Now, an *ammonite* shell offers us at least the following: first, the shell alone shows an abundance of differences in shape and combinations of characters; there are thin disks and thick disks, some with lower, some with higher apertures of the most varied silhouettes, with broad or narrow, shallow or deep umbilica, with tightly wound spirals or spirals that barely touch or have even become completely free (figs. 3.32, 3.41, 3.152). And there is every conceivable transitional form and the most varied kinds of linkage between characters.

As further criteria, we have the growth lines, which correspond to former aperture openings, reflect the incremental growth of the shell and, once again, show an abundance of the most varied shapes, from simple straight lines to complicated crescent-shaped undulations. They, like the final aperture, may have on the ventral side a backward sinus or a projection, from which inferences can be drawn regarding the structure of the funnel. Further, the shell is sculptured: there are ribs, nodes, and spines, parabolic ribs, constrictions, keels and furrows, multifarious kinds of spiral sculpture showing regular transformational trends during ontogeny, all of which can also be used in determining lineages.

Fig. 3.14. Suture line of *Pinacoceras metternichi* (v. Hau.) Upper Triassic of Hallstatt (Salzkam-mergut, Upper Austria). Much reduced. (After F. von Hauer.) The suture line is very complicated owing to its extremely fine crenulation, but only relatively slightly differentiated, for the individual lobes and saddles show a positively schematic uniformity among themselves. Further, since the crenulation starts during ontogeny from only one pole—from the lobe (as will be mentioned later), these Triassic ammonoids are at a lower evolutionary level than Jurassic and Cretaceous ammonites, which have bipolar crenulation.

And then, most important of all, there is the suture line, the morphologically extremely significant line corresponding to the point of attachment of the chamber walls to the external shell, where they are secured by finely branched muscle endings. We are not only able to observe sutures in the infinite variety of form of the mature state (fig. 3.14) but can also follow every particular of their development, from the first, still very simple stage on, for with the shell, we hold the entire ontogeny of the hard parts in our hand (figs. 3.37–39, 3.162–64). Extensive differences can be seen in the manner in which the lobes and saddles originate, as well as in their further elaboration.

We recognize that there are fundamentally different ways to divide the suture lines into their main elements and, further, into their secondary notching, and that superficially similar mature structures are really only homeomorphic; they have arisen in various ways and are similar in appearance only. In suture lines, we have a very sensitive instrument for determining kinship. The chambering itself may be broad or narrow; the terminal portion of the shell—the latest living chamber, which is left open—can be short or long; these characters can be applied both taxonomically and phylogenetically.

And then we have the siphuncle, a tube that runs through the entire chambered portion of the shell; the position of this structure, its course of development, the various ways in which the septal necks (fig. 3.36) are constructed, and so on, provide extremely valuable criteria for identification. Next in line of importance is the type of protoconch and its anterior margin, which is formed by the prosuture; this character varies among species in fundamental ways. Further, we have the impressions left by the attachment muscles and the aptychus and the anaptychus, platelike structures whose function has not yet been completely explained, but which nonetheless offer information useful for systematics.

If we stop to consider how many combinations of these characters would be possible, the number would by far exceed the million mark. This large number of interactive characteristic features offers a reliable guarantee of the correctness of the lineages we have established. If we occasionally run into trouble by incorrectly associating two given forms or lineages on the basis of some identical features, time after time we ultimately come upon other important diagnostic characters that reveal the mistake and set us on the correct path. Moreover, the chronological arrangement of the material eliminates certain possibilities of association right from the start, limiting the selection within which associations are possible.

Discoveries about the structure of most ammonite soft parts are indeed impossible for paleontologists and will probably always be so. For example, we cannot assess whether ammonites had *four* gills, which would make them identical in this regard to their only living Recent relative, *Nautilus*. For *Nautilus,* zoologists use the number of gills as a basis for the subclass Tetrabranchiata, contrasting it with the Dibranchiata (squids and octopuses), which have *two* gills.

Unfortunately, only relatively little information that would contribute to an understanding of fossil ammonites can be gained from the living *Nautilus*. For example, when we are not in a position to come to a satisfactory understanding of the importance of the siphuncle, the mechanism of submerging and rising in the water, and so on, the reason is that thus far, these things have been insufficiently investigated in Recent forms, too. In this instance, zoology is hardly further along than paleozoology. The clarification of these matters, a valuable contribution, surely awaits us.

But in matters of ammonite taxonomy and phylogenetic assessment, paleontology is autonomous. Paleontology has at its disposal, in contrast to the single contemporary relict of this group of cephalopods (which, moreover, is only very distantly related to ammonites), such an overwhelming quantity of material in such a great abundance of forms that it must apply its own standards and methods of classification without regard for the taxonomic principles used for Recent cephalopods. We hope to have demonstrated that the results are valid.

In addition, we can get ideas about the mode of life of ammonites from the structure of their shells, the associated fauna, and the facies of the rock in which they are embedded; here, the zoologist's direct observation (for the most part) of his subject matter must be replaced by detective work, a search for clues. In this roundabout way, inferences about the length of time it takes for a shell to grow can be drawn, a subject about which there are no immediate criteria and analogues with living relatives available. Here is an example:

Not infrequently, Jurassic ammonites have a growth on the shell caused by a serpula (a worm that lives in a calcareous tube), which has colonized the flank of the shell, then turned to the perimeter and grown along it (pl. 16B, *a*). Specimens such as the one shown in plate 16B, *b,* where the last whorl of the ammonite coils on top of and beyond the first portion of the worm's tube, show that the colonization took place while the ammonite was alive, and based on certain other considerations, it follows that the growth of the two shells must have been roughly in step. From the known growth rate of Recent serpulid tubes, we can figure out the unknown rate of the one on the ammonite shell. It turns out that the construction of one whorl of our ammonite took about one half to one year. Since the whorl includes about twenty-five septa and chambers, the construction time for a single chamber or for the corresponding section of the tube would be set at about a week to a month.

This is indeed only an approximate measurement with quite variable limits; nonetheless, it has some significance for certain biological conclusions. An older hypothesis holding that only female ammonites had shells, and that the forward movement within the shell and the formation of septa was caused by the accumulation of eggs during reproductive cycles, is hereby unequivocally laid to rest.

In *nautiloids,* the phylogenetic precursors of ammonites and close relatives of the living *Nautilus,* some of the characters listed above are lacking. But there are others that appear in their stead, for example, the particularly complex construction of the siphuncle, showing the development of extremely complicated internal canal systems (pl. 17). The high degree of differentiation among the multiplicity of forms in these structures, which, however, are constant within certain groups, leads to the idea that they arose repeatedly and erratically. They provide, therefore, a reliable index character for elucidating kinship. In addition, there are within the chambers calcareous deposits of the most varied kinds and shapes; for example, in one group of forms, there may be calcareous beaks. These, too, constitute a large body of important characters and guides.

However, there are differences among cephalopods in their usefulness for phylogeny. *Belemnites* (figs. 2.42–44), relatives of today's squids, are less suitable. Recently, efforts have been made to draw far-reaching phylogenetic conclusions just from the shape and ontogenetic transformation of the rostrum, a cylindrical or conical calcareous body secreted internally by the animal, but the grounds for doing so seem somewhat narrow. Yet many Jurassic squids yield a considerable amount of information about the organization of their soft parts: the mantle, portions of the arms and sucker disks, the ink sac, stomach, intestine, fins, and so on, are discernible.

* * *

It is the same with *bivalved mollusks* (clams, oysters, mussels) as with other mollusk shells—the shell is by no means just a simple integument, an external container for the animal, but rather an integral component of the organism. The two-valved shell is typical of the structure of this form, and there are a multitude of correlations between the shells and the soft parts, in particular with regard to the gills, the mantle fold, and the musculature. Further, the valves yield information about the number and size of the adductor muscles, the edge of the mantle, and many other features of the soft parts. According to D. L. Frizzell, at least fifty major morphological characters are at our disposal on the bivalve shell. If we take only ten different manifestations of any one of those characters, there are 10^{50} possible combinations that can be used in taxonomy and phylogeny!

The taxonomic classification of bivalves has been carried out on various grounds. The zoological system (P. Pelseneer and others) is based on the structure of the gills, a character that is inaccessible to the paleontologist. On the paleontological side (Melchior Neumayr and others), a different system, one based only on the shell, and particularly on the structure of the hinge, has been devised, making it possible to classify fossil material accordingly. Remarkably enough, these two independent systems of classification, constructed on different bases, are in broad agreement—convincing proof that the use of shells for this purpose is valid and that the relationships so established are legitimate.

*   *   *

The least profitable of the mollusks for phylogenetic purposes would seem to be the fossil *snails*, or gastropods. For classifying Recent snails, the anatomy of the soft parts (nervous system, gills, radula) has proved to be the most suitable basis, and these characters are almost entirely unavailable to the paleontologist. One might think at first glance that the snail shell, with its relatively few characters, would offer only inadequate grounds upon which to base an explanation of natural kinship and individual evolutionary series. Yet it gives pause for thought when an outstanding expert on Recent mollusks such as J. Thiele, in his well-known *Handbuch der systematischen Weichtierkunde* [Guide to the systematics of the mollusks], ascribes to the shell a high and often even critical significance, regarding it as a diagnostic structural element of the molluscan type.

In this regard, the investigations conducted by F. A. Schilder on the form-rich Cypraeacea group of snails (pl. 18C) is very informative and of fundamental importance. He proposed a classification of Recent and fossil forms that included two superfamilies, five families, ten subfamilies, and ninety-two genera, based purely on the morphology of the shell. He was later able to conduct anatomical investigations of the soft parts of sixty-one of the genera, in particular, of the mantle and the radula, and *remarkably, these yielded extensive confirmation of the systematics he had previously worked out independently, based purely on conchological evidence!* Thus, even with the seemingly unsuitable snails, a careful analysis of the shells and their individual features led to incontestable results.

Recently, W. Wenz was very successful in his attempts to elucidate the origin and early phylogeny of gastropods based on fossil material. In the Early Paleozoic, in addition to the two large groups Pleurotomariacea and Bellerophontacea, there was a third, which Wenz recognized as a separate group, calling it Tryblidiacea. These were snails with bowl-shaped to conical shells that had formerly been included with the Docoglossa, family Patellacea, the still-living group of limpets and cap snails.

However, Wenz separated them from the Patellacea, for these had only begun to develop long after the Tryblidiacea had become extinct. Further, according to Wenz, the symmetrical bowl-shaped form of the tryblidiacid shell, together with the separate, symmetrically paired attachment-muscle impressions (pl. 19A) were indicative of a *primary* bilateral structure and an original orthoneury, that is, a parallel, uncrossed arrangement of the nerve commissures. All other gastropods with bowl-shaped to cap-shaped shells, among them also the Patellacea, are, in contrast, *secondarily* bowl-shaped, deriving from forms with spiral shells and, like those, chiastoic, having crossed nerve commissures and a torsion of the visceral hump.

This makes the Tryblidiacea the most primitive of all gastropods thus far known and is entirely consistent with the ideas that were formed about the protoform based on comparative anatomical studies. The impressions left by paired attachment muscles—a primitive character—leads to the conclusion that the ancestors were segmented. There is a parallel here with the Loricata (the chitons), which suggests a common origin for the two groups. (Remarkably, in both the tryblidiacids and the loricates, the number of pairs of attachment muscles is never greater than eight, but can indeed be less, owing to fusion.) However, whereas in the loricates segmentation of the shell was maintained, in the tryblidiacids the individual segments of the shell fused tightly to form a uniform shell, and in some cases, the attachment muscle impressions have joined.

As the visceral hump was elevated, forms with higher, conical shells arose from the originally flat, disklike forms; then, for biomechanical reasons, the shell began to coil, developing a bilaterally symmetrical spiral in the bellerophonts

Fig. 3.15. *Pleurotomaria subscalaris* Desl. Middle Jurassic from Bayeux (France). ½ ×. (After K. A. von Zittel 1924.) The shell of the Pleurotomariacea is an asymmetrical spiral, top-shaped to more or less pyramidal, characterized by a slit in the aperture and a connecting slit band, which corresponds to an anal sinus of the mantle. Pleurotomarians flourished during the Paleozoic and Mesozoic. In the Cretaceous, they decreased sharply in frequency and richness of forms and from then on proceeded slowly toward extinction. Today, in the seas of the West Indies, the Moluccas, and Japan there are still a few species of the genus *Pleurotomaria,* which have survived unchanged from the Lower Triassic to the present day, representing true "living fossils." In both shell structure and anatomy (two double-filamentous gills, primitive rhipidoglossal jaw [radula] with a central plate and extremely numerous narrow marginal plates, and so on), the pleurotomarians are the oldest of all living gastropods, all of which apparently derive from them.

(pl. 19A, *b*) and an asymmetrical spiral in the pleurotomarians (fig. 3.15). The symmetrical shell coiling of the bellerophonts leads to the supposition that these forms retained bilateral symmetry and primary orthoneury in their anatomy.

This supposition would be confirmed if we could find in bellerophonts, too, symmetrically arranged paired attachment-muscle impressions instead of the muscle spindles common to the rest of the gastropods. Such evidence has now, in fact, been produced, crowning the at first purely theoretical deduction. This confirmation was the result of a search guided by such a prediction: in a lower North American genus of bellerophonts, three pairs of attachment-muscle impressions have been discovered, arranged similarly to those found in the Tryblidiacea (pl. 19A, *b*).

On the other hand, in the few surviving species of the pleurotomarian group, signs of descent from bilaterally symmetrical forms can be discerned in individual organ groups, thereby establishing that these animals represent the oldest asymmetrical gastropods; their anatomical traits are thus quite consistent with a derivation from the tryblidiacids.

This is *convincing proof for the reliability of evolutionary conclusions drawn from the often underappreciated gastropod shells,* and for the complete accord between assumptions based on comparative anatomy and the observations of actual demonstrable evidence in the fossils. Thus, even the shapes of gastropod shells allow substantiated conclusions to be drawn as to the anatomy of the soft parts: the musculature, which is not directly accessible; the structure of the nervous system; and according to recent investigations by G. Delpey, perhaps also the number of gills. Moreover, we also have, although only in individual instances, fossil radulas—tongues covered with fine teeth—which occupy an important place in zoological systematics.

An Evaluation of Vertebrates and Invertebrates

We do not want to go into discussions of other groups of animals here. The ones already presented should suffice to establish, first, the *fundamental suitability of paleontological material,* and then, especially, the *importance of fossil invertebrates for phylogeny.* It is indeed indisputable that the criteria available to vertebrate paleontology are, for the most part, more deeply rooted in internal organization than those obtained from the external skeletons of invertebrates. Nevertheless, many paleontologists consciously prefer the latter, for they have a number of critical advantages over vertebrates.

1. Of fundamental importance is their far *greater frequency and distribution* and the correspondingly much larger body of material available for study. Under certain circumstances, vertebrate carcasses and skeletal remains can also accumulate in considerable masses, as, for example, in the Californian asphalt pits of Rancho la Brea, in the German Kupfershiefer, in Ice Age cave deposits, in the carcass fields in the lignite of the Geiseltal, and so on. These are always, however, only isolated occurrences or horizons, which usually lie far apart in space

and geologic time. Thus, the possibility of following the evolutionary course of any particular group from system to system or even throughout an entire age, layer by layer, and with abundant documentation, as is the basic situation with invertebrates, is usually denied to the vertebrate paleontologist.

2. It should be emphasized that the remains of invertebrates are in a *far more complete state of preservation.* The valves of brachiopods, the exoskeletons of trilobites, and the shells of ammonites all represent the complete skeleton of the animal, whereas complete skeletons of vertebrates are rarely found. The possibilities for error of interpretation are certainly considerably greater for a single tooth or a limb bone found in isolation than for a complete invertebrate exoskeleton. Purely anatomical features and the law of correlation do not go far enough to allow absolutely reliable statements to be made about very fragmentary individual remains, as many recent incorrect interpretations prove. Preservational conditions for terrestrial vertebrates, in particular, are unfortunately much more unfavorable than those for marine invertebrates.

3. The most important advantage is one already touched on, that with invertebrates, in many, many instances, the *entire course of the ontogeny* of the hard parts has been preserved in the fossil record. An ammonite or snail shell, the corallite of a stony coral, the test of a foraminiferan—all these incorporate not only the mature form of the organism but contain the complete sequence of every individual developmental step, from the first secretion of the exoskeleton to its final state. Thus, the coming into being and the gradual transformations taking place during ontogeny can be followed step by step.

What this possibility means for the elucidation of the course of evolution does not have to be explained in any more detail here. It is enough to refer to the so-called biogenetic law and to the fact that all phylogenetic change takes place and manifests itself during the ontogeny of each individual. Thus, in certain ontogenetic stages, we read the phylogenetic event directly, *in statu nascendi,* as it were.

Observations of this kind are, in general, denied to vertebrate paleontologists. In exceptional cases, the vertebrate paleontologist may indeed have the skeletons of juvenile animals at hand (for which, however, the preservational situation of the more fragile bones and, in particular, of the important joint surfaces, is even less favorable than for the more resistant adult skeletons); he can sometimes compare the milk dentition with the definitive form; as a rule, however, reliable proof of the species to which such remains belong is missing. It is naturally impossible to follow the different developmental stages in the same individual. Many times, the taxonomic position of juvenile or larval forms cannot be determined reliably at all. For example, do the Phyllospondyli represent the larval forms of other stegocephalians (primitive amphibians), as Romer tried to show was probable, and if so, which ones?

Such uncertainties can hardly appear in the realm of invertebrates. Indeed, here too, it is usually not possible to interpret the isolated embryonic shell of a snail or the shell of a larval ammonite with absolute certainty. Given the massive quantities of material, however, such an unidentifiable specimen can easily be set aside; juvenile stages are always found in abundance in specimens that have been preserved in totality, where they are indisputably associated with the mature form to which they belong.

4. The final point to emphasize is that the *organization of invertebrates is simpler and more easily observable in its totality* than that of vertebrates. But when it is a question of discovering general laws of evolution, it is obvious that one proceeds from that which is simple and easier to analyze to that which is more complex.

For these reasons, it follows that *the interpretation of invertebrates,* apart from its practical importance for stratigraphy, *is quite on a level with purely biological and, particularly, evolutionary vertebrate research and, in many respects, even superior to it.*

If the hard parts of fossil invertebrates often do not permit investigation and consideration of the *same* anatomical features that underlie zoological studies of the corresponding forms, they still offer such a wealth of diagnostically useful characters that erroneous interpretations are ruled out or kept to a minimum. In place of zoological principles of classification, others must be used that are more suited to the fossil material and have even been developed from it. In cases in which the results obtained by the paleontological and the zoological methods have been accessible to verification, it has turned out that there is broad agreement between them; this is excellent proof that the analysis of hard parts does indeed yield reliable results.

It goes without saying, however, that such research must be carried out with the aid of the entire arsenal of *modern biological methodology,* that the endeavor is always directed toward elucidating the ontogeny, architecture, and internal structure of the shell or other type of exoskeleton to the greatest possible extent. Today, this work is carried out with refined technological aids. If one thinks that invertebrate paleontology, or the whole of paleontology, can still be regarded with a certain condescension, then one should take a closer look at the modern procedures used and not measure the science against a stage of development, long since surpassed, in which the goal was only to make a somewhat sketchy description of fossils based on characters of external form and to name them without asking any biological questions.

Plants

In conclusion, a quick glance at paleobotany. It could perhaps be said that the preservational state of the objects studied is insufficient for taxonomic and phylogenetic purposes. But investigatory procedures used to study the morphology and comparative anatomy of living plants are basically applicable to fossil material, too. Research is not limited to external relationships of form; proper maceration procedures permit the preparation of thin sections and slides, which make it possible to elucidate all kinds of anatomical and histological details, the structure of the leaf, stem, flower, and fruit, the structures of individual cells and tissues, of the diagnostically important epidermis, the stomata, the spores, pollen, and so on (fig. 2.76; pls. 20 and 21).

Nonetheless, paleobotany suffers from two disadvantages that are similar in many respects to those applicable to vertebrate paleontology. First, plant remains are not common except in coal deposits and their associated rocks; and since our sedimentary rocks are primarily marine in origin, there is no possibility of obtaining large amounts of material from each individual horizon and following the uninterrupted development of the continental plant world.

The second drawback is that only rarely is an entire plant preserved with all its parts in natural association. Usually only separate components of plants are found, which makes it difficult to determine with certainty to what plant an isolated leaf, bit of wood, fruit, and so on, belongs; often, it is impossible to make a positive identification. But this drawback is more a temporary one and will not endure as a fundamental aspect of paleobotany. The gap in knowledge will steadily diminish as chance finds of plant parts in their original association turn up; for this reason, the perceived deficiency does not present an insurmountable limitation to knowledge.

GAPS IN THE FOSSIL RECORD

The problem of the incompleteness of the fossil record has been touched on repeatedly in previous chapters; it is closely intertwined with questions of the preservational quality of fossil material. Since, however, the expression "gaps in the fossil record" has become a cliché, used on every suitable, and more frequently, unsuitable, occasion to point out the deficiencies in paleontological data, it is appropriate to devote some special attention to this subject.

The reference to the fragmentary or incomplete preservation of fossils comes out of the early days of evolutionary theory, when it became clear that paleontology could only partially fulfill and corroborate the expectations that accompanied the conceptual basis of the science. As we all know, Darwin's theory of evolutionary descent asserts that organisms evolve slowly and very gradually through the smallest of individual steps, through the accumulation of an infinite number of small transformations. Consequently, the fossil organic world would have to consist of an uninterrupted, undivided continuum of forms; as Darwin himself said, geological strata must be filled with the remains of every conceivable transitional form between taxonomic groups, between types of organizations and structural designs of differing magnitudes.

Discontinuities between Structural Designs

Fossil material did not then and, based on the present state of our knowledge, does not today meet this challenge, not by a long shot. It is true that we know of countless lineages with continuous transformation, in as uninterrupted a sequence as could be desired. However, each time we go back to the beginning of these consistent, abundantly documented series, we stand before an unbridgeable gulf. The series break off and do not lead beyond the boundaries of their own particular structural type. The link connecting them is not discernible; the individual structural designs stand apart, beside one another or in sequence, without true transitional forms.

Even the initial joyous satisfaction that once greeted, for example, the discovery of the famous ancestral bird *Archaeopteryx* did not prove to be justified. Despite all its similarities to reptiles, *Archaeopteryx* is a true bird; the boundary between the reptile type and the bird type has not yet been bridged by a continuous, uninterrupted linking series.

In this state of affairs, there are two logical possibilities: (1) the concepts of evolutionary events are not true, or at least not without reservations; or (2) the fossil material is too incomplete to prove actual evolutionary pathways. The first question was not explored; convinced of the unquestionable rightness of evolutionary concepts, people chose the second possibility out of hand and spoke thenceforth of the gaps in the fossil record; this sharply reduced the documentary value of paleontological material. For a while, even in paleontological circles, too much weight was attached to this second interpretation, thereby adding grist to the mills of the doubters in other camps.

Obviously, there are gaps, even considerable gaps, in the preservation of fossil creatures; it would be silly to deny that. The criticism applies primarily to groups of soft-bodied animals that had no hard parts susceptible to fossilization. For example, for sea anemones and jellyfishes, many groups of "worms," slugs, holothurians, tunicates, and so on, we cannot count on finding remains. Yet those we do very occasionally find, widely separated from one another both in time and space, are due to extremely favorable conditions of fossilization, conditions that are realized only as the exception, and these remains are sometimes in an astonishingly good state of preservation.

On the other hand, we can naturally study the evolutionary process only on forms that we have and that are suitable for the purpose; we cannot use those we do not have, or do not have enough of. And we have at our disposal, thanks to mass collecting along the profile of a great many groups, such an abundant, consistent, chronologically well-ordered material that we are entirely capable of arriving at binding evolutionary assertions, even by stringently critical standards. This is the case with shelled cephalopods, stony corals, and so on; and in working out the evolution of these animals, it does not concern us much that the tunicates, perhaps, or the annelids, show large gaps in their particular fossil records.

Investigations of this kind, therefore, are by no means based only on completely isolated, chance finds, as is sometimes assumed by those who do not understand the strict paleontological methodology. Modern paleontology goes about its work systematically with hammers, picks, dynamite, and shovels, collecting from coherent, fossil-rich stratigraphic series, accumulating massive amounts of material that usually contains countless specimens of individual species.

Then, when the fossil contents of two successive massive beds are compared, we are confronted with three different kinds of situations with regard to individual forms:

1. We observe species that have not undergone transformation; they pass from the older to the younger strata unchanged.

2. Other forms from younger horizons differ from those of older ones, but the modifications are insignificant in nature. The forms link up with the preceding species so closely that they can be interpreted as being direct descendants that have undergone transformation in small, individual steps. This transformation usually continues in subsequent strata, and what we have, then, is a closed, uninterrupted series showing gradual, smooth transformation.

3. In addition, but much less frequently, we come across forms here and there that are quite different from any other form previously present, forms that are not connected in an unbroken line with previous ones but rather appear suddenly as new designs.

And these are by no means just isolated occurrences; these strange new forms are usually also represented by large numbers of individuals. Nonetheless, there is no connecting link with the stock from which they derive. The continuity of the other species gives us no reason to suspect interruptions in the deposition of the layers, or subsequent destruction of layers already deposited, which, furthermore, would be revealed by other geological criteria. Nothing is missing here, and even drastic changes in living conditions are excluded, for the facies remains the same.

Further, when we see this situation repeated in all stratigraphic sequences of the same time period all over the world—we have described an example of this in chapter 2—we cannot resort to attributing this phenomenon to immigration of the new type from areas not yet investigated, where perhaps a gradual, slowly progressing evolution had taken place. *What we have here must be primary dis-*

continuities, natural evolutionary leaps, and not circumstantial accidents of discovery and gaps in the fossil record.

The Fossil Record and Evolution

Countering this assertion, the objection has been repeatedly raised that the paleontologist is very far from knowing the totality of all individuals that have populated the earth during geologic time, nor will he ever know it. This, in itself, is true but as an argument is completely inappropriate. The same situation is true for zoologists and geneticists with regard to the world of living animals. No one would ever demand of the science of genetics that it analyze all living specimens of its favorite research subject, *Drosophila,* the fruit fly, before arriving at binding results. Likewise, it is enough for the paleontologist to make a precise, thorough study of a number of selected examples for which sufficient raw material is available. It should not be forgotten that the documented evolutionary process is *not a genealogy,* which is based on the knowledge of every individual member of a strain.

The effect of these limitations of the fossil record and of particular finds can be likened somewhat to a kind of *material- and time-lapse* photography: At any given time we have at hand only samples taken from the vast number of generations of a species and from the enormous abundance of species themselves. Thus, the entire sequence of generations appears compressed and abbreviated. But it is exactly this that constitutes the essence of phylogeny and that is completely sufficient for its work. Within this context, it is important to establish that in accordance with the statistical laws of randomization the proportions are not changed as the material is condensed and that the selections taken from the fossil record extend evenly throughout the entire lineage from beginning to end.

When, therefore, the preserved material is sufficient to substantiate continuous evolutionary lineages *within* the individual structural designs, it should follow, if the assumption of a gradual bridging of the type boundaries by means of small developmental steps is correct, that the same situation applies *between* them. Moreover, in view of the significant differences we see among the organizations of the individual types, a connecting series of the sort just referred to would not even have to have been very long or composed of many members. However, there are no special conditions whatsoever in the fossil record that would indicate that gaps had repeatedly affected only the sequences that connect structural designs. The gaps that exist in the continuity of forms, which we always encounter at those very points, are not to be blamed on the fossil record; they are not illusions, but the expression of a *natural, primary absence of transitional forms.*

To avoid this conclusion, one favorite device is to pin all hopes on future discoveries and on the exploration of regions heretofore untapped. Since fossil material is inexhaustible in its abundance, each day brings new discoveries and descriptions of new species and genera. It should make us stop and think, however, that among them, in spite of tireless search, the hoped-for series of connecting forms—the "missing links" of the cliché—have never been found.

Even the recent large expeditions to lands previously unexplored have turned up nothing that is basically surprising and hardly anything that is fundamentally new. Great quantities of specimens have ended up in museums, some of them indeed considerably augmenting and filling out knowledge we already had; but this material has not trespassed on the general rule described, that the closed evolutionary lineages we have before us regularly break off as we near their roots. Nothing in the future will change this.

In a later connection, after we have become acquainted with examples of a few specific phylogenies, we shall return to this question with a new perspective and realize that in very many instances, for purely morphological and biological reasons, transitional forms between the different structural designs and organizations are not even possible or conceivable. In many ways, we have been searching for things that cannot possibly exist. And to show, moreover, how thoughtlessly the cliché about the gaps in the fossil record is sometimes used, I shall now briefly discuss two examples.

Two Examples

In the older Paleozoic, from the Ordovician to the Middle Devonian or the lowest Upper Devonian, the group of corals known as the *heliolitids* (fig. 3.16) are found with great frequency worldwide. These forms have a certain superficial resemblance to the *helioporids,* representatives of the alcyonarians (octocorals), which begin in the Upper Cretaceous and extend to the present (fig. 3.17).

This similarity has caused various authors to assume a phylogenetic relationship between the two very different ancient groups, to place the heliolitids with the alcyonarians and to ascribe the absence of intermediate forms in that enormously vast period of time between the Upper Devonian and the Lower Cretaceous to gaps in the fossil record. This view, like so many fundamental errors, clings tenaciously to life and is presented again and again, primarily in text books, even though it has been repeatedly and inconstestably refuted.

Important factors in the assessment are that both the heliolitids and the helioporids build massive, strongly calcified colonies, which, accordingly, are capable of being included in the fossil record even under unfavorable conditions of preservation; and as already stated, that they are extremely common in the layers in which they do occur. Therefore, there is clearly no plausible reason at all why the intermediate forms, if they had really been present, should not be found in the intervening rock layers. Only one sensible conclusion is possible, that the heliolitids died out in the Devonian—that they really do give out at the point where their last representatives are observed.

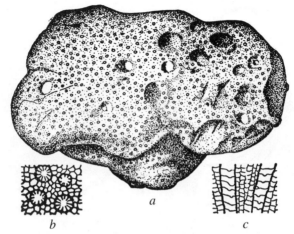

Fig. 3.16. *Heliolites porosus* Goldf. Middle Devonian, from the Eifel [Germany]. *a.* A corallum, consisting of individual corallites bound securely together by a tubular mesogloea (coenenchyme) to form a massive colony. ⅓ ×. In the Eifel, the coralla of this species attain a diameter of more than one-quarter of a meter; they appear as massive mushroom- or cushion-shaped calcareous lumps, which are extremely resistant to destruction. (The weight of the specimen in the illustration in its fossilized state is almost 5 kilograms!) *b.* Cross section through a few corallites and the coenenchyme surrounding them. Enlarged. *c.* Longitudinal section. Tubes of corallites with widely spaced bases and the narrow coenenchyme tubes with closely spaced bases. Enlarged.

Upon closer examination, going beyond the consideration of the apparently insignificant superficial similarity of form, it also appears indisputable that morphologically and structurally the heliolitids cannot be allied with the alcyonarians but rather that they are descendants of the Tabulata of the hexacoral group, having differentiated in a particular direction. In this case, therefore, the notion of "gaps" is called upon to support erroneous phylogenetic concepts, acquiring in the process an intellectual authority that borders on the miraculous.

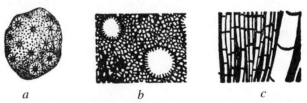

Fig. 3.17. Corallum (*a*), cross section (*b*), and longitudinal section (*c*) of different species of *Polytremacis,* from the Upper Cretaceous; the genus belongs to the order Helioporacea. *a* is natural size; *b* and *c* enlarged. (After E. Stromer von Reichenbach 1909.) The coralla of this species attain a considerably larger size than the fragment shown here; because of their size and, further, because of their massive structure and strong calcification, they are very resistant to agents of destruction.

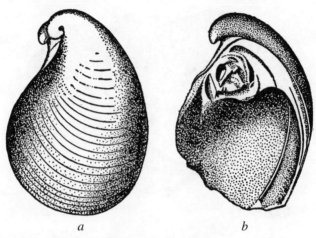

a   *b*

Fig. 3.18. *Megalodon abbreviatus* (Schl.). Upper Middle Devonian, from Paffrath, near Cologne (Germany). 1 ×. *a.* The entire shell. Many specimens considerably exceed the size of the one illustrated, attaining a height of 10 centimeters. *b.* Interior of the right valve showing the thick hinge plate, which fills in the beak region with its mass and bears powerful hinge teeth.

* * *

For the other example, we mention a plump, very thick-shelled clam from the Middle and early Upper Devonian, which, because of the sturdy hinge teeth on extremely massive hinge plates, has received the descriptive name *Megalodon* (fig. 3.18). In the Upper Triassic, there appears in enormous, often almost rock-forming masses, a similar but larger and heavier clam that has also been assigned to the genus *Megalodon,* or included in it under its own name, *Neomegalodon* (fig. 3.19).

Even recently it has again been stated that the two have a direct phylogenetic connection. Since, however, links from the intervening period of time are not known, the idea of gaps in the fossil record is once more held responsible, which, in this case, is absolutely absurd. Such a conclusion would only be justified if, for example, the two occurrences were separated by a series of rock completely devoid of fossils. In reality, however, between them lie the rocks of the Upper Devonian, the entire Carboniferous, the Permian, and the early Triassic, which have yielded an enormous abundance of forms from the most varied phyla of the plant and animal kingdoms, among them objects of the utmost delicacy and fragility.

Now, if anything is capable of being preserved it is certainly these very plump, thick-shelled clams, which do, in fact, appear massively in the Middle Devonian and the Upper Triassic. Therefore, their absence during the long intervening period can only be interpreted as meaning *that the presumed intermediate forms*

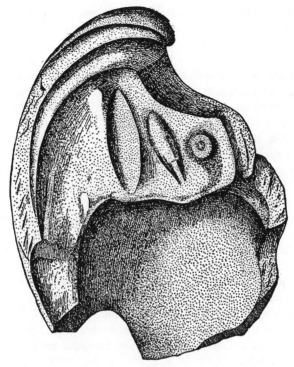

Fig. 3.19. *Neomegalodon ampezzanus* Hoern. Upper Triassic of St. Cassian (Upper Austria). This enormous bivalve grew to more than 20 centimeters. The illustration shows the beak region of the left valve with its thick hinge plate. 1 ×. (After F. Frech.)

never lived, that the two types are not in any way related but rather only resemble each other superficially, providing an example of convergent evolution.

Gaps in the Material Studied by Paleontology and Neontology

In summary, it should be stated that *the cliché about the incompleteness of the fossil record has been seriously abused.* When all the unjustified exaggerations and erroneous assertions are traced back to the actual facts of the matter, it can be said that the gaps in the material available to the paleontologist certainly do not exist to the extent true for the zoologist, who endeavors to construct a natural system and to reconstruct the course of evolution based only on animals alive today.

There is probably no single group of plants or animals that contains living representatives of all of the types and manifestations of all of the structural designs it once held, to say nothing of those extreme instances, such as, for example, the cephalopods, brachiopods, pelmatozoans, reptiles, vascular cryptogams, cycads, and so on, of which we have today only the scantiest remains of the enormous abundance of forms that once existed.

With regard to brachiopods, we have already given a few figures showing the numerical relationship between fossil and living forms. Likewise, we have also mentioned in another context the fact that of the thirty-four orders of reptiles known to have existed, only four (!) extend to the present, and one of these (the rhynchocephalians) is represented by only a single living genus and species, a mere scrap from a once great diversity of forms. Furthermore, to say just a bit more, the number of species of fossil sea urchins known to date exceeds that of living sea urchins by at least a factor of five. Among the fossils, many now-extinct families and orders with remarkable, revealing structures are represented.

Even with mammals, a relatively young group richly represented in the present, the percentage of extinct forms predominates. According to recent calculations (Simpson 1945), of the around three thousand valid genera, two-thirds are known only as fossils, and this figure is continually being revised upward. Moreover, there is fossil evidence of all of the thirty-two orders of mammals recognized today. Fourteen of these orders have died out completely, and of the eighteen extending on into the present, fifteen contain extinct families, which have no representatives among present-day fauna.

In the plant world, the twenty-five species of living horsetails, which are all quite similar, stand in contrast to the almost ten times greater number of extremely form-rich fossil Articulatae. In the present, the Dasycladacea (calcareous algae) are represented by ten genera; in contrast, fifty-eight fossil genera are known. It is obvious that this abundance of fossil forms should not be disregarded, that only by taking it completely into account will a well-rounded picture of morphology and phylogeny emerge.

Added to this is the fact that zoologists and botanists have no real clues at all to *the temporal factor.* But purely morphological deductions are often misleading, as a control by chronologically arranged fossil material often shows. It makes no sense, however, to emphasize the advantages one branch of research has over another and to play off one against another. The possibilities for investigation and the results achieved in one discipline supplement those of another in extremely valuable ways; not dogmatism and arrogance but trustful cooperation lead to the common goal of elucidating the course of organic evolution and phylogeny.

PATTERNS IN EVOLUTION

TWO INTRODUCTORY EXAMPLES

If we want to gain insight into the general processes and principles of phylogeny, it is essential first to create an *empirical basis.* This could be done by presenting several individual examples from widely differing groups of animals. This way of proceeding would draw together a number of interesting partial observations and would undoubtedly be the most entertaining and stimulating for the reader.

Nevertheless, we think that another path should be chosen, one that is more profound: that of a relatively complete analysis of the entire evolutionary picture of two selected groups of animals. In so doing, it is unavoidable that several details not in themselves very engrossing will be mentioned. But by proceeding in this way, we shall gain a rounded overview of the way evolution proceeds within in a few major stocks; we can then fall back on this information later as we make general observations.

The Unfolding of the Cephalopods

Cephalopods (literally, with "legs" attached to the head), the evolutionary course of which we shall pursue here in somewhat more detail, have already been touched on several times in this book. We recall that they are a particular class of mollusks; and that they have a shell which, in contrast to that of clams and snails, is tubular, regularly divided into transverse chambers by partitions called septa, and run through lengthwise by a siphuncle, a tube that perforates the septa (fig. 3.20).

The primordial, most ancient group of cephalopods is the *Nautiloidea*, which brought forth an enormous abundance of forms in the geologic past and which

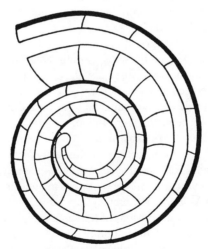

Fig. 3.20. Medial longitudinal section through the initial chamber and first whorls of the ammonite *Polyptychites*, from the Lower Cretaceous of northern Germany, showing the internal structure of the cephalopod shell. 30 ×. (From Schindewolf 1933.) In these early whorls, the siphuncle does not yet lie directly against the outer wall, as is the case in the later whorls of ammonite shells. It begins in a bladder-shaped distension, which protrudes into the relatively large, centrally located initial chamber.

today would be extinct if it were not for a single living genus, *Nautilus,* found in the Indo-Pacific realm. Nautiloids form the root stock of the vast host of extinct *ammonites,* and further, of the ten- and eight-armed *squids and octopuses,* which extend into the present. In a certain sense, then, the Nautiloidea are the backbone of the entire cephalopod line.

Within this line, the relationships and continuities of individual groups are to some extent clear, but the actual origin of the line is shrouded in darkness. Without a doubt, the roots of cephalopods are to be found in some primitive mollusk; yet we have discovered nothing that would point in a particular direction. *We know of no connecting, transitional link to other molluscan types.* All of the primitive forms we have found to date show the finished, typical cephalopod characteristics, at least with regard to the shell, which cannot help but constitute a valid distinguishing feature of the basic cephalopod structural design.

Nautiloids

THE EARLY LINEAGES

The oldest remains that, with regard to form, must be assigned to the nautiloids are the *volborthellans* (pl. 18A and B). These are small, slender, conical shells about half a centimeter long, which are found in enormous quantities here and there in the Lower Cambrian of the Baltic region, Scandinavia, Poland, and Canada. They have very closely spaced chambers and a narrow, centrally located siphon; unfortunately, the preservational state is not such that all the details of structure, of the diagnostically important siphuncular necks and rings, among other features, can be clarified.

Consequently, there is a degree of uncertainty in the interpretation of the volborthellans. They are completely isolated stratigraphically and fit only imperfectly into the groups of later forms, with which they probably have no immediate phylogenetic connection; rather, volborthellans may represent a kind of unsuccessful prototype and experimental form from the same roots that later produced the consistent cephalopod evolution.

Better preserved and interpretable finds do not occur until the Upper Cambrian of East Asia. There, the genus *Plectronoceras* (*Plectr.* in fig. 3.21) appears, which typically includes very small species only about one centimeter long with very closely spaced septa (three to four partitions per millimeter). They are more broadly conical than *Volborthella,* are slightly curved, and have a marginal siphuncle. The septal necks, protrusions of the septa at the place where the siphon passes through, are short and curve slightly downward. The siphuncular rings, which form the actual wall of the siphuncle and connect the neck of one septum with that of the previous one, bulge sharply outward. This gives the siphuncle the appearance of a string of beads. The interior of this structure is empty, devoid of any contents that might be subject to preservation.

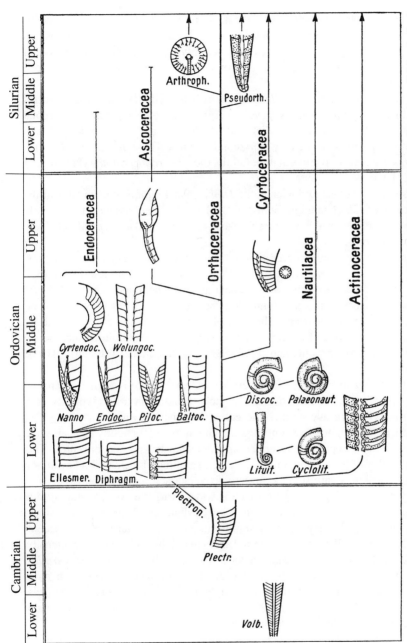

Fig. 3.21. Unfolding of the Nautiloidea in the early Paleozoic, shown diagrammatically. The individual forms are shown in medial longitudinal section with the exception of the arthrophyllids (cross section) and the representative of the Nautilacea, which is shown three-dimensionally. The names in italics are genera, those in roman type are families, and those in bold-face roman are the phyla. (From Schindewolf.)

Whereas cephalopods are rare at the upper limit of the Cambrian, we encounter representatives in great abundance in the layers of the earliest Ordovician lying immediately above, an abundance not only of individuals and numbers of species but also of the very different organizational types that are developed here.

In the lowest Ordovician, *Plectronoceras* is immediately followed by forms that, like the plectronocerans, have a marginal, narrow siphuncle, short septal necks, and closely spaced chambers but which show in the siphuncle thick calcareous deposits separated by horizontal sutures. These forms and *Plectronoceras* are placed together in one family, the *Plectronoceratidae* (Plectron. in fig. 3.21.)

Also in the earliest Ordovician, the *Diphragmoceratidae,* descendants of the plectronoceratids (Diphragm. in fig. 3.21) develop. These forms, either slightly curved or straight, also have closely spaced chambers and a marginal, relatively narrow siphuncle.

However, the diphragmoceratids developed two thoroughly new features: longer septal necks and true diaphragms in the siphuncle. In these forms, each neck extends from one septum to the next and even continues a bit farther into the previous neck. Thus, the siphuncle is here completely enveloped by extensions of the septa—the necks. This type of structure is extremely diagnostic for the Endoceracea, one of the first major lineages of the nautiloids, which derive from the plectronoceratids and to which the diphragmoceratids must also be assigned.

In these root forms of the Endoceracea, however, there is at first no consistency at all in the structure of the siphuncle. In addition to the typically long septal necks, there appear in some cases short necks, which then are either straight, collarlike, or folded outward and laid more or less tightly up against the lower side of the septum. Further, the siphuncle may be marginal or central. This *great lability* of the mixture of forms at the beginning of the nautiloid evolution stands in striking contrast to the strongly fixed characters in the lineages derived from them. Even the shape of the shell is widely variable; we find side by side straight shells and slightly curved ones, long cylindrical shapes to short conical ones—in short, in every respect a significant tendency to variability is evident, a tendency that will not be encountered again in anything like the same intensity.

The diphragmoceratids are succeeded immediately by the *ellesmereoceratids,* forms with the long septal necks typical of the lineage but with even more closely spaced chambers and a narrow siphuncle without siphuncular deposits. Issuing from this group, or perhaps from the Diphragmoceratidae, are the many forms of the actual *endoceratids* and *piloceratids.*

The superfamily of the endoceratids contrasts with previous forms in that it is characterized by more widely spaced chambering and correspondingly very long septal necks, which may extend almost the length of two air chambers. In typical representatives, the siphuncle is marginal, extremely wide and has in its interior a new diagnostic feature—telescoping, funnel-shaped walls, or endocones, the tips of which continue as short tubes. The spaces between the partitions are filled with biogenic calcareous deposits.

The endoceratids are one of the first groups of cephalopods to proliferate massively. The order is very rich in forms and exhibits basic differences in, among other things, the shape of the initial chamber as well as of the initial segment of the siphuncle that lies within it (*Nanno, Suecoceras, Endoceras*); consequently, the order Endoceratida splits into a great number of families and genera.

At the beginning of its evolution, this group of forms is still quite *labile*. In addition to the typical forms with marginal siphuncles, there are some with siphuncles centrally situated (*Wolungoceras*); and beyond the preponderance of straight shapes there are also curved ones (*Cyrtendoceras*), and so on. All these forms appear in the Lower Ordovician *explosively,* during a burst of extreme evolutionary intensity. The space available in our figure 3.21 is not adequate to display the abundance of types developed during this time period *side by side,* with the result that the phase of frenzied evolution of forms appears here considerably more strung out than it actually was.

Nor could the natural chronology of these forms, which appeared simultaneously and in parallel, or in rapid succession at short intervals, be adequately depicted. Connecting transitional forms among the different types of endoceratids with regard to initial chambers, and so on, are not known; the different types stand side by side, *unmediated by transitional forms,* and evidently arose through large, single steps in, one might say, a headlong rush of evolution.

During the further evolution of the endoceratids, a *consolidation* takes place, partly through stabilization and solidification of the typical traits of the forms and partly through the extinction of extreme types. During this portion of the unfolding, which extended on into the Gothlandian (Silurian), we observe a *gradual* elaboration of the forms in small steps that do not further alter the now-established basic organization. Continued specialization brings forth, among other things, the *gigantic forms;* before their extinction, many endoceratids attained a length of up to five meters, a size about five hundred times that of the initial primitive nautiloid forms (fig. 3.141*a, b*)!

The piloceratids mentioned above are considerably smaller; they were short and conical, and the siphuncle was unusual in that it took up more than half the diameter of the shell. (Our illustration shows the entire shell of a piloceran reduced in size, whereas the endoceran next to it is represented by only the initial portion of its shell!)

* * *

Another lineage that separates from the plectronoceratid complex during the Lower Ordovician is that of the Orthoceracea, relatively simply structured forms compared with the Endoceracea. They are primarily straight and have a narrow siphuncle located either at or near the center of the shell; in general, the siphuncle contains no biogenic calcareous deposits. The septal necks are short and straight, collarlike, and the rings are simple cylinders and mostly quite long, for the chambers of these forms are usually widely spaced. Long septal necks completely enclosing the siphon itself, as are typical for endocerans, do not appear in these forms.

Within this group, too, the oldest representatives still show considerable *variability of form.* The septal necks may appear not only in the typical straight design but also more or less sharply folded outward, and the rings, correspondingly, bulge outward. A sharp difference between these two types of designs, which are of great taxonomic importance for the younger members and the descendants of the Orthoceracea, is impossible to make in Lower Ordovician forms, for there, these characters have not yet been integrated in a stable way.

Only later do *stable lineages* emerge from the initial richly varied confusion of forms. *The stable lineages, with their basic structures now fixed, extend on into the Upper Triassic, through a large number of systems, whereas the labile phase of this line lasted only a very short time.*

*   *   *

Another independent, sharply delineated group appears in the Lower Ordovician. This is the Actinoceracea, forms with more or less marginal, wide siphuncles that resemble a string of beads and are filled with very complicated structures composed of internal secretions. In the vicinity of the septal necks, here always folded back and lying against the underside of the septa, ring-shaped structures are secreted, each of which consists of an endocone surrounded by a wreath of radiating canals (pl. 17). This is a highly differentiated vascular system within the siphuncle; when the animal was alive, this system was embedded in mineralizing connective tissue and thus has been preserved. Furthermore, in the air chambers there are biogenetic calcareous deposits, indicated in figure 3.21 by stippling.

The actinoceres are either immediate descendants of the plectronoceratid root group or of the initial stages of the orthoceran line—the distinction is not really meaningful given the lability of these early forms.

This line, too, has a long life span, up to the Upper Carboniferous. *During that time, the basic organization of the siphuncle, evidently developed very quickly during the Lower Ordovician, does not change again;* only with regard to the width of the siphuncle and its more or less marginal position, to the greater or lesser degree of slenderness of the shell, to the elaboration of the shell sculpture, and to the deposits inside the chambers do *insignificant modifications appear in continuous, slow evolution.*

*   *   *

Another fundamentally divergent kind of differentiation of the siphuncle characterized the Cyrtoceracea, which also split off from the Orthoceracea in the Lower Ordovician. In the Cyrtoceracea, radial longitudinal lamellae developed inside the siphuncle, which, in cross section (see the figure next to the longitudinal section of the cytoceran in figure 3.21) thus resembles the interior of a coral. Here, too, as in the actinocerans, the siphuncle looks like a string of beads, and the septal necks are folded back. But the cyrtoceral shell always shows a more or less pronounced curvature, whereas that of the actinocerans is predominantly straight.

The Cyrtoceracea extend at least to the end of the Devonian. Here, too, once the basic structure of the siphuncle characteristic for this lineage has been established, it is maintained unaltered; only the position of the siphuncle, the degree of curvature, the cross section of the shell, and the sculpture undergo insignificant variation.

*   *   *

We emphasized earlier that the initial forms of the nautiloid line were very labile with regard to the length of the septal necks and the external silhouette of the siphuncle and, in a certain sense, still contained all the possibilities for later evolution. In this respect, the roots of the individual lineages appear to be members of a common collective group.

However, with regard to the internal structure of the siphuncle, which is clearly of great significance for the organization of the animals, the forms are distinctly separate from one another. Smooth transitions between the conical endosiphonal sheath of the endocerans, the ring-shaped calcareous pads of the actinocerans with the enclosed, complex vascular system, and further, the radial lamellae of the cyrtocerans have never been found despite the enormous quantities of material available, and because the characters are antithetical, it is also not anticipated that they will be. This can only be an example of *saltational, unmediated changes in organization accomplished in relatively large, individual steps,* which are already manifest *in the early juvenile stages,* at the very beginning of the secretion of the shell.

Even starting with the initial chamber, many far-reaching differences appear. The initial chamber of the actinocerans, for example, is characterized by special features and differs considerably from the same structure in the orthocerans, which is formed in a completely different way. Likewise, we have seen that within the endocerans, very different types of initial chambers and first segments of siphuncles appear, which means that *at the very beginning,* the shell embarked on a divergent evolutionary path.

*   *   *

Another lineage of short duration issuing from the Orthoceracea in the Ordovician is the peculiarly differentiated Ascoceracea. Its shell is dimorphic. Up to a certain point, the shell looks like a slender, slightly curved orthoceran shell with its cylindrical siphuncle. Then, however, the shape changes, becoming sharply distended, pear-shaped; in this portion, the most recent living chamber of the animal (the anterior unchambered section of the shell) and the air chambers (abandoned former living chambers), which normally lie *behind* one another, are to a certain extent pushed *inside* one another.

From a particular growth stage on, the living chamber remains stationary; unlike the situation found in all other cephalopods with external shells, it no longer moves forward as a result of the secretion of the typical septa at its posterior end. The normally chambered posterior portion of the shell, which resembles the orthoceran shell, is probably shed, and from then on, new, vestigial air chambers develop on the dorsal side of the living chamber, rising gradually, step by step, above it. Only the sharply narrowed marginal extensions of the air chambers and septa, and the siphuncle, now distinctly resembling a string of beads, are still formed at the base of the living chamber.

Evidently, these diagnostic, unusual characters of ascocerans also developed *suddenly and discontinuously;* at least, *no transitional forms of any kind* are yet known. However, once this new evolutionary direction has been introduced into the basic features by means of a radically new structuring, progressive differentiation of the septa and the siphuncle continues consistently right up to the last representatives at the end of the Middle Silurian (or better, the Middle Gothlandian). Here, then, we have a *continuous unfolding in small evolutionary steps,* the separate stages of which can be rather well substantiated in an unbroken line.

STRUCTURE AND MODE OF LIFE OF THE PRIMITIVE NAUTILOIDS
When one tries to visualize the organization and mode of life of the cephalopods discussed thus far, one must construe them as creeping bottom dwellers of shallow seas. We have pointed out that the chambering of all of the old representatives is extremely closely spaced. The air chambers are really just narrow cracks, and the amount of gases they were able to hold was certainly insufficient to balance the weight of the soft parts and the shell, which was relatively heavy owing to the numerous septa. Thus, buoyancy through the shell was still lacking. Therefore, it is very likely, as Karl Beurlen has indicated, that these forms, like the rest of the mollusks, still had a foot; this structure had not yet been modified into the funnel that is diagnostic for younger cephalopods and that was used to expel respiratory water and in locomotion (fig. 3.22).

We have no direct information about the number of gills in primitive nautiloids. Since they form the roots of the Recent *Nautilus* (tetrabranchiates, forms with four gills), as well as of squids and octopuses (dibranchiates, which have two gills), there are two possibilities: either they had two gills, and the second pair of gills of the living *Nautilus* is an innovation, or four gills were present in the primitive forms, and the presence of only two in the dibranchiates is due to

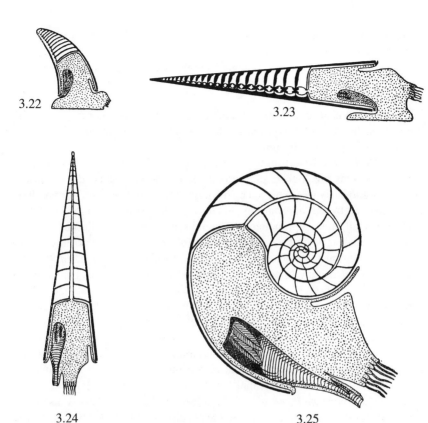

3.22

3.23

3.24   3.25

Fig. 3.22–25. Posture of the shell in various types of nautiloids, shown schematically in longitudinal section.

3.22. The oldest, smallest forms with very closely spaced shell chambering of the type *Plectronoceras* had no buoyancy and could not hover. The benthic-living animals carried their shells much as a snail does and probably still had a snail-like foot. The organization of the body would not have been much different from that of primitive mollusks.

3.23. For shells like those of the *actinocerans,* with extensive, heavy calcareous deposits in the siphuncle and chambers, especially in the ventral portions of the chambers, we may assume a horizontal position. These forms probably also still had a snail-like foot.

3.24. For *orthocerans,* with their widely spaced septa and high, gas-filled chambers without internal calcareous deposits, we can assume the ability to hover and a perpendicular position for the shell; these forms probably also had a funnel.

3.25. In the *Nautilacea,* the shell was carried with the air chambers upward and the living chamber pointing downward.

Reliable evidence for the organization of the soft parts of Paleozoic forms is lacking. Therefore, they are shown much simplified and idealized in the illustrations: For a medial section through the soft parts, only the outline and the mantle cavity, or funnel, is given. The postulation of four gills, two on each side, is hypothetical. All much reduced in size, with the exception of the shell of *Plectronoceras,* which is about 2 ×. (fig. 3.22).

degeneration, analogous to the evolutionary gill reduction in snails. When we combine taxonomically the Paleozoic lineages described and the Recent *Nautilus,* the emphasis is not on the zoological criterium of number of gills but on the similarity of shell type and its relation to the soft parts, and to the unquestionable phylogenetic context.

These oldest nautiloids would still have lacked the sturdy, horny jaws of the living *Nautilus;* we only find fossils of these organs in association with shells of the nautiloid-type spiral from the Permian on.

In contrast to the first, primitive forms, the chambers in typical endocerans and actinocerans are higher; they might cause one to assume that a certain amount of buoyancy was achieved by the gas-filled shell. However, this possibility is obviated by the multifarious biogenetic calcareous secretions inside the shell. The shell shows a pronounced bilateral symmetry in all of its parts.

The siphuncle is never located in the center of the shell but rather more or less close to the ventral side, sometimes lying directly against that side. Further, the siphuncle is extensively filled with internal calcareous secretions beginning at the tip of the shell; these, too, are typically thicker on the ventral side. When combined with the great width of the siphuncle, this adds considerable weight to the shell and works against buoyancy, particularly at the tip.

In addition, in the air chambers there are usually still very thick biogenetic calcareous deposits, which also reach their largest extent on the ventral side and in the region of the tip; often, the ventral portion of the chamber is completely filled. There may be some potential for slight buoyancy remaining, enough to keep the shell *horizontal* and up off the substrate (fig. 3.23). This is a safe assumption, for the animal was hardly in the position to drag a shell that was often one meter or even several meters in length along the bottom behind it.

Validation for the assumption that the position of the shell in life was horizontal is provided by the occasional preservation of traces of color and pattern, which are limited to the dorsal side of the shell, whereas the ventral side, directed toward the bottom, is faintly colored or not at all. These forms would not have been good swimmers and probably remained mostly on the bottom. Nonetheless, they had wide distribution and are found in all Early Paleozoic marine deposits.

From a biological point of view, the Orthoceracea are to be assessed differently. In them, the siphuncle is located at or near the middle of the shell, is narrow, and has no calcareous deposits. These are generally also lacking in the air chambers, or, if present, are very insignificant and deposited evenly around the circumference of the chamber. Unlike the others, this type of shell does not show a distinctly bilateral structure and an extra weighting of the ventral side. Here, the relation of the shell to the environment is the same in all radii; it must have had a *vertical* position in life. The tip of the shell would have pointed upward, for the buoyancy was greatest there; the living chamber, heavy with the weight of the soft parts, hung down (fig. 3.24).

The orthocerans were planktonic forms, with perhaps some active swimmers, in which the foot was beginning to roll to form the funnel. This is true at least for the vast majority of the forms; a few of the collateral lineages, however, returned to life on the bottom.

The Ascoceracea, which in their juvenile stage had an *Orthoceras*-like shell, must have had a mode of life analogous to that of the orthocerans. When the animal matured and shed its shell, it became secondarily bottom dwelling. The heavy living chamber, filled with the soft parts, with its insignificant laterally appendant reduced air chambers, made a swimming mode of life impossible. Here, too, the mature shell typically shows distinct bilateralism.

Only in the orthocerans did the shell attain importance as a mechanism for hovering, after having developed originally purely as a protective structure for the soft parts. Here, the structural design of the shell with its air chambers took on the function consistent with its organization and for which it seemed, in a certain sense, to be destined, with the result that, from that time on, a state of equilibrium was established. The long life span of these forms is probably related to this development.

Apart from that, in other groups of cephalopods we are also able to establish that the freely moving forms had an advantage over the bottom-living, creeping forms and had a broader temporal range. This is true, for example, for a branch of the actinocerans, usually tied to marine bottoms, called the sactoceratids, in which the siphuncle is considerably narrower than that of its relatives and contains only slight amounts of calcareous secretion as ballast. These forms were undoubtedly more mobile and have the longest life spans of all the actinocerans.

Particularly obvious is the progress made by the squids and octopuses as they acquired the capability for active swimming. This mobility gave the group a distinct advantage over the nautiloids, and it was more successful in the long run: it is present in a considerable abundance of forms in modern seas, whereas the once so flourishing nautiloids have all become extinct, with the exception of three closely related species of the single genus *Nautilus*. And we observe the reverse in numerous aberrant ammonite variants that made the transition to a secondarily bottom-dwelling mode of life—their life span is considerably shorter.

But the elongate, straight, rod-shaped orthoceran shell with the downward hanging soft parts certainly does not yet represent a particularly advantageous biological solution. Aside from ascent and descent, actual moving from place to place was probably possible only to a limited extent; lateral navigation was lacking, for the funnel (probably still only incompletely developed), because of its vertical position, could as yet hardly be set into horizontal motion.

THE NAUTILACEA

A much more suitable position for shell and animal was attained through the spiraling of the shell; we see, in contrast to the stalled attempts by many other nautiloid lines, the successful realization of this position in the Nautilacea, the last large group that remains to be discussed here.

In these planospirally coiled forms, the center of buoyancy, which in their straight-shelled orthoceran ancestors lay at one end of the shell, has been transferred to the center of the shell. The underside of the living chamber, with its ballast of soft parts, enclosed the chambered spiral; the aperture was directed toward the side (fig. 3.25); in this position, the funnel was then able to effect lateral, more active movement from place to place, a considerable improvement in the animal's performance. The robust development of the funnel in these forms can be inferred from the hyponomic sinus always clearly in evidence on the ventral side of the shell aperture; in earlier types, this feature is either lacking or, if present, not as deeply incised.

Furthermore, the shell has now become more compact and integrated; the air chambers of the initial portion of the shell are better protected against injury from external sources than they were in the cumbersome primitive forms. The biological advantages of this new, integrated type of shell extend in several directions and have as their consequence the fact that in their competition with the tenacious Orthoceracea, the Nautilacea are finally victorious: they and the freely moving squids and octopuses replace the orthocerans from the Triassic on.

The development of coiled shell types evidently took place several times and from root forms of orthocerans that differed somewhat in particulars of temporal occurrence and morphology. Some of these evolutionary lines may have come to a dead end; one, or perhaps even several, however, were successful and led to further improvement. We have models for this coiling process in the *Discoceras-Palaeonautilus* group (figs. 3.26, 3.27) and, particularly informatively, in the *Lituites* series, various members of which are shown in more detail in figures 3.28–31.

The starting point of this evolutionary series is a form (fig. 3.28) that is basically still straight and rodlike, showing a slight curvature only in the initial portion of the shell, at its tip. Thus, it is thoroughly connected to the orthocerans, which are straight even in their juvenile stage. From that initial form, others derive as the coiling progresses from the tip onward (figs. 3.29, 3.30) until finally a form is achieved (fig. 3.31) in which the spiral makes up the preponderance of the shell. Here, no straight arm such as is seen in the preceding types is formed; from this point on, even the final portion of the shell is curved in a sickle-shape. Only the last segment, the living chamber, departs somewhat from the spiral pattern and does not lie directly upon the previous whorl.

After the basic form of coiled shell has been acquired in the way described above, the unfolding of the Nautilacea proceeds in several parallel lines; this same type of development is also seen in other stocks: collateral lines and different kinds of specialization issue now and then from the main stock and disappear again after a more or less brief life span without having deviated further from the norm or exceeded the boundaries of the structural design of the type.

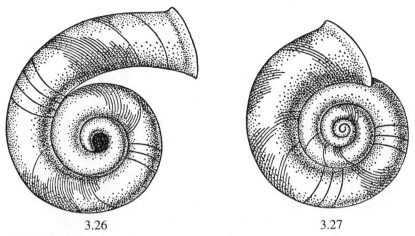

3.26   3.27

Figs. 3.26–27. Two developmental steps in the coiling of the shell in the Nautilacea from the Lower Ordovician of northern Europe.
3.26. *Discoceras.*
3.27. *Palaeonautilus.*
The tightly spaced cross lines, convex toward the front and running parallel to the aperture represent growth lines; the more widely spaced concave lines indicate sutures and are on those parts of the shell that are lacking shelly matter; in other words, they are present on the steinkern.
(After Schindewolf 1936.)

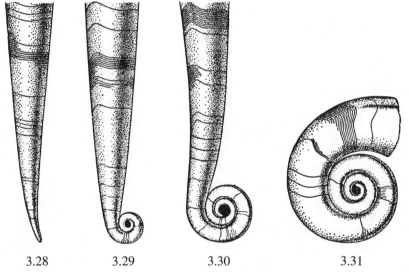

3.28   3.29   3.30   3.31

Figs. 3.28–31. Course of development of the shell coiling in the lituites from the Scandinavian Lower Ordovician, slightly schematic.
3.28. *Rhynchorthoceras.*
3.29. *Ancistroceras.*
3.30. *Lituites.*
3.31. *Cyclolituites.*
(After Schindewolf 1936.)

A widespread crisis in the evolution of the Nautilacea occurred at the boundary between the Triassic and the Jurassic. At the end of the Triassic, the form experienced a period of great diversity, during which it proliferated extravagantly. Richly sculptured forms appeared, some growing to enormous sizes and others developing highly complex sutures; in their shell differentiation they exceed by far the modern *Nautilus*. This entire diversity of forms died off without descendants at the boundary between the Triassic and the Jurassic—except for a *single* line that had not differentiated, one that consisted of small, smooth forms with simple arcuate suture lines (see the diagram in fig. 3.32). This is the only form to carry on past the end of the Triassic; in the Lower Jurassic it became the starting point of a new unfolding of the Nautilacea line.

During the course of further evolution, there were no more sweeping modifications of structure. The line is narrowly specialized and largely stabilized; it has nothing like the flexibility shown by its initial forms at the outset of nautiloid evolution. From at least as far back as the Permian, the organization of the soft parts of nautiloids differs hardly at all from that of the living *Nautilus,* their last descendant.

In contrast, the old conservative line of the Orthoceracea, which had remained simple, had also maintained a greater plasticity, and in the Triassic, through a radical modification, it brought forth a completely new type of form—the squids and octopuses. We do not need to consider here some of the smaller groups shown in figure 3.21, such as the arthrophylls and the pseudorthocerans, which separated from the orthoceran line during the Gothlandian (Silurian) and really represent only parallel evolutionary lines within the orthoceran framework.

CONCLUSIONS

What we wanted to demonstrate by the preceding remarks is summarized as follows: In the unfolding of the nautiloids, the behavior in the initial stage is thoroughly different from that in the further course of the phylogeny. The Ordovician, particularly in its earliest period, is a time of *frenzied evolution* and the sudden appearance of almost all of the types of shells that were ever to appear; these types then proved to be constant and stable over a long period of evolution.

At the very beginning of the evolution of the cephalopod line, the special structural types of the individual subgroups took shape from the mass of labile forms, the various trends being established *in large evolutionary steps, without connecting transitional links.* Innovation is so *sudden* and *volatile* that it is often difficult to decide with certainty which forms are primordial and which derived—which is the ancestor and which the descendant.

Once the basic forms of the individual structural designs had been established, however, a *continuous, slow evolution set in,* which can be clearly followed thanks to the presence of *closed series of gradually changing forms.* This contrast between the brief creative period of the Ordovician, which provided so many forms, and the long period of continued evolution of the nautiloid lineages thus established, which extended through several to many systems and entailed only elaboration and specialization, can be seen clearly in the diagram in figure 3.32.

Ammonoids

Moreover, the old, long-lived central stock of the Orthoceracea is the root of a very important group of cephalopods, one that differs thoroughly from all nautiloids in a great many characters and is considered an independent entity. This is the well-known group of the ammonites and their precursors, which are placed together in the order Ammonoidea.

ORGANIZATION

The typical ammonites, or Ammon's horns, which developed an enormous abundance of forms during the Mesozoic and are of great importance for chronology and stratigraphy, resemble the Nautilacea in that they have planispirally coiled shells. Aside from that, the two types of forms cannot be confused.

An important and easily recognizable distinguishing feature is the *suture*. This, as we recall, is the line along which the transverse cameral partition attaches to the inside of the shell wall. In specimens consisting of only the internal filling of the shell (the actual shell wall is lacking)—the so-called steinkerns— the sutures are exposed to view (pl. 7); in specimens with shells, the sutures can be made visible by grinding or etching the shell away.

In all the nautiloids, these sutures are always very simple, and may run in a completely straight line around the shell tube, may be slightly arcuate, or if divided, may show only a very few forward or backward curvatures. The portions of the line that undulate convexly toward the aperture of the shell are called *saddles,* and those that are convex toward the rear, *lobes.* The sutures of nautiloids, even at their most highly developed, have only a *few* such lobes and saddles; in contrast, the ammonite suture is divided into a great many elements.

In addition, in ammonites, the lobes and saddles exhibit a fine *crenulation,* a breaking-up of the line into minute notches and teeth, whereas in the nautiloids, those undulations of the line are always smooth and unnotched (fig. 3.32). Further on, we shall return to the details and the biological significance of this sometimes extremely exaggerated frilling of the septal margins.

In contrast to this high degree of differentiation in the suture line and the multifarious forms it takes are the great simplicity and uniformity in the evolution of the *siphuncle.* It is always slender and tubular (fig. 3.20); ammonoids do not have wide siphuncles, or siphuncles with the appearance of a string of beads, or any kind of internal siphuncular structures, vascular systems, and so on, as are so frequent in the nautiloids. Further, the ammonoid siphuncle is always very close to the margin, predominantly on the ventral, convex, external side of the shell; in only one small group does it lie on the concave, dorsal side of the whorls.

The *initial chamber,* the protoconch, is also differently shaped; we shall have more to say about this. Ammonoids lack the horny jaws of the Nautilacea; instead, they have the platelike structures of the *aptychus* and the *anaptychus,* which are lacking in the nautiloids. In short, there are a great many far-reaching differences in the organization of the two groups of forms.

Easy as it may be to distinguish a differentiated ammonite from the representatives of the nautiloids, it is sometimes difficult to do the same for the oldest representatives of the line, the primitive goniatites, which were not much different from their orthoceran roots. Nevertheless, *one* set of characters emerges that is diagnostic for all ammonoids and also serves as an easy way to distinguish their simplest, as yet unevolved representatives from the orthocerans: the marginal placement of the siphuncle on the shell wall, together with the ubiquitous siphuncular lobe, a lobe that is coupled with the siphuncle and its position. There is no such consistent combination of characters in the nautiloids.

PALEOZOIC AMMONOIDS

We find the oldest forms having this typical combination of characters in the Lower Ordovician of Bohemia (*Eobactr.,* fig. 3.32). The shell is still straight and rod-shaped, like that of *Orthoceras.* Whether these Ordovician types are directly connected phylogenetically with the ammonoids, which did not appear until later and did not proliferate to produce an abundance of forms until the Lower and, especially, Middle Devonian, is uncertain. Thus far, not enough attention has been focused on those forms that differ only slightly and superficially from the orthocerans; thus, there are still considerable gaps in our knowledge here.

There is a certain striking parallel with the situation at the root of the nautiloid line, namely, with the isolated, sudden appearance of the genus *Volborthella,* which we have described. Similarly, it may have been that *Eobactrites* was also a form that anticipated later evolution, a prototype that was unsuccessful, however, and went no further.

In any event, the actual, consistent, rapidly progressing evolution of the most primitive ammonoids, the Goniatitacea, did not begin until the Lower Devonian and early Middle Devonian. There, at the onset of the evolution, there are small, rod-shaped forms with elongate-ovate protoconchs (*Bactr.,* figs. 3.32, 3.34*a*), which still completely bear the stamp of *Orthoceras,* but which, like *Eobactrites,* distinguish themselves by having a marginal siphuncle and a siphuncular lobe. Apart from this one small lobe, the suture line is still almost completely straight (fig. 3.33*a*).

One derivative genus (*Lobobactrites*) shows advancement over *Bactrites* in

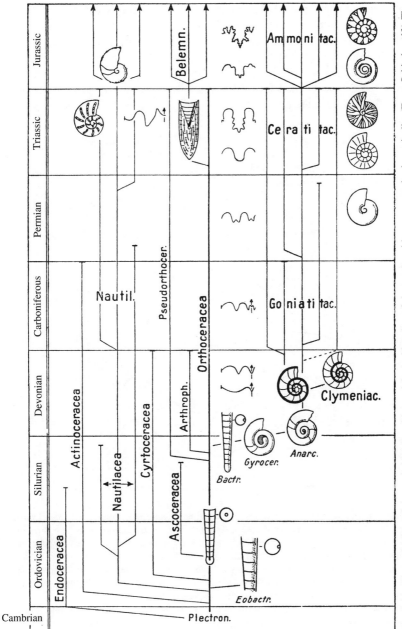

Fig. 3.32. The unfolding of the cephalopods, with special consideration of the ammonoids, represented schematically. (From Schindewolf.) The development of the nautiloids in the early Paleozoic is shown in more detail and with illustrations in figure 3.21. The meaning of the different type faces is the same in both figures.

that the cross section of the shell, originally round, has become flattened laterally and is elliptical; further, on each of the flat sides a broad, arcuate lateral lobe has developed, separated from its counterpart by a high, antisiphonal saddle (so called because it lies on the side of the shell opposite the one where the siphuncle is located, fig. 3.33*b*). The growth lines, too, have taken on a specialized form.

From these types, coiled forms developed very rapidly. Transitional forms such as the lituites at the root of the Nautilacea (figs. 3.28–31) are not present;

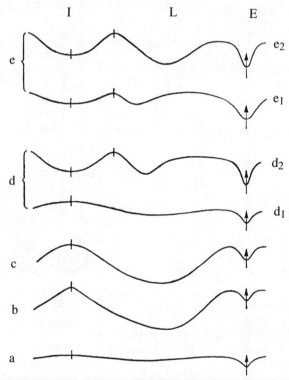

Fig. 3.33. Development of the suture line in the primitive goniatites from the German Lower and Middle Devonian. *a. Bactrites schlotheimii* (Qu.). *b. Lobobactrites ellipticus* (Frech.). *c. Gyroceratites gracilis* Bronn. *d. Anarcestes (Anarcestes) lateseptatus* (Beyr.): d_1 = one of the first sutures with an antisiphonal saddle in the region of the umbilicus without a concave coiling zone; d_2 = a later suture line with an internal *lobe* in place of the antisiphuncular *saddle*. *e. Werneroceras ruppachense* (Kays.): e_1 = First suture line (primary suture), already showing an internal lobe; e_2 = Suture line at maturity. (After Schindewolf 1933.)

In these and all later illustrations of the suture line, the portion representing the inner, antisiphuncular side—the concave zone of the whorl—is at the left, separated from the outer portion of the suture line by a short vertical line, symbolizing the seam of the whorl. The center line of the antisiphuncular saddle—the inner lobe (I)—is also represented by a vertical line; that of the outer lobe (E), is represented by an arrow.

the modification here must have been much more sudden, the result of *large single steps* affecting the entire shell at once.

Lobobactrites is immediately followed by spirally coiled forms (*Gyrocer.,* figs. 3.32, 3.34*b*), in which, however, the descent from straight-shelled ancestors is clearly apparent. In these forms, coiling is at first quite incomplete. The initial portion of the shell does not show a tight, closed spiral; rather, there is a relatively large umbilical perforation, an opening that goes right through the center of the shell. This has to do with the fact that the relatively large, elongate-ovate protoconch, inherited from the bactrites, has not yet become modified to meet the new demands of coiling; at first, its axis is only slightly curved.

The rest of the shell, however, from the first air chamber attached to the protoconch on, shows spiral coiling, however loose it may be at first, in the sense that the individual whorls only touch and do not enclose one another as is typical for all ammonoids. In ammonoids, each whorl is mostly—and sometimes even completely—enveloped by the subsequent one, and consequently, there is a deep, concave impression on the inner sides of the whorls, which is still lacking on the whorls of the primitive *Gyroceratites.*

Apart from the coiling, this genus is identical with *Lobobactrites* in all characters of the siphuncle (fig. 3.34*b*), the suture line (fig. 3.33*c*), and the growth lines. The convex outer side of the shell, where the siphuncle is situated, corresponds to the ventral side of the animal, as can be inferred from an indentation in the aperture and the growth lines and from the slot for the ventrally situated funnel.

Further coiling of the shell proceeded very quickly and in several parallel lines at once. In subsequent forms (for example, in *Anarcestes,* figs. 3.32, 3.34*c*), the protoconch is more compressed and curves more sharply. In connection with this, its aperture becomes more erect and the umbilical perforation grows smaller.

In the region of the umbilical perforation, the suture line still shows the same form as in *Lobobactrites* and *Gyroceratites:* there, on the inner side of the whorl, it has an antisiphuncular saddle (fig. 3.33d_1). The later whorls, however, enclose one another and form on their inner walls a concave zone for reception of the convex outer side of the previous whorl; in the center of this concave zone, at the place formerly occupied by the original antisiphuncular *saddle,* an inner *lobe* develops (fig. 3.33d_2).

Thus, in the ontogeny of these forms, the basic elements of the suture line that are characteristic of all younger ammonoids take shape for the first time: here, the internal lobe (I) appears, supplementing the already present siphuncular, or external, lobe (E) and side, or lateral, lobe (L). In all later ammonoids, the internal lobe develops right from the beginning, from the first suture, in accelerated, abbreviated development (fig. 3.33e_1).

The final stage of evolution, marked by a spirally twisted protoconch with a

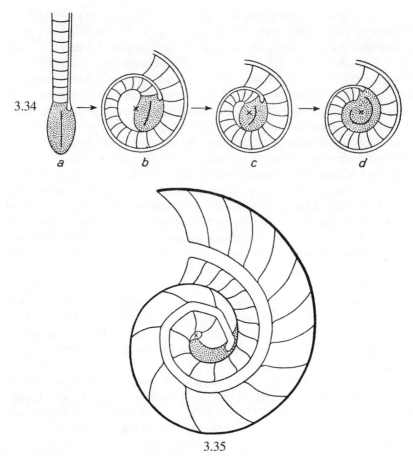

3.35

Fig. 3.34. Development of the primitive goniatites from rod-shaped orthocerans, shown in slightly schematic medial sections through the protoconch. The main steps are as follows: the straight *Bactrites* or *Lobobactrites* (*a*); the Lower Devonian *Gyroceratites,* with more extensive umbilical perforation (*b*); the not yet completely coiled lower Middle Devonian shell type of *Anarcestes* or *Mimagoniatites* (*c*); and finally, the completely coiled upper Middle Devonian forms, without any umbilical perforation, of *Werneroceras* and *Agoniatites* (*d*). The protoconchs are stippled, their longitudinal axes indicated by a heavy dark line. The cross shows the approximate location of the axis of coiling. All 9 ×.

Fig. 3.35. For comparison, a cross section through the nautilid *Discoceras teres* (Eichw.) from the Lower Ordovician of Scandinavia, five times enlarged. In a comparison with the coiled goniatites shown in figure 3.34*b–d,* considerable differences are seen in the position of the axis of coiling, the shape of the protoconch, the size and shape of the plump, very rapidly growing first whorl, and the different position of the siphuncle. Thus, there is no question of the goniatites having descended from the coiled types of the Nautilacea. (From Schindewolf.)

vertical aperture to which the first whorl, which encloses the protoconch completely, attaches (fig. 3.34*d*), has already been attained by the early Middle Devonian.

Moreover, there is a notable difference between this type of coiling process and that of the Nautilacea, in which the space problem posed by the coiling at the beginning of the shell is solved in a fundamentally different way. Here, the originally elongate-ovate or conical protoconch modifies to become a broad, shallow, bowl-like structure, sickle-shaped in medial section, which is very much like the shape of the subsequent air chambers and actually represents only a kind of end cap for the first whorl (fig. 3.35).

Further, in the Nautilacea, the protoconch shifts *to one side* of the axis of spiraling, which is situated within a small, persistent umbilical perforation. In contrast, in the ammonoids the twisted protoconch itself is situated in the center of the shell, and the axis of spiraling passes *through its center* (fig. 3.34*d*).

By the time the first fully spiral ammonoids arose *in a relatively brief period of time* by the process just described, a design had been created that, except for a few aberrant types, underwent *no further basic alterations during the subsequent long evolutionary period,* which extended from the late Middle Devonian to the Upper Cretaceous. Immediately after the final form had taken shape, however, there was a powerful impulse toward splitting into multiple evolutionary directions marked by different ways of structuring the siphuncle and the sutures, changes in the sculpturing, and special kinds of shell construction within the framework of the spiral architecture.

The first significant event is the separating off of the Clymeniacea, which are distinguished from the Goniatitacea, with their external (ventral) siphuncle, by having a consistently internal (dorsal) siphon (fig. 3.32). Clymenians are incontestably derived from the goniatites; nevertheless, in spite of the abundance of material available and the existence of continuous profiles, there are no forms that show a gradual transition between the two different types of organization; *the transformation must have taken place very rapidly and abruptly during early stages of ontogeny.*

Both the goniatites and the clymenians split up into a number of different evolutionary lines, which evolve in parallel. From the Goniatitacea (perhaps also from certain of the Clymeniacea) come the Ceratitacea, in the Triassic, and from those, in turn, come the Ammonitacea, in the Jurassic; the two groups proliferated, producing an enormous wealth of forms (fig. 3.32).

General Evolutionary Trends

If we look first at the ammonoid line in its entirety, we see that it is dominated by certain gradual evolutionary tendencies, similar in kind but by no means always simultaneous, that take place in a great many parallel lineages. One example that should be cited is the gradual reversal of the *curvature of the septa.*

In the most primitive forms, goniatites of the Lower and Middle Devonian and some of the Upper Devonian, a medial section (fig. 3.36*a*) shows that the septa are *concave* forward, toward the aperture, and in this respect are identical with the situation found in their ancestral forms, the orthocerans, and with the nautiloids quite generally (figs. 3.21–25, 3.35).

In the rest of the more advanced Upper Devonian goniatites, the septa are concave only in the first whorl, after which they gradually reverse themselves; in the mature stage, they are *convex* forward (fig. 3.36*b*). As the lineage continues, convexity appears at ever earlier ontogenetic stages (fig. 3.36*c–e*). From the Carboniferous on, the first septum usually bulges forward, but there and even on into the Permian and the Triassic, there are isolated forms, evidently members of more slowly changing lineages, in which the first septum or septa are concave, and only the later ones show the flattening and then the forward bulging. Finally, in all Jurassic and Cretaceous ammonites, even the first septum is thoroughly convex (fig. 3.36*f*).

The same evolution takes place independently in the Clymeniacea line: the septa in the oldest forms are concave forward, and those in the most recent forms are convex.

The attainment of convex septa represents a biological advance. On one hand, the resistance to water pressure afforded by this type of domed construction is greater than in septa bulging in the opposite direction, subject to tension. For one thing, convexity allows for the regulation—the reducing or expanding—of the quantity of gases enclosed between the septum and the septal membrane attached to it, a space that functions as a "swim bladder."

*   *   *

Further, the ammonoid stock shows an observable, gradual, step-by-step *reversal of the septal necks,* those ring-shaped or tubular extensions of the septa through which the siphon passes. In Devonian and Carboniferous goniatites and in clymenians, as well, they are directed *backward* (fig. 3.36*a–c*), once more in agreement with the nautiloid mode of construction (figs. 3.21–25). Even in some Permian and Triassic ammonoids, the septa remain pointing toward the rear for the life of the animal.

But in the dominant, phylogenetically more advanced forms of those groups, this situation is still found only in the initial stages, and then gradually, at ever earlier stages of ontogeny, the septal necks reverse (fig. 3.36*d, e*). The original septal extension, directed backward, is progressively shortened, and in its place, at the point where the septum is perforated, a *forward-pointing* collar develops, short at first, becoming gradually longer. In a certain sense, the septal necks are pushed through the septa from the rear to the front.

In Permian ammonoids (fig. 3.36*d*), this reversal takes place during the formation of the third to sixth whorls; in Triassic ammonoids (fig. 3.36*e*), in accel-

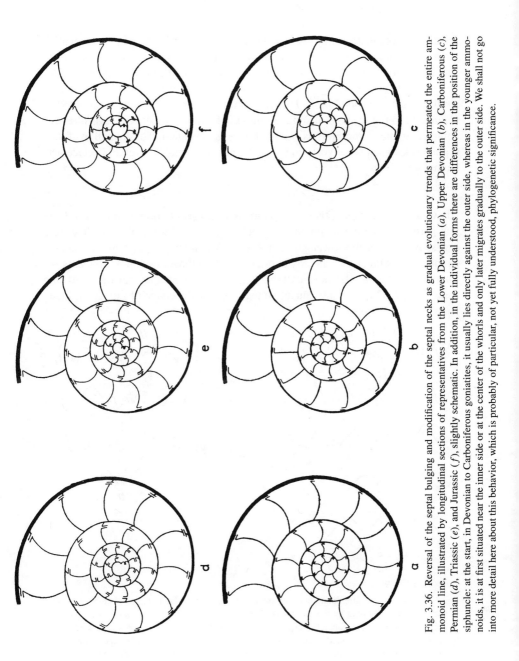

Fig. 3.36. Reversal of the septal bulging and modification of the septal necks as gradual evolutionary trends that permeated the entire ammonoid line, illustrated by longitudinal sections of representatives from the Lower Devonian (*a*), Upper Devonian (*b*), Carboniferous (*c*), Permian (*d*), Triassic (*e*), and Jurassic (*f*), slightly schematic. In addition, in the individual forms there are differences in the position of the siphuncle: at the start, in Devonian to Carboniferous goniatites, it usually lies directly against the outer side, whereas in the younger ammonoids, it is at first situated near the inner side or at the center of the whorls and only later migrates gradually to the outer side. We shall not go into more detail here about this behavior, which is probably of particular, not yet fully understood, phylogenetic significance.

erated development, during the second and third; and in the phylogenetically youngest representatives, the ammonites from the Jurassic and Cretaceous (fig. 3.36*f*), the primitive form of septal necks directed backward no longer develops at all—instead, right from the outset, a transitional form is emplaced similar to the one found in the second and third whorls of Triassic ammonoids. Compared to the reversal of septal doming, which in some instances is already complete in the Carboniferous, the phylogenetic change in the orientation of the septal necks takes place very slowly and is not yet complete even in Cretaceous ammonites.

*   *   *

Above all else, however, there is a progressive *differentiation of the suture line* that permeates the entire ammonoid stock with its multitude of individual lineages. At first, in the lower goniatites and clymenians, sutures are extremely simple, consisting, as we already know, of only the three basic lobal elements— internal (I), lateral (L), and external (E)—and the saddles separating them (fig. 3.33*e*, 3.37*a*$_1$). In subsequent younger forms, the shape of the lateral lobe

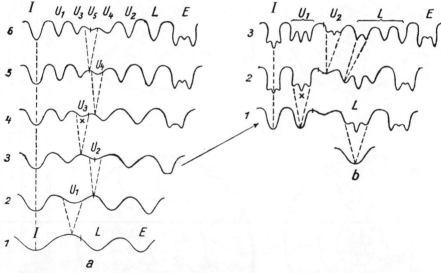

Fig. 3.37. Progressive differentiation of the suture through an increase in the number of lobes and saddles. *a. Saddle-splitting* by the process of alternating (ventropartite) division of the saddle and its subelements (umbilical lobes U_1–U_5), which are located between the inner (I) and lateral (L) lobes. *b. Lobe-splitting* by saddles arising in lateral lobe L in the two umbilical lobes U_1 and U_2. The ontogeny of the suture line in a Permian adrianitid (*a*) and in a Permian stacheoceratid (*b*) forms the basis for these illustrations. For each of these developmental stages there are corresponding phylogenetic stages of forms showing the illustrated degree of suture differentiation as a mature feature. (From Schindewolf.)

is transformed from shallow and curved to deep and pointed, and the number of lobes and saddles increases.

This progressive subdivision of the suture proceeds in a very regular, precise way. First, the internal saddle, located between I and L, divides; at its apex, an umbilical lobe (U_1) appears, shallow at first, then growing deeper (fig. $3.36a_2$). Many forms remain at this stage for the rest of their existence; in the phylogenetically younger forms, however, an analogous lobe, termed U_2, appears in the saddle U_1/L (fig. $3.37a_3$). This splitting of the saddle located between I and L may continue with the development of additional umbilical lobes (U_3, U_4, U_5, and so on—fig. $3.37a_{4-6}$), and in the process, moreover, several different kinds of sequences and their further development can be distinguished.

Very interesting insights are revealed if the progressive differentiation of the suture is followed in detail phylogenetically in its relation to ontogeny (fig. 3.37, right). The ontogenetically first suture, or *primary suture,* of primitive ammonoids (1 in the Devonian series in figure 3.37), which is easily exposed by removing the individual whorls back to the initial chamber, shows the familiar three elements of the primary developmental mode. To the extent that there is in *Devonian* representatives any further development at all, the elaboration of a new lobe, U_1, normally takes place between the twentieth and the twenty-fifth suture. In *Carboniferous* forms, the primary suture is the same as in Devonian forms; however, lobe U_1 now appears in accelerated, abbreviated development in the second or third suture, after having moved up to this early developmental stage gradually in certain transitional forms.

In *Triassic* ammonoids, further abbreviation of the developmental process results in the same kind of differential stage being emplaced immediately in the primary suture, and then, somewhere between the twelfth and sixteenth sutures, a second umbilical lobe (U_2) forms. Further, the external lobe is divided by a low medial saddle, which then appears regularly in all younger ammonoids from the Carboniferous on (but in the Carboniferous, not emplaced until an advanced developmental stage).

Finally, *Jurassic* and *Cretaceous* ammonites, dispensing with the previous course of development, start out in the very first suture with such a developmental stage (two umbilical lobes and a medial saddle in the external lobe), and then, somewhere between the twelfth and fifteenth sutures, develop further by emplacing another umbilical lobe (U_3). This manner of suture development provides a very impressive, directly quantifiable example of *accelerated development,* of the skipping over of older stages of differentiation and the moving up of newer ones to ever earlier stages of ontogeny.

Now, in ontogeny there is yet another suture that precedes the primary suture—that of the septum that closes off the initial chamber. In form, it is completely different from the primary suture and is contrasted with it as an embryonic suture, or *prosuture.*

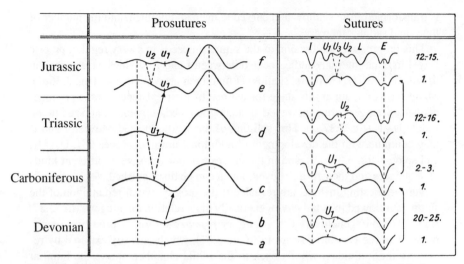

Fig. 3.38. Phylogenetic development of the prosuture and the primary suture in ammonoids. *Prosutures: a, b.* Asellate type without a well-defined saddle. *c.* Latisellate A-type with broad inner and outer saddles. *d.* Latisellate B-type with the inner saddle divided by a small lobe (u₁). *e.* Angustisellate A-type with tongue-shaped narrowing of the outer saddle and displacement of the original inner lobe u₁ toward the seam. *f.* Angustisellate B-type divided by another small lobe (u₂) in the inner saddle. (Since this lobe is laid down dorsal to u₁, it is not homologous with umbilical lobe U₂, which arises ventral from U₁, of the subsequent true suture.) *Sutures:* The illustrations show a pair of sutures from a typical representative from each segment of evolution (systems). In 1, the primary suture is shown; added to it in each case is the younger stage at which the number of lobes increased. The arrows refer to those ancestral stages that are taken on by the primary sutures. (From Schindewolf.)

Instead of having, as the primary suture does, an internal and an external *lobe*—the most important typical features of every true ammonoid suture—the prosuture develops an inner and an outer *saddle;* in younger representatives, the forward arching of these saddles is ever more pronounced, accentuating the divergence of the prosuture from the primary suture. Further, the sequence in which newly arising elements (u₁ and u₂) are emplaced in the course of phylogenetic unfolding is different from that followed in the primary suture and the sutures developing subsequent to it. This situation can be seen in the illustration in figure 3.37, left; this is not the place to describe it in more detail.

It should only be pointed out that it appears impossible to trace the primary suture back to the prosuture because of the large morphological gap between them and because of their completely different modes of elaboration. The primary suture, however, is closely tied to subsequent sutures with regard to form and furnishes the point of departure for their always gradual further development.

In each individual ammonite, the primary suture is connected phylogenetically to the primary suture of its ancestors (or to one of its subsequent, more advanced stages of differentiation) by skipping over the prosuture, which always appears only singly and which is inserted into this developmental process as a completely alien element and is itself connected to the prosuture of predecessors. Within individuals, this development of the prosuture does not continue; the prosuture represents a developmental dead end and does not have anything to do with the formation of actual true sutures. Structures corresponding to the prosuture never appear in the suture lines of adult individuals.

In ammonoids, then, between the secretion of the prosuture and the primary suture the organism most have undergone a kind of metamorphosis, an abrupt, radical modification. This makes the prosuture a typical larval structure, thus providing a rare instance of a fossil in which we can directly observe *the relationship between the evolution of larval or embryonic elements and that of the definitive organs.* This is the reason that we mention, at least briefly, this somewhat special situation.

Taking this exposition further, we observe that in certain ammonoid lineages, instead of or along with the splitting of the *inner* saddle there is also a corresponding division of the *outer* saddle. There, also in regular succession, carrying the evolutionary trend already embarked upon forward in a consistent manner, new lobes, termed adventitious lobes (A in figure 3.39), arise. Here, too, we observe the same behavior as in the derivatives of the inner saddle, namely, that the new lobe is first emplaced at an advanced stage of ontogenetic development and then, in descendants, moves up step by step to appear in ever earlier sutures. Subsequently, in late ontogeny a new lobe is once again emplaced, which, in descendants, follows the same developmental path.

In contrast, other ammonoids drop the principle of *saddle* splitting, described above, back in early juvenile stages and introduce *lobe* division as a new way of increasing the number of lobal elements: starting in the lateral lobe (L) and often also in the two umbilical lobes (U_1 and U_2), two saddles arch forward from each, breaking up the originally uniform lobe into three segments, which quickly become independent and are in some cases again divided by the arching of a saddle (fig. 3.37b).

Obviously, there can be no linking transition of any kind between these two fundamentally different ways of elaborating the suture line; the two paths toward differentiation diverge at very early juvenile stages, and there are no transitions between these early sutures and the adult sutures, which are composed in a completely different way (fig. 3.37a_6 and b_3). In itself, it would be entirely conceivable that the different ways in which the splitting of lobes and saddles was accomplished were first established late in ontogeny and then moved up into early juvenile stages. Nevertheless, we have no proof of any kind for such a gradual process; even the oldest representatives of the principle of lobal splitting embark upon this path at a *very early developmental stage.*

From the Permian on, and in isolated cases even in the Carboniferous, a

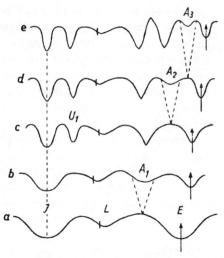

Fig. 3.39. Progressive differentiation of the suture through unilateral ventropartite splitting of the saddle and its subelements located between the lateral (L) and external (E) lobes. *a, b.* Two stages of the ontogeny of a primitive representative of the genus *Cheiloceras,* an early Upper Devonian goniatite. *c.* Suture of a highly differentiated, younger species of *Cheiloceras,* in which the umbilical lobe has been added and, further, the rest of the lobes have become deeper, narrower, and in some instances, pointed. *d.* Suture of the genus *Sporadoceras,* which evolved from *Cheiloceras,* with two adventitious lobes. *e.* Suture of the late Upper Devonian *Discoclymenia* (which, in spite of its confusing name, is a goniatite), with three adventitious lobes. In the ontogeny of *Discoclymenia,* the preceding series of phylogenetic steps of the suture is repeated.

crenulation of the lobes appears that is to a large extent diagnostic for younger representatives of the ammonoid stock and itself proceeds in a progressive, regular way (figs. 3.32 and 3.152). Here, the bases of the lobes do not remain entire or simply pointed, as in the primitive forms already presented, but are notched to produce two or three points or are even finely saw-toothed, a mode of elaboration that is very diagnostic for many representatives of Triassic Ceratitacea (pl. 7 and fig. 3.39*a*).

In other forms from the Triassic system, notching is not limited to the bases of the lobes but rises up the sides of the saddles and then either leaves the apex of the saddles looking like smooth-edged loops or goes on to notch these, too. After that, a *secondary* notching sets in, dividing the primary teeth into even smaller units and leading ultimately to sutures as complex as the example shown in figure 3.14.

A basic advance in notching is introduced in Jurassic ammonites in that it proceeds not only from *one* pole, the base of the lobe, but from *two* poles simultaneously, from the lobe *and* from the apex of the saddle, thus working against itself (bipolar notching of Rudolf Wedekind, see figs. 3.32, 3.152, 3.162–64).

Finally, in many of the most recent ammonites from the Cretaceous, a reduction of the notching takes place. There are forms that show a fine, even serration of the base of the lobes, which makes them resemble to a large extent the considerably older ceratites of the Triassic system (fig. 3.40a, b). In spite of this, there is no phylogenetic connection, nor does the same form recur in identical fashion. Individual saddles of these forms show notching at their apices, proving that the animals in question are degenerate descendants of bipolarly notched forms and, therefore, that they are at a higher phylogenetic level than the Triassic ceratites.

It used to be that this circumstance went unnoticed, and a direct connection with the superficially similar Triassic forms was carelessly assumed. It did not seem to matter much that among the millions of ammonites from that vast period of time between the Triassic and the Upper Cretaceous no transitional links have ever become known: gaps in the fossil record! G. Steinmann postulated that transitional forms may have lived in the area where the Indo-Pacific Ocean is located today. Since then, however, an abundant ammonite fauna from the Jurassic of that geographic area has been described, and among it not a single ceratitic form has been found. Here, too, the cliché about gaps has proved untenable, especially since we now know that the phylogenetic relationships lie in a completely different direction.

The strict pattern according to which suture lines developed, the different ways of splitting into lobes and saddles, and the multifarious forms of notching and combinations of notching are extremely important for the taxonomic and evolutionary understanding of ammonites. But the progressive differentiation of the suture plays an important functional role as well.

For one thing, the intensive frilling of the septal margin, the result of which is, of course, the suture line, increased the solidity and rigidity of the shell. For another, it graphically demonstrates the increasing differentiation of the muscles of the septal membrane, which attach to the curves and notches of the suture. The suture, in turn, is extremely closely related to the regulation of the air chamber, which is enclosed between the septum and the septal membrane and is of critical importance to the animal in diving and swimming. For this reason, and because of the forward bulging of the septa and the thinner shell, ammonoids, as

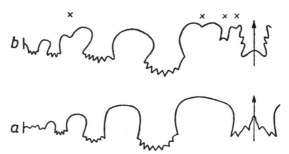

Fig. 3.40. Comparison of the sutures of a "true" ceratite (a) from the Triassic (Upper Muschelkalk) and the "Cretaceous ceratite" *Tissotia* (b), from the Upper Cretaceous (Emscher).

a whole, must have been considerably more mobile, agile swimmers and divers than the Nautilacea, with their otherwise similar shell form.

MESOZOIC AMMONOIDS

Although the evolution of the suture, as we have said, permeates the entire extent of ammonoid stock, from the Devonian on into the Upper Cretaceous, its unfolding is by no means completely uniform. Rather, it is subdivided into different levels, or cycles, which seem to be surprisingly similar. We have already discussed the older cycle of the goniatites and clymenians; now we shall turn to the behavior of Mesozoic ammonoids. Within the overall framework of the directional elaboration of the suture there are evolutionary processes of shorter duration affecting the *sculpture*.

One such level of unfolding occurs in the *Triassic*. The most primitive, simplest representatives of the Ceratitacea dominant there are smooth forms without distinct ribbing. Soon, simple, radial, unsplit ribs appear (fig. 3.32). These sculptured forms are then followed by others, whose ribs bifurcate near the external side. In the course of further evolution, the point where bifurcation occurs moves forward toward the umbilicus, and the branches of each individual rib split again toward the outside. (There is not enough space in figure 3.32 to show all the separate stages of the sculptural evolution; therefore, only one simple, early form and the final result are given. A more complete presentation is seen in figure 3.152.)

This progressive evolution from smooth shells to those with multibifurcate ribs takes place within a whole cluster of parallel lineages, not always absolutely simultaneously and in many conservative or prematurely extinct lines also not attaining the highest stage of sculpturing, but we can still speak, in general, of a directional, steady progression of sculpture. All these lines then died off conspicuously, after having reached their highest degree of sculptural specialization, at the boundary between the Triassic and the Jurassic—all except one, which remained smooth and undifferentiated and survived this critical juncture to become the starting point of a new cycle of ammonoid evolution, of a renewed and profuse proliferation.

Exactly the same course of sculptural evolution that we see in the Triassic is repeated in the *Jurassic-Cretaceous* era in the Ammonitacea: at the beginning, there are once more smooth, unsculptured forms; these go on to develop simple ribs, then bifurcate ribs, then multibifurcate ribs (figs. 3.32 and 3.152). Thus, we encounter types with split ribs at the lowest level, in the goniatites and clymenians; at the midlevel, in the ceratites; and at the upper level, in the ammonites. However, these levels differ in the way evolution progresses and in the correspondingly varying degrees of differentiation of the suture.

Since external environmental factors do not play a role in the extinction of diverse forms of Triassic ammonoids at the Jurassic boundary, the reasons must be sought in the animals' internal organization. We have seen that Jurassic ammonites show, in addition to the gradual progressive evolution of numerous features inherited from their ancestors, a newly acquired character: the notching of the suture proceeds not only from the base of the lobe but also from the apex of the saddle. It cannot and should not be maintained that this modification *by itself* is decisive and that the Triassic lineages that were denied this advance were *therefore* condemned to extinction.

Since the notching of the suture accommodates the attachment of the septal membrane musculature, the possibility so created of establishing forward points of attachment at the apex of the saddle, thereby improving the diving mechanism, may have been of at least some significance. We should probably assume, however, that there were, in addition, still other and probably more important internal modifications of the organism that we cannot read from the hard parts that have been preserved. The character we see—the notching of the suture—may therefore be only the visible, abbreviated expression of organizational changes that occurred.

Such a structural modification was impossible for the mass of Triassic ammonoids. They died out without descendants—and as we have just said, it was precisely *those lineages that were the most highly differentiated* with regard to sculpture and other features that disappeared. The extinction was preceded by *overspecialization* and *symptoms of decline* of the most various kinds. The basic type—a flat, spiral shell, so consistently and tenaciously adhered to, degenerated. Shortly before extinction, forms appeared in which the final whorl became detached from the rest of the spiral (fig. 3.41*a*); in other forms, the shell became secondarily straight (fig. 3.41*b*), and in still others, the shell was transformed into a snailshell-like spire (fig. 3.41*c*). In addition, reductions in the sutures and other characteristics of degeneration are in evidence.

The changeover from unipolar to bipolar notching of the suture happened *abruptly during early ontogeny;* whether or not a notching pole was to be established at the apex of the saddle was already settled in the minute stages of the shell. This transformation (or the change of organization connected with it) took place in only one main group—the phylloceratids—which, simple and unspecialized, had retained its smooth form; this line continued to evolve past the end of the Triassic.

Once the new basic organization was established, another intensive radiation set in immediately, *in large evolutionary steps and at a rapid evolutionary tempo;* the stock broke up into a great many different models and separate lineages, which we encounter in particular in the lowest Lias of the Mediterranean area (fig. 3.32). Later on, within each lineage, evolution proceeded *gradually and continuously in small, individual steps.* Here, we observe many transitional, intermediate forms; new characters, usually quantitative in nature, affecting the prominence and number of individual structures, are acquired toward the end of

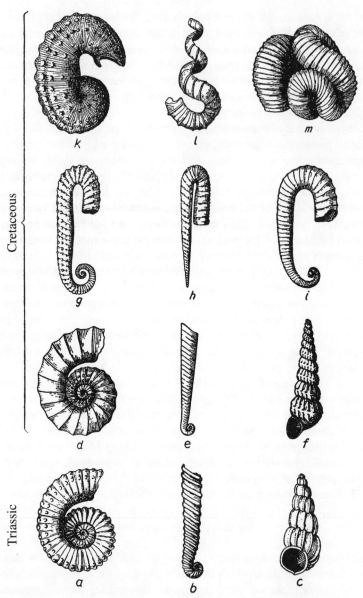

Cretaceous

Triassic

Fig. 3.41. Aberrant shell types of Upper Triassic and Cretaceous ammonoids showing the dissolution of form within a stock facing extinction, in some instances with broad similarities of form. *a. Choristoceras. b. Rhabdoceras. c. Cochloceras. d. Ammonitoceras. e. Baculites. f. Turrilites. g. Ancyloceras. h. Hamulina. i. Heteroceras. k. Scaphites. l. Hyphantoceras. m. Nipponites.* Most considerably reduced. (After Janensch, d'Orbigny, Roman, von Zittel, and others.)

ontogeny and then, in descendants, are advanced toward ever earlier stages of development.

* * *

The more recent evolutionary cycle of ammonites, which, beginning in the Lower Jurassic and extending on into the Upper Cretaceous, encompasses a considerably longer period of time than the Triassic cycle,[3] provides an exact copy of the previous evolutionary phase in many respects. As we have already said, it is dominated by a progressive evolution of the *sculpture*—implanted in individual series and through which each one passes independently; this mode of evolution also occurred in the Triassic Ceratitacea. Within the individual lineages, shorter evolutionary trends appear once more, with regard to *shell* type. We see forms supplanting one another over and over—shells with a broad or narrow umbilicus; shells with a high or low aperture; shells that are broader or narrower or, in the external portion of the whorl, more pointed than curved in cross section; with or without keels; and on and on (fig. 3.152).

These different forms of shells, with more or less pronounced sculpture, sometimes combined with sturdy nodes and long spines, are the actual *adaptive forms,* intimately related to the mode of life of the species in question and decisively influencing water resistance and, consequently, swimming capability. The same types of sculpture and shells seen in Triassic ceratites are repeated in Jurassic and Cretaceous ammonites, so that for almost every single older form there is a corresponding more recent counterpart, which differs only in the formation and degree of evolution of the sutures.

Further, more recent ammonites, after having experienced a *phase of explosive evolution* and a long *period of gradual progressive specialization,* arrive at a *phase of overspecialization and decadence of form* completely analogous to the one the Triassic representatives experienced. The closer we get to the end of the Upper Cretaceous, to the final extinction of the ammonite stock, the more frequently we encounter degenerate forms: with the final whorl detached and showing many differences with regard to details of shape (fig. 3.41*d,g,i,k*); with secondarily straight shells (fig. 3.41*e,h*); with snail-type spirals (fig. 3.41*f,l*); and finally, with shells all coiled together in a completely irregular tangle (fig. 3.41*m*).

No linking, transitional forms whatsoever between the different types of shells—between the normal planospirally coiled shell and either the corkscrew snail-type spiral or the tangled coil—are known, and owing to the nature of things, they are not anticipated. The transformation in the mode of coiling must

3. Since the Jurassic and the Cretaceous together lasted about 115 million years, and the Triassic can be set at 25 million years, the more recent Jurassic and Cretaceous evolutionary stages lasted four to five times longer than the older Triassic.

have taken place *discontinuously, in early ontogenetic stages,* at the beginning of shell formation.

In shell shape, the straight-shelled Cretaceous ammonites repeat the corresponding Triassic types, as do the bactrites and the orthocerans, which mark the beginning of ammonoid evolution. Just as we have seen with Cretaceous "ceratites," however, here too it is not a matter of a complete reversal of evolution, not a reappearance of completely identical forms, for the individual types differ sharply in the extent to which the sutures have developed. Moreover, the secondarily straight shells with their spirally twisted initial chamber and the coiling of the first whorl still show clear evidence of their origin from coiled forms, whereas in the primarily straight shells, the initial chamber is ovate and not coiled.

Furthermore, toward the end of the Cretaceous, overspecialization is evidenced by the presence of *gigantic forms* with shells of up to 2.50 meters in diameter (fig. 3.14*c*), which doubtless exceeded by far the size proportions that were optimal for ammonites and contrast sharply with primitive ammonoids, which have an average size of only a few centimeters (fig. 3.141*d*). All these overspecialized, degenerate forms were incapable of further evolution, and since there was no vigorous revival of ammonite stock at the Cretaceous-Tertiary boundary, their fate was sealed.

A Glance at the Dibranchiates

For the sake of completeness, the last group of cephalopods—the dibranchiates [Coleoidea], or squids, cuttlefishes, and octopuses—get at least a few words as an afterthought. They developed during the Triassic from orthocerans, thus going back to the group that gave rise to so many different lines of cephalopods.

In these forms, the increase of swimming capability followed a path different from the one we have seen in the nautiloids and ammonoids. Whereas in the latter, an ecologically more practical shell form was attained through the coiling of an originally long, straight shell, in the dibranchiates the straight form of the calcareous skeleton is retained; however, the soft parts grow backward, enclosing the chambered shell and depositing on it a thick calcareous sheath (figs. 3.42–44; fig. 3.32, Belemn.).

In the process, the buoyancy of the tip of the shell was offset, causing the equilibrium of the shell to shift from the vertical, which was disadvantageous, to the horizontal. The funnel changed as the two lobes fused to form a consummate organ for swimming; the animals became active swimmers, shifting to a purely nektonic mode of life.

This better swimming capability based on a perfected organization of the soft parts finally made the shell superfluous as a structure for support and buoyancy. The chambered portion of the shell was increasingly reduced and replaced by a muscular mantle; at first, the calcareous sheath enclosing the shell remained as an internal support structure, and finally, it, too, disappeared. This evolutionary process can be clearly followed in every detail in the decapodes, or ten-armed squids—the Belemnoidea, the Teuthoidea, and the Sepioidea. According to facts pointed out by P. Schulze, these differences between the squids and octopuses, on one hand, and the tetrabranchiates, with their external shells, on the other, go back to a *simple modification during early ontogeny,* namely, to the fact that the shell gland did not fill out but closed up instead.

It has often been postulated that ammonoids, too, underwent a similar kind of shell reduction and that they did not actually die out but lived on as modern shell-less forms, as octopodes, or eight-armed octopuses. This concept has very little plausibility. Given the good capacity for preservation of the shells and the amount of material available to us, there would have to be some observable trace of the reduction and modification of the shell, just as there is for the decapod squids, for which the evidence is indeed present in as much detail as could be desired.

Most unlikely, moreover, is the notion that the living genus *Argonauta* (p. 16C)—the familiar "paper boat"—which has also been found as a fossil from the Late Tertiary on, is a descendant of the ammonites. The spiral shell secreted by *Argonauta* females does indeed offer a certain superficial resemblance to the ammonite shell, but in morphological particulars (shell structure, lack of chambers and siphon, and so on), there is no comparison at all. Further, as a structure produced by the epidermal glands of the arms, the shell is not in the slightest homologous to the shell secreted by the mantle epidermis. Finally, the embryonic stage of *Argonauta* shows a small shell sack as a vestige of a rudimentary internal shell, which indicates a descent from forms whose shells had already been internalized. There are no grounds at all to support the view that this was originally the shell of an ammonite.

CONCLUSIONS

When we summarize the evolutionary events described in this survey of the unfolding of the ammonoids, we must see that their evolutionary process does not consist of a uniform, smoothly progressive modification of species but rather that it is decidedly *phasic with quantum transitions.* The stock unfolds stepwise, the different levels distinctly separated by critical turning points, and exhibits an unmistakably *periodicity.* Just as an hourglass narrows at the center, so is the breadth of the stock sharply and repeatedly constricted.

Further, the individual characters—hierarchically arranged and superimposed upon one another—clearly behave in various ways. The modification of several sets of characters—septal bulging, septal necks, sutures—permeates the stock throughout its extent. These are *unique, orthogenetic,* linear progressions, which, the suture development in particular, are the main indicators of the phylogenetic position of the individual forms and the determiners of their chronological classification. As the sutures and the sculpture evolve, three successive levels of phylogeny can be distinguished—Paleozoic, Triassic, and Jurassic-Cretaceous—which, in outward appearance, show broad similarities.

At the beginning of each of these three levels there is a *thorough reorganization* of structure, carried by only a single line of unspecialized forms, one slender shoot from the former abundance. This reorganization is *abrupt, discontinuous, and takes place in early stages of ontogeny;* each time it happens, it creates a new, general, basic form, which immediately, in a rush of evolution and with large transformational steps, splits into different designs. Each *phase of explosive form-building* inaugurates a new radiation, and soon the stock, decimated by mass extinction at the end of the previous level, is restored to its former breadth.

The first phase of heightened evolutionary intensity then gives way to a *period of calmer, more gradual transformation,* which proceeds *continuously, in small, separate steps,* usually in *late stages of ontogeny.* Its essence consists of the *differentiating elaboration* of a previously created structural basis. At the end of the evolutionary cycle, we encounter once more an extinction of all the specialized lineages in conjunction with the most varied *manifestations of degeneration.* (Even at the end of the lowest level, we find shell abnormalities of a most significant kind, but we shall not go into more detail about them here; see p. 216ff.)

Within the framework of the directional, expansive evolutionary course taken by the *characters of the first rank* already named, evolution, likewise clearly directional, of the sculpture also takes place; it is narrower in scope than that of the sutures and, as an *evolution of characters of the second rank,* is limited to a particular level. It is definitely rhythmic and periodic in nature in that the sequence of types of ribbing is repeated three times, beginning anew in each cycle with the reappearance of the same forms at a higher level.

Just as the evolution of characters of the first rank affects the entire stock, so this second-ranking evolutionary course permeates the entire level, holding sway equally over all the parallel lineages included in it. Thus, the degree of sculptural evolution establishes the phylogenetic position of a given form within the framework of the category to which it belongs.

Subordinate in turn are *characters of the lowest rank* and least temporal extension, such as the shape of the shell, the evolutionary course of which marks shorter segments of the lineages and is repeated several times during the course of a single sculptural cycle. Only these features have a direct bearing on the mode of life and are indeed much more critical for swimming and mobility than is the general evolutionary level of the stock to which the bearer of a particular shell belongs.

If this phylogenetic pattern is also exhibited by other stocks and is thus a universal phenomenon, it must be of primary importance for the interpretation of evolutionary processes.

The Unfolding of the Stony Corals

The oldest verified stony corals (Madreporaria) occur in the upper Ordovician. Corals have indeed supposedly been found as early as the Cambrian, but these are extremely doubtful. For the time being, it must be assumed that the first secretion of a calcareous skeleton by a coral took place during the Ordovician, permitting us to speak of true stony corals. Their predecessors are thought to have been soft-bodied organisms that had not yet developed fossilizable hard parts.

Significant gaps in the fossil record of true stony corals, from the time they secreted a skeleton on, are not to be assumed; their first observed appearance coincides largely with the beginning of their existence. This is inferred from the fact that we already know of structures that resemble coral reefs in the Cambrian, but these were not constructed by stony corals, as in later times, but by archaeocyathids (fig. 2.8). This group of forms used to be combined with the corals, but recent research has shown that the Cambrian reefs were built by poriferans, in fact, by a very peculiar group of extinct sponges. From the Ordovician on, we have an uninterrupted record of the evolution of stony corals provided by calcareous skeletons in an excellent state of preservation.

Immediately upon their first appearance, corals, like cephalopods, break up into a great many divergent trends of differentiation. Right from the beginning, we have side by side the two main groups that typify the entire Paleozoic: the pterocorals [Rugosa] and the tabulates.

The latter form compound coralla composed of extremely slender-tubed individuals and are characterized by the appearance in these tubes of floorlike

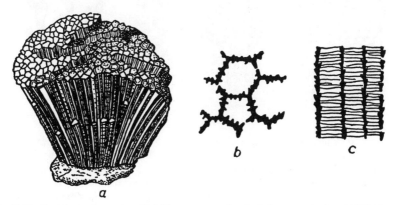

Fig. 3.42. *Favosites polymorphus* (Goldf.) as an example of tabulate organization. Middle Devonian, from the Eifel. *a*. Corallum natural size. *b*. Enlarged cross section through several corallites; walls with uneven thorns (septa) and isolated pores. *c*. Enlarged longitudinal section showing closely spaced tabulae. (After K. A. von Zittel 1924.)

plates, or *tabulae,* the structures for which the group is named (fig. 3.41). The tabulae are secreted by the posterior end of the body and seal off each living chamber, or cup, from the rest of the tube posterior to it, which is no longer inhabited. These forms either lack septa entirely or develop them only as spines or even faint ridges.

In contrast to these "corals with floors," the pterocorals can be called "corals with walls, or septa." These are solitary (figs. 2.15, 2.16, 3.49*a–g*) or colonial forms (pl. 5C; fig. 3.49*h,i*) in which the individuals are always considerably larger than tabulate individuals. These forms also develop tabulae to close off the older portion of their cylindrical, or conical calcarious containers behind them, but the dominant elements here are the *septa.* In corals, the term septa (sclerosepta) refers to radial calcareous membranes, or partitions, which attach to the outer wall and divide the shell lengthwise into individual chambers.

These septa are entirely different from the previously discussed skeletal structures of cephalopods. From a purely functional point of view, cephalopod septa correspond to the tabulae in corals in that both effect a transverse chambering of the shell tube. But since the two structural designs are so completely different, no further comparison is possible.

Both the tabulates and the pterocorals develop all their different types of forms immediately upon their appearance and in rapid succession; in the course of their evolution, variations appear only in details and are limited in scope. For example, among the oldest tabulates are the *favositids* (fig. 3.42), with a honeycomblike corallum composed of many separate tubes, each with septa and mural pores already present; the *chaetetids,* without septa and pores; the *halysitids*

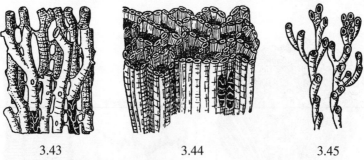

3.43   3.44   3.45

Figs. 3.43–45. Different types of tabulates. All actual size.
3.43. *Syringopora ramulosa* Goldf. Lower Carboniferous.
3.44. *Halysites catenularius* (L.). Gothlandian.
3.45. *Aulopora tubaeformis* Goldf. Middle Devonian.
   The species and genera illustrated differ not only in the different growth form of the corallum (*Aulopora* forms creeping colonies, the entire undersides of which were attached to a foreign substrate; this cannot be discerned from the picture), but also, fundamentally, in the development of the septa and the tabulae, with regard to the type of budding, and so on. For *Syringopora* and *Halysites,* a few of the tubes are broken open to show the tabulae diagrammatically.

(fig. 3.44), with the individuals arranged in chains; the massive *tetradiids,* with four septalike structures; and the *heliolitids* (fig. 3.16), which developed a spongy coenenchyme between the individual tubes. Immediately thereafter come the *syringoporids* (fig. 3.43), the *auloporids* (fig. 3.45), and other groups, all of which continue on through the entire Paleozoic with only slight changes. Here, too, the contrast between the *initial rush of evolution,* the sudden bringing forth of all of the different types of forms, and their *later very slow transformation,* during which almost nothing fundamentally new arises, is particularly striking.

Pterocorals [Rugosa]—Cyclocorals [Scleractinia]

The pterocorals also make their first appearance with almost all of their important forms in place; the breaking up into large groups with different evolutionary trends took place very early. As far back as the Ordovician, we already find side by side solitary corals and colonial forms, some without a columella and some with an advanced columella (the columella is a solid axis situated at the center of the skeleton), types that have developed an inner wall, types with varying tabular structures, and all kinds of other differentiations. There are also perforate types, whose calcareous skeletons are not uniform, like those of most other corals, but full of holes; these types appear very early, in the Gothlandian.

Taxonomically speaking, therefore, the suborder of the pterocorals breaks up immediately into a great many superfamilies and families. They have in common a particular type of septal apparatus, one that is diagnostic for pterocorals; we shall now examine this structure more closely.

STRUCTURAL DESIGNS OF THE SEPTAL APPARATUS

With corals, too, if preservational conditions have been favorable we are fortunate to have access to the entire developmental history of the skeletal elements of a single individual, from its first appearance on. We can take a series of cross sections from the calcareous corallite and, using them, follow in every detail the origin and transformation of the septa, which are of decisive importance for the morphology of the coral animals.

In the *pterocorals,* the secretion of the septa begins with the emplacement of an axial septum, a lamella that passes through the axis of the corallite, dividing its cavity into right and left halves (1,1 of the lower row of cross sections in fig. 3.46). It is composed of two individual septa, the cardinal septum (C) and the counter septum (C'), which are directly opposite each other and, as development continues, usually separate.

Very soon, two pairs of lateral septa appear, the first usually the pair next to the cardinal septum (2,2 and S,S in the illustrations), followed by the other pair, called counterlateral septa, next to the counter septum (3,3 and S',S'). At their beginnings, both pairs are small and peripheral, exhibiting first a bilaterally symmetrical orientation, but they soon increase in length and shift to a radial position, with the result that the corallite is now divided evenly into six parts.

Remarkably enough, however, as development continues, this six-part radial structure is not retained. The counterlateral septa move closer to the counter septum; the two large radial pockets that they enclosed are thus reduced in size or develop no further, so that only four large spaces remain, which are emphasized in our illustration by the heavy lines marking the six original septa.

Then, in these four sectors more septa are secreted; these are called *metasepta,* to contrast them with the six original septa, or *protosepta.* They appear in each of the four sectors, or quadrants, in unilateral succession, on the side of the septum facing the cardinal septum, as the numerals in figure 3.46*a,* right, show. In accordance with this serial mode of origin, the metasepta develop in graduated lengths, at least in the early stages of formation, and are usually pinnately arranged. This typical behavior is responsible for the name of Paleozoic corals—pterocorals.

At maturity, however, after septal formation has ceased, the septa of advanced forms lose their pinnate assemblage. Also at that time, the most recently formed, shortest ones attain their full length, coming to resemble the older septa, and take on a radial arrangement (fig. 3.46*b*). In addition to these long proto- and metasepta, which, as major septa, can extend to the center of the corallite or almost so, there is usually a circlet of shorter septa, which, as *minor septa,* insert between each two major septa. Metasepta do not appear in the two reduced spaces on either side of the counter septum. Only one minor septum appears in each, which, indeed, may occasionally exhibit considerable length (fig. 4.46*c,* right).

The main features that we can single out from this developmental process are the following: the primordia of the septal structures of pterocorals are emplaced *bilaterally;* the six protosepta develop in a pronounced *bilateral symmetry.* Not until these are fully developed does a *radially symmetrical form* appear, and this form is *hexamerous,* or six-sided.

The radial arrangement, however, is restricted to only a temporary stage. As the sectors immediately to the left and right of the counter septum diminish in size and, further, as the metasepta arise in sequence in the remaining four sectors, purely *bilateral tendencies* reappear; throughout this entire course of development, there is always only one plane of symmetry, which opposes two parts, mirror images of each other.

Not until construction of the major septa is finished does a certain radial arrangement reappear, at least superficially; however, it is only an *appearance of radial symmetry,* if we take into consideration the different sizes of the septa as reflected by their sequence of origin. Even at this stage, however, the minor septa still produce bilateralism. These septa appear in unilateral sequence just as the metasepta do, but always somewhat later than the major septa *to which they are assigned.*

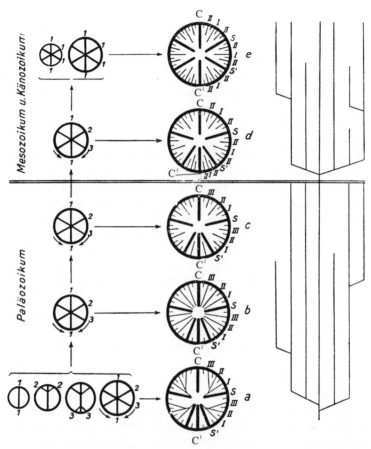

Fig. 3.46. Development of the septal structures of stony corals, from the original bilaterality in the pterocorals [Rugasa] (Paleozoic) to the six-part radiality of the cyclocorals [Scleractinia] (Mesozoic and Cenozoic); shown diagrammatically, in cross sections. In the left column of the illustration, the vertical arrows indicate the course of development of the early juvenile protosepta stage. The horizontal arrows represent the ontogenies that lead from the protoseptal stages—which are morphologically alike everywhere but differ in the way they are emplaced and in their potential—to the mature stages (center). On the right, a very general diagram showing the division into two cycles as the stock unfolds. (From Schindewolf.)

* * *

If we compare with this the development of the septal structures in *living stony corals* [Scleractinia], the latter appear to have a completely different structural design, one that bears no relation to that of the pterocorals. In modern stony corals, the six protosepta do not arise as sequences of pairs but rather *simultaneously, as an integrated circlet* or cycle (fig. 3.46e). The hexameral partitioning of the space thus established is retained unchanged; there is *no subsequent degeneration of any sector at all.*

Consistent with this, in every one of the six sectors metasepta are secreted, always in *radial-cyclical* form. Thus, the first to appear is a circlet of six metasepta (I in fig. 3.46*e,* right), one in the middle of each sextant, dividing it exactly in half. (In many cases, accelerated development results in this first cycle of metasepta appearing simultaneously with the protosepta, resulting in the presence of twelve septa right from the start.) Then, a second circlet of metasepta (II), again twelve in number, is emplaced, each one inserting between a protoseptum and a metaseptum from the first cycle. A third cycle follows, dividing the newly created segments in the same way, and so on.

Thus, the septal apparatus of Recent stony corals (madreporarians) is characterized by a distinctly *hexameral radiality* at all stages, both with regard to the first emplacement of the structures and to their further development. Consequently, within the comprehensive unit of the corals, these forms are assigned to the six-radiate *Hexacorallia* [=Scleractinia].

Formerly, the Paleozoic pterocorals were referred to as the "Tetracorallia" to contrast them with the six-radiate forms. This term was arrived at based on metasepta being secreted into only the four quadrants, which had led to the assumption that all of the development was quadripartite and that there were only four protosepta. Recent phylogenetic research has refuted this decisively, showing that pterocorals as well as living madreporians have *six* protosepta.

TRANSFORMATION OF THE TWO STRUCTURAL DESIGNS

In addition, certain other similarities of form between the two groups of corals with regard to the rest of the characters, no matter how different they appear at first glance to be, can be observed; we shall discuss these now. Anticipating the results, we assert that the phylogenetic relationship between the two groups will be proven indisputably: Paleozoic pterocorals are the stock from which living stony corals arose and should be joined with them in the *order Madreporaria* [Zoantharia], which used to be limited to only the Recent group of forms.

Within this comprehensive group, we place the younger type with cyclical septal insertion in a suborder, Cyclocorallia, to contrast it with the extinct suborder Pterocorallia. In geologic time, the two groups succeed one another directly: the pterocorals extend into the Permian, with perhaps a few stragglers on into the Triassic; the oldest representatives of the cyclocorals appear in the Triassic, replacing the pterocorals.

We must now explain how the transformation from the original bilateral plan to the radial one took place. Relatively insignificant is the *change in the way the six protosepta are laid down—from bilaterally to radially-cyclically.* The changeover is not complete until the cyclocoral group; we still observe in many Triassic representatives of this group (fig. 3.46*d,* left) and even in many Jurassic ones that, just as in Paleozoic types, the pairs of protosepta are secreted in sequence. This character is thus not limited to the structural design of the ptero-corals but spreads to that of the cyclocorals as well.

All it takes here is a slight shift in the relative growth rate or an acceleration of ontogeny, for the originally separate, sequential appearance of the protosepta to become compressed into a single simultaneous cycle and to make it so that the two lateral pairs of septa look like the first pairs of protosepta—the cardinal and the counter septa—in both length and position right from the start. This transformation, important for the construction of a completely radially symmetrical form, took place at *early ontogenetic* stages and, indeed, *with a small leap,* for there can be no smooth transition between the two alternatives of sequence and simultaneity.

It is worth noting that even in Recent stony corals, the soft parts, which relate to the septa, still bear distinct traces of the former sequential development of the protosepta. The calcareous septa of the corals develop in radial pockets of the soft tissue. Each pocket is composed of two membranes, fleshy septa, or mesenteries, which enclose the calcareous septum and precede it in development. Even in Recent forms, however, the mesenteries do not form simultaneously but in sequence. The details of this sequential origin are not of interest here. It should only be pointed out that two opposite pairs of mesenteries—the directive mesenteries—develop first, and that these are the very pairs that enclose the first septa (the cardinal and the counter septa) emplaced in fossil forms.

In its particulars, the homology is such that the dorsal pair of directive mesenteries corresponds to the cardinal septum, the ventral pair to the counter septum. A chain of observations and conclusions that cannot be discussed here in detail puts us in a position to establish an indisputable relationship between the septal apparatus of Paleozoic pterocorals and that of living forms and to show the homologues among the individual septa.

Thus, the temporal sequence in the formation of the protomesenteries of Recent representatives mirrors the sequential appearance of the protosepta in Paleozoic pterocorals and in the older cyclocorals. In this respect, the mesenteries have retained a phylogenetically primitive behavior; in living cyclocorals, indications of a former bilateral structure are restricted to this developmental stage only.

In contrast, the simultaneous emplacement of the six protosepta marks this mode of development as derived, the result of accelerated development, the combination of individual formative events into a single event. This acceleration has proceeded even further in the forms already mentioned, in which even the first six metasepta are emplaced simultaneously with the protosepta.

* * *

The most striking difference between the pterocorals and the younger madreporarians is that in the latter, *all six of the primary radial sectors are developed* and provided with metasepta, whereas in the pterocorals, only *four* of them follow this pattern. At first, evolution with regard to the size of the sectors that lie on either side of the counter septum is quite gradual.

In *early Paleozoic* stony corals, these sectors usually degenerate completely during ontogeny to the size of a simple interseptal space (fig. 3.46*a*, right). In many *late Paleozoic* forms, however, the reduction in size does not go that far; the affected sectors remain a certain size and in each a secondary septum develops, appearing early and growing to a significant length (fig. 3.46*c*, right).

The *Triassic* descendants of stony corals show a reduction of both of the radial fields to half the size of the rest of the sectors (fig. 3.46*d*, right). True metasepta develop in them, exactly half as many as in the four normally developed sectors. Even in the Jurassic, corals still appear that belong to a lineage that has remained primitive; in these forms, the two sectors remain smaller and do not show the secretion of metasepta in later cycles. These forms are neither four- nor six-radiate; they are a distinct intermediate form between the typical pterocorals and the normal type of younger madreporarians.

Even among *Recent* cyclocorals there are still representatives in which the two ventral radial sectors are less developed than the rest. They are smaller, the emplacement of their metasepta is delayed, and they often do not attain the full number of cycles developed in the other sectors.

This gradual atrophy of the sectors as ontogeny proceeds, regulated by the duration of ontogeny, signifies *quantitative character differences* and, correspondingly, indicates *smooth transition.* The situation here is one of an extensive, slow evolutionary process dominating the entire stony coral stock orthogenetically.

On the other hand, there is a *basic qualitative difference* in that in pterocorals, metasepta do not form in the two sectors in question whereas in the cyclocorals, they do. Indeed, there is a certain morphological similarity inasmuch as in many Paleozoic forms, the sector in question already has a long secondary septum, which superficially looks exactly like a primary septum.

Nevertheless, the secondary septa have a developmentally different origin from that of the primary septa; secondary septa represent exosepta, that is, they are secreted *between* the individual pairs of mesenteries of the primary septa; in contrast, the primary septa are endosepta, appearing *within* their own mesenterial pocket. The construction of such individual mesenterial pockets in the sectors in question constitutes a fundamentally new character in the cyclocorals, which appears *abruptly, without transition, in early stages of ontogeny.*

*   *   *

The transformation to the *radial-cyclical emplacement of the metasepta* typical of the cyclocorals also took place suddenly in early juvenile stages. It is obvious that there can be no smooth transition between the unilateral, serial insertion of metasepta in the old madreporarians and the cyclical process that occurs in the younger forms. *Either* the first metaseptum arises on one side in the individual sectors, immediately next to its respective lateral septum, and the subsequent ones are emplaced *unilaterally* dorsal to it (fig. 3.46*a–c,* right) *or* the first meta-septum appears at the center of the radial section and the later ones are secreted *on both sides of it* (fig. 3.46*d–e,* right).

The choice between these alternatives occurs in early juvenile developmental stages; subsequent ontogeny cannot alter the development once it has embarked on one of the two paths and thus cannot establish the slightest kind of transitional form reflecting gradual, step-by-step development. Through a *sudden transformation in early ontogenetic stages,* the old, previously adhered-to organization interfaces directly with a new one.

Characteristic similarities of form between the two structural designs with regard to the mode of insertion of the metasepta are also observed, providing further evidence of the existing relationship. In certain of the most recent representatives of the pterocorals, isolated in the Carboniferous and more widespread in the Permian, a transformation in the development of the length of the meta-septa within the individual quadrants takes place. We have seen that in the older forms, the metasepta usually either decrease in length the later they are developed (fig. 3.46*a,* right) or can all be the same length, the latest septa to form continuing to grow (fig. 3.46*b,* right).

The younger types cited—they are called plerophylls—deviate from this pattern in that each septum that appears in the center of a sector reaches a particular length and thickness (fig. 3.46*c,* right). Then, the septa on both sides of it are shorter. Cross sections through the adult stages of such forms show broad similarities to those taken from cyclocorals. If the secondary septa are included, the impression is given that in this case, three cycles of metasepta developed (fig. 3.46*c* and *d,* right).

This impression, however, is a superficial morphological illusion; in terms of evolution, there is still a purely unilateral, serial sequence of origin. The long central septum does not correspond to a primarily deposited metaseptum of a corresponding cycle but is rather a later arising metaseptum that has simply grown longer than the others. The sequence of emplacement of the individual septa is thus still completely in agreement with the typical pterocoral sequence, as the numerals in figure 3.46*c,* right, show.

Nonetheless, it is unmistakable that this refashioning of the way length is graduated points to the direction taken up by the Mesozoic cyclocorals: *a changeover from pseudocyclomerism to true evolutionary cyclomerism effected by an abrupt transformation in the sequence of emplacement,* which can also be construed as a change in relative growth rates.

Evolutionary Behavior

In summary, it can be said that the whole infinitely long evolution of the stony corals is permeated by a structural trend that dominates all others, namely, a *replacement of the original bilaterality of the septal apparatus by a hexamerous radiality.* This evolutionary course embarked upon in the Ordovician has not yet come to completion, for even in Recent corals, many bilateral features have been retained: the emplacement sequence of the mesenteries; the development of the directive mesenteries; the often not purely cyclic appearance of the metasepta; the occasional retardation of the two ventral sectors; the lateral compression of the mouth and gullet, and so on. What we see, then, is an agelong, exceedingly slow rate of evolutionary change.

This entire transformational course can be divided into several separate processes that work in concert. The reshaping from the older to the younger structural design of the Madreporia [Zoantharia] is thus complex in nature, the subprocesses exhibiting different behavior. Sometimes they proceed *continuously and gradually,* as, for example, in the appearance of the protosepta, originally sequential and then combined into a single cycle; and further, the slowdown in the reduction of the two ventral sectors. These evolutionary courses follow the same direction in a great many parallel lineages and *extend beyond the confines of the two structural designs,* sometimes bridging them and, along with other morphological similarities, giving evidence of the phylogenetic relationship.

On the other hand, there are *abrupt, discontinuous restructurings,* which appear *at the divide between the two structural designs;* these changes are related to the *same* characters that form the decisive organizational features of the cyclocorals, namely, the filling of each of the six sectors with metasepta and their cyclical emplacement. This far-reaching transformation takes place *without transition, in early ontogenetic stages.*

A gradual, smooth bridging of these qualitative differences is not possible. Nor can the peculiarities of the septal apparatus of the cyclocorals be derived from the differentiated septal structural design of adult pterocorals, for this is firmly established unilaterally in a particular direction. Rather, the parting of the ways of the two different developmental courses takes place at an early juvenile stage, that of the formation of the six protosepta; up to that point, development was the same. This developmental stage forms the only common, unifying element in the chain of ensuing, transformed ontogenies.

The abrupt transformation of the pterocorals into cyclocorals took place as the Paleozoic gave way to the Mesozoic, indeed, in a single lineage, the group of the plerophylls (family Polycoelidae). This is the group mentioned earlier in connection with a purely morphological similarity to the cyclical septal apparatus of younger madreporarians (fig. 3.46*c*, right). The behavior of the coral stock at this divide between the two structural designs corresponds completely to the situation we encountered in the evolution of the ammonoid stock at the Triassic-Jurassic boundary. We have here exactly the same process of a *massive die-off,* an extinction of all families and individual lineages except for a single one, which managed to leap beyond the confines of its structural design (see the diagram at the far right in figure 3.46).

This significant event proves that it is by no means an intellectual construct, an invention of the human mind, to speak of a new structural design or type of organization. Quite the contrary, there is here a completely *real, natural interruption of evolutionary processes,* the termination of an old principle and the emergence of a new one.

Furthermore, here again it is an *unspecialized, conservative line* that alone maintained the plasticity necessary for a far-reaching transformation while all the differentiated coral lineages around it became extinct. The plerophylls are offshoots of the zaphrentoids, an ancient group of corals that remained simple, having a massive calcareous skeleton without columellae (columns), with undifferentiated tabular structure, without a peripheral zone of vesicles (dissepimentarium), and with extremely simple, microscopic septal structures.

As soon as the new basic organization of the cyclocorals is in place, there is once again, in large individual steps, a splitting into widely disparate groups and evolutionary trends; the stock once more enters a *phase of explosive development of forms.* Thus, almost all of the coral families that extend to modern times are already established in the Triassic, and during the long interval, they have undergone no further appreciable, radical transformations. Even here, of course, evolution does not cease, but it consists only of a *gradual, continuous elaboration and progressive specialization* of characters already in place.

*   *   *

Within the broad evolutionary course of the stony corals as here described there are *evolutionary phases of a second rank,* which relate to other characters and run their course seemingly independently, apart from the processes determining symmetry, which are superimposed on them and take place in every individual lineage. And in these secondary phases there are also close parallels with the unfolding of the cephalopods, in which, as we have said, at every single developmental stage of the suture, at each new plateau, or at the various levels, we saw other kinds of differential trends that were narrower in scope.

In corals, these shorter evolutionary series affect the structuring of the tabulate and vesicular zones, the internal and external walls, the elements of the axis, and so on, features that are the criteria for identifying the individual families and give the different time periods their distinctive character. For example, the broad distribution and advanced evolution of columns is typical for the Carboniferous (fig. 3.47).

In connection with these secondary evolutionary trends, forms are frequently repeated within the individual levels, with copies of older forms appearing. This multiple *convergence* and *homeomorphy* can easily be confusing and has often led to seeing phylogenetic relationships between forms that resemble one another superficially, but only superficially, when in fact such a relationship does not exist.

Going back to the Carboniferous column-bearing corals already mentioned, it

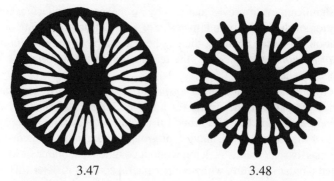

3.47   3.48

Figs. 3.47–48. Homeomorphy between a Carboniferous pterocoral and a Tertiary cyclocoral with regard to the structure of a central column and a certain superficial similarity in the way the septal apparatus is structured.
3.47. *Cyathaxonia cornu* Mich. Lowest Lower Carboniferous (Lower Tournaisian) from Drewer (Rhenish Mountains). 8 ×.
3.48. *Turbinolia sulcata* Lam. Lower Tertiary (Middle Eocene) from Grignon (France). 15 ×. (From Schindewolf 1942.)

should be pointed out that one of these forms, the genus *Cyathaxonia* (fig. 3.47) is extensively similar to the genus *Turbinolia* (fig. 3.48), which appears in the Tertiary and extends on into the Recent. Because of this, *Cyathaxonia* was formerly assigned to the family Turbinolidae, and even today, there are those who regard Tertiary (!) turbinolids as being the immediate descendants of Carboniferous (!) cyathaxonians. The lack of connecting forms during the long interval is once more ascribed to "gaps in the fossil record"!

However, a comparison of the septal structures shows immediately that the two genera are in reality fundamentally different: *Cyathaxonia* is a typical pterocoral, and *Turbinolia* a true cyclocoral whose roots are to be sought within this group. The similarities in the styliform columns and a few other superficial characteristics are based only on analogy and not on immediate phylogenetic relationship.

Thus, in each evolutionary phase of the second rank there appear in frequent alternation and succession the most varied forms of corallites—large and small, high and wide, cylindrical, conical, discoid, and slipper-shaped, straight and curved (fig. 3.49)—all forms directly related to the mode of life, to the locality, to ocean currents, and so on. In contrast to the shell form of free-living cephalopods, the shape of the corallite in sessile corals shows great variability. No particular evolutionary trend is recognizable; the forms repeat again and again within the individual families and genera and bear no direct relationship to the dominant trend of the stock, which proceeds along its orthogenetic path independently of the zigzag course followed by the external characters.

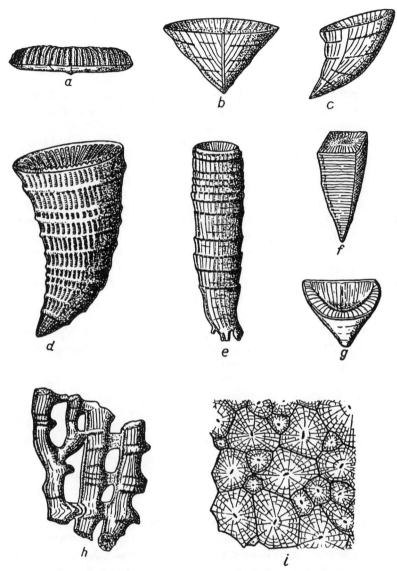

Fig. 3.49. Different forms of corallites in pterocorals and cyclocorals. The upper illustrations show solitary corals; the lower two are colonial types. *a.* Discoid. *b.* Erect turbinate. *c.* Curved turbinate or short conical. *d.* Long conical or horn-shaped. *e.* Cylindrical. *f.* Four-sided pyramidal. *g.* Slipper-shaped. *h.* Portion of a corallum composed of loose bundles. *i.* Cross section through a massive corallum, the individual corallites in a honeycomb arrangement. (After D. Hill, somewhat changed.)

Pterocorals—Heterocorals

In addition to the cyclocorals, there is another group that derives from the ptero-corals—that of the very remarkable heterocorals, which constitute a separate suborder within the madreporarian group.

They appear suddenly in a particular horizon of the Lower Carboniferous, attain wide distribution throughout all of Europe and East Asia, and then die out without descendants after a brief life span. Their calcareous corallite is a slender cylinder of considerable length; all we usually find of them are short fragments of the kind shown in plate 19B.

COMPARISON OF THE SEPTAL APPARATUSES
A cross section through the simplest form of heterocorals is presented schemati-cally in figure 3.50. It shows a thick outer wall and an internal septal apparatus consisting of four septa arranged in the form of a cross and joined at the center. Toward the periphery, the two lateral septa bifurcate. In addition, between the septa, a section through a tabula is visible. In these forms, the tabulae tip at an angle toward the rear, or the apex, and are therefore cut through by any cross section. These features are of no further interest here.

The primitive representatives of our group, which are lumped together in the genus *Hexaphyllia,* remain at this simple evolutionary level of septa apparatus for their entire life span. In other, succeeding forms (genus *Heterophyllia*), the number of septa increases considerably (figs. 3.51 and 3.54).

The guiding principle can be seen in the diagram in figure 3.55*b.* Here, a system of concentric circles encloses the series of sequential ontogenetic devel-opmental stages. The inner circle shows a developmental level with four septa arranged perpendicular to one another, as is to be expected in the juvenile stage of *Hexaphyllia* and *Heterophyllia.* The next larger circle circumscribes a state in which the two lateral septa have split toward the periphery, just as is seen in mature stages of *Hexaphyllia* and in juvenile stages of *Heterophyllia.*

Then there is a splitting in the upper and lower septa that is exactly analogous to that which already appeared in the lateral septa, and simultaneously, within the forks of the latter and arising from them at an angle, new septa (metasepta) are secreted. The same process takes place again at a later stage in the forks of the upper and lower septa. Further, between the branches of the two lateral septa, which are usually one step ahead of the other two septa in their development, metasepta of the second order are emplaced in the cyclic fashion depicted in the illustration.

Here we encounter a design principle for the septal apparatus that differs fun-damentally in several respects from the one explained for pterocorals.

1. The pterocorals form *six* basic septa, or protosepta (fig. 3.55*c* and *d*), but our heterocorals only *four;* in them, protoseptal pair III does not form.

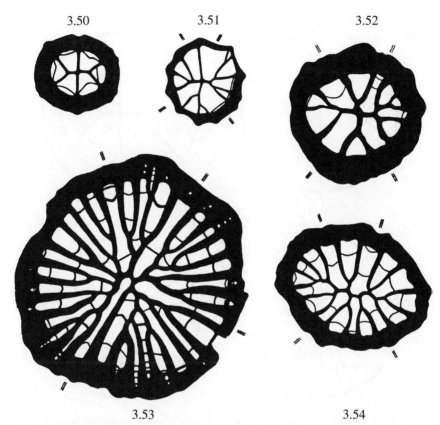

3.50   3.51   3.52

3.53   3.54

Figs. 3.50–54. Cross sections through *Hexaphyllia* and through juvenile and adult stages of various species and subgenera of *Heterophyllia*. Lower Carboniferous (Upper Visean, D_2).

3.50. *Hexaphyllia mirabilis* (Dunc.), from Altwasser, in Lower Silesia [Poland]. 12 ×.

3.51. Juvenile stage of *Heterophyllia* (*Heterophyllia*); cf. *parva* Schdwf. Same as above. 15 ×.

3.52. Juvenile stage of *Heterophyllia* (*Heterophylloides*) *reducta* Schdwf., from Elsoff, near Battenberg (Rhenish Mountains). 13 ×.

3.53. Mature stage of *Heterophyllia* (*Heterophylloides*) *reducta* Schdwf., from the Segen Gottes Mine, near Altwasser. 8 ×.

3.54. Adult stage of *Heterophyllia* (*Heterophyllia*) *grandis* McCoy, from Rothwaltersdorf, in Lower Silesia. 9 ×.

(From Schindewolf.)

2. A splitting of protosepta, or more precisely, their replacement in later separate developmental stages by two septal branches of equal length, does not take place in the pterocorals.

3. A certain similarity of structure can, however, be seen in that in both instances further development of the septal apparatus takes place in *four* radial sectors, or quadrants. We recall that in the pterocorals, of the six sectors origi-

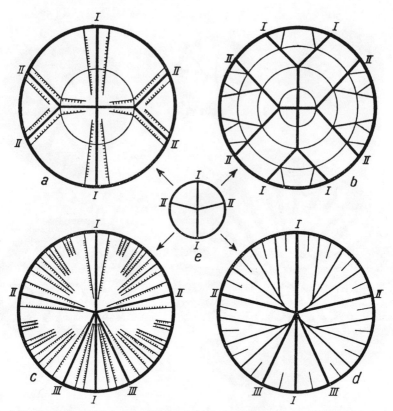

Fig. 3.55. Comparison of the structural designs of the septal apparatus in the heterocorals (*a, b*) and the pterocorals (*c, d*). *e* shows the early juvenile developmental stage common to the two, from which the very different mature forms of the septal apparatus diverge. In *a* and *c*, the mesenteries, which enclose the calcareous septa, are reconstructed diagrammatically.

nally set up, the two lying on each side of the counter septum atrophy and no metasepta are secreted into them (fig. 3.55*d*).

But even in this point, the similarity is only superficial; in reality, there are far-reaching differences here, too: In the pterocorals, the metasepta are emplaced in the spaces *between* the protosepta (fig. 3.55*d*). In the heterocorals, however, these spaces remain free of septa and usually stand out clearly as an "X" consisting of septa-free zones with parallel borders (figs. 3.51–54, 3.55*b*).

Based on an analogy with Recent stony corals, there is no difficulty in obtaining a picture of the relationship between the calcareous septa and the soft parts in pterocorals (fig. 3.55*c*). The protosepta lie enclosed within a pair of long mesenteries, and the rest of the septa are emplaced *outside* of these in the sectors between the mesenteries. In the heterocorals, in contrast, there is a peripheral subdividing of the four original mesenterial pockets (fig. 3.55a), and the new septa arise subsequently *within* these pockets, whereas the spaces separating them contain neither mesenteries nor septa.

In other words, the quadrants of pterocorals and heterocorals are not homologous. At the most, a similar structural tendency can be discerned, which in the heterocorals, however, is realized in a different way by different means.

4. Further, the metasepta of the two groups are emplaced in completely different ways: in the pterocorals, bilaterally-serially in unilateral succession, and in the heterocorals, radially-cyclically in a coherent circlet.

THE DERIVATION OF THE HETEROCORALS

All of these are fundamental, qualitative differences, which preclude tracing the heterocorals back to differentiated mature stages of pterocorals. Rather, the starting point of their organization must be sought in an *early ontogenetic developmental stage* of a pterocoral, one in which the possibility of developing in either of the two different directions remained open. Such a developmental stage is shown in figure 3.35e. We already know it as a very early stage in the ontogeny of pterocorals: the septal apparatus exhibits only the cardinal and counter septa joined to form an axial septum (I,I), and one pair of lateral septa (II,II).

At this point, the two divergent, irreconcilable paths separate as follows: (a) in the pterocorals, a third pair of protosepta forms, but in the heterocorals it does not; (b) in the heterocorals, the four protosepta divide, but in pterocorals, they do not; (c) the metasepta of the heterocorals are emplaced as a circlet within the branches of the split protosepta or pairs of protomesenteries, but in the pterocorals, they arise sequentially outside of the mesenterial pockets of the protosepta.

What we have here are two fundamentally opposing, alternative developmental plans; *the emergence of the heterocorals from the mature stages of the pterocorals through many small changes along the path of continuous race and species formation is therefore inconceivable.* The only possibility is a *sudden restructuring early in ontogeny,* at the stage of differentiation in pterocoral ontogeny *just prior* to the one in which the change in course embarked upon by the heterocorals was shut off and no longer possible.

This explains the *lack of transitional forms* between *Hexaphyllia,* the first to incorporate our new type of organization, and any other representative of the pterocorals. Since the transformation had to have taken place in early juvenile stages of a few individuals, thereby directly interfacing the new and the old, such intermediate links are hardly to be anticipated; thus, the fact that there are none has nothing to do with that old standby but long-since overdrawn account—"the gaps in the fossil record."

Similarly, *Hexaphyllia* and *Heterophyllia,* as the two different developmental stages of the heterocoral structural design, stand in sharp contrast, unconnected with one another. In itself, a stepped transition would be conceivable here. One could imagine something like a stepwise transition with a splitting of *both* pairs of protosepta, yet still without the formation of metasepta; thus far, however, this is not known as an adult stage. Throughout the broad distributional range of these forms, we have only representatives of the two independent genera *Hexaphyllia* and *Heterophyllia,* and these by the thousands; the two forms are so thoroughly different that one would have to assign them to different families even if they were the only member.

Therefore, it is extremely likely that the new septal design not only arose or was introduced abruptly but also that its subsequent elaboration and improvement took place *without any transition, once again in one large transformational step.* Those who do not acknowledge that data accumulated thus far are sufficient proof of this may content themselves by hoping that intermediate forms will yet be found. In any event, however, this much is certain, that even further evolution of the new septal type must have proceeded extremely rapidly, since *Hexaphyllia* and *Heterophyllia* appear side by side in the very same bed and have only a very narrow vertical range.

In view of the abruptness of the restructuring of the type and of its having taken place in early ontogenetic stages, *gradual selectional processes could hardly have participated,* especially since the new structural design of the heterocorals is clearly inferior to the long-lived, durable organizational type of the pterocorals. Moreover, it should also be pointed out here that the thoroughly new organization of the septal apparatus is on a different plane from that of characters bearing any relation to the mode of life, upon which selection operates.

In the new organization, the entire basic structure has been transformed; within the new framework, the continuous chain of species differ from one another in insignificant individual characters such as the size and shape of the corallite, the thickness and sculpture of the outer wall, the changing number of metasepta formed by the genus *Heterophyllia,* and so on. This is yet another situation in which there is a clear *separation between basic structural characters, which arose discontinuously,* introduced the new type, and became the common property of all of the members of that type, and the subordinate *special, or specific, characters,* which affect only certain individual groups and exhibit *gradual change in small individual steps.*

*   *   *

The examples presented here from the phylogeny of cephalopods and corals have familiarized us with a number of commonalities and regularities; it is essential now to discuss these in more detail, to corroborate and explain them through further examples.

GENERAL RESULTS

Evolution or Creation?

Thus far, our arguments have been based on the theory of evolutionary descent, accepting it as correct without closer examination. Like the vast majority of biologists, we are convinced that the objective material before us is absolutely compelling, clear-cut evidence for historical organic change and for the actual descent of individual groups of plants and animals from one another.

The countless observations of small changes in form that actually take place before our eyes, the pattern of the graded steps in the multiplicity of forms, which is expressed in the natural system, the temporal and spatial order in the appearance of the individual stocks, and much else speak convincingly in favor of the theory of evolutionary descent; *it is only from the comprehensive viewpoint of this theory that all of these individual phenomena find a satisfactory, consistent interpretation.* The concept of descent in its most general form, the assertion that organisms have evolved and exist in an actual, natural relationship based on descent, thus appears to us to be *an irrefutable logical necessity.*

Nevertheless, it cannot be denied that even today there are still isolated opponents of the theory of descent who do not share the view represented here. Occasionally, especially in newspapers and popular magazines, there are attacks on these basic biological concepts, which are presented—usually out of some kind of ideological prejudice—as doubtful or even as a doctrine considered outmoded by science. This creates some uncertainty in the minds of people not familiar with the subject, and therefore, it is not entirely useless to examine here a few facts and concepts that must be held essential as *compelling evidentiary proof of the theory of evolutionary descent* and to explain why it must be accepted as beyond doubt.

The Chronological Order of the Types

The most important evidence obtained from paleontology is probably the circumstance that the individual phyla, classes, orders, and so on, of organisms appear in a temporal-historical succession corresponding to their increasing organization and level of evolution.

A graphic example for many organisms is shown in the chronological table (fig. 3.56) of the appearance of the individual vertebrate phyla. We see that the most primitive vertebrates, the jawless agnathans, appear in the Ordovician, and the placoderms—cartilaginous fishes (Chondrichthyes or Elasmobranchii)—and bony fishes (Osteichthyes or Teloeostomi) appear one after the other from

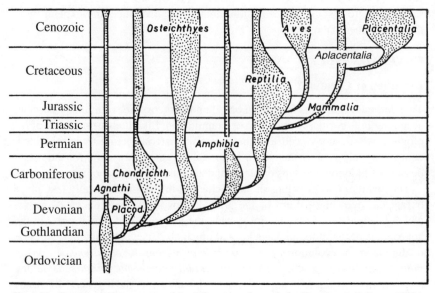

Fig. 3.56. Chronological appearance of the different vertebrate phyla. (Based on A. S. Romer 1933.) There is a strict parallelism between the relative levels of organization of the phyla and the time of their appearance in the geological past. The width of the individual branches should provide a rough idea of the increasing and decreasing range and abundance of forms in the groups in question.

the Gothlandian [Silurian] to the Lower Devonian. The next higher vertebrate type, the amphibians, is found evidentially for the first time in the Upper Devonian; the reptiles begin in the Upper Carboniferous, the primitive, aplacental mammals in the Triassic, and finally, the most highly differentiated of vertebrates, the placental mammals, in the Upper Cretaceous. In exactly the same way, we observe in the plant kingdom a regular temporal succession in the appearance of the algae, lower and higher vascular cryptogams, gymnosperms, and angiosperms.

The main types of the plant and animal world appear in exactly the same sequence historically as they do in the natural system of progressively more highly organized forms. The same thing is true for types of lesser magnitude, whose increasing levels of organization within the framework of their comprehensive structural designs also follow a strict temporal sequence. Since these relationships are repeated at every level of magnitude with insistent regularity, we must recognize in them *compelling statistical proof for an evolutionary connection between types that succeed one another in time.*

If there were no such evolutionary connection, if, rather, the individual groups of organisms had been created independently of one another, it would be completely incomprehensible that they always appear temporally and spatially at the very places in the geological record where their appearance is anticipated given their respective levels of organization. The various entities would then have been created in exactly the same order *as if* they had evolved from one another. But this shows that, in fact, the new, more highly organized types could have arisen from their more primitively organized temporal forebears *only by means of evolution;* otherwise, there would be no intrinsic reason at all for this *regular parallelism between structure and historical succession.*

* * *

One could perhaps make the objection that in reality, there is no improvement or higher evolution of organisms at all and, correspondingly, also no orderly sequence from *lower* to *higher;* that such an interpretation is an unwarranted evaluation from the human perspective. According to such a view, a protozoan, a snail, or a fish is as complete in its way and as well-adapted a being as the *highest* vertebrates, the mammals. All these creatures are fit for life, entirely capable of fulfilling their place in nature, which means that we have no right at all to speak of progressive evolution. This assertion is partially correct, in that the aptitude for life of individual types of plants and animals and their particular adaptations to their environments represent specific incommensurables, which cannot be peremptorily evaluated in a relative way.

But this by no means excludes our being able to determine that in the plant and animal kingdoms there are completely *objective differences in the level of organization* or complexity of structure that qualify the organisms for extremely varied levels of performance, from lower to higher. Thus, there can be no doubt that multicellular organisms, with their extensive division of labor among the individual cells resulting in an expanded functional utility, must necessarily be regarded as more highly organized than unicellular organisms; or that amphibians, whose anatomy allows them to live both in water and on land, have come to occupy a broader habitat than the fishes, thereby proving themselves to be at a greater advantage. Ecologically speaking, there is an unmistakable, analogous ascent in the plant kingdom: whereas the oldest plant life was tied exclusively to water, younger groups of plants, through their improving organization, became increasingly independent of water, thereby ultimately being in a position to colonize all the continental areas previously closed to them.

Mammals are another example: through live birth and brood care, which offer improved security for the offspring; through the improvement of the circulatory system and of metabolism; through the extensive division of labor in the dentition, and so on, they have achieved considerable morphological-physiological advantage over reptiles. Furthermore, we observe a progressive evolution of the central nervous system, of the more advanced functions of the brain and the sensory organs, an increased independence from the environment—risen to its greatest degree in humans—and so forth.

To this extent, therefore, we are completely justified in speaking of *improvement, increase in performance, or increase in advantage of individual groups of organisms in proportion* as the groups succeed one another in both time and the natural system. The only relevant criterion for this conclusion is the increasing general degree of organization, which provides for increased mastery of the environment and independence from its changes and not, however, a particular organic differentiation and adaptation to special life circumstances that have developed within the framework of the individual types of organization.

Moreover, the most important thing is that *the next higher group assumes the organization of the lower one preceding it and builds on it.* Thus, a multicellular stage is necessarily preceded by a unicellular stage, as we see repeated in the ontogenetic development of every multicellular creature. The organization of the amphibians is connected to that of the fishes and is thus characterized as being derivative; the organization of the mammals builds on that of the reptiles, making the reptiles a preliminary stage of mammals.

And yet it is not that these organizational types are completely alien to one another, standing in unrelated contrast. Rather, they exhibit many traits in common; this is expressed taxonomically when we join the amphibians, reptiles, birds and mammals to form one comprehensive type, the tetrapods, and subsume these in turn, along with the agnathans and fishes, in the superordinate structural design of the vertebrates. Therefore, each type takes from its predecessor a large number, in fact, the great preponderance of bodily organs and then distinguishes itself only through the modification of individual complexes of characters such as the new acquisition of certain organs that themselves provide the peculiarities of the type in question and lead to its being set apart within the framework of the superordinate structural design—the "umbrella organization"—to which it belongs.

The Absence of Gradual Transitions

The material we study in paleontology has led us to the realization that, contrary to the classic theory of evolutionary descent, the individual types of structural designs are not smoothly connected by a long chain of transitional forms linked by small transformational steps, that the features peculiar to them are not smoothly bridged, but that they appear in contrast with one another, set apart by large discontinuities. Opponents of the theory of descent have advanced this situation, the absence of the links that, according to Darwin, are to be expected, as proof that the more comprehensive typal entities could not have descended one from another, thereby calling into question the justification for the entire theory of descent. The argument centers on these structural designs of a higher order of magnitude; continuous evolutionary change on the scale of the lower types, of genus or family, for example, is generally conceded (except by a few extreme doubters); it is also not easy to explain away.

Nonetheless, it is clear that the gaps that exist between the major type organizations by no means disprove the concept of descent as such; the facts of the matter say only this much, that the concept of the origin of the types held thus far, that is, the development of the radical differences between types by way of a gradual, fluctuating modification of species, is not correct. Certainly, the types do differ among themselves with regard to particular fundamental characteristics—otherwise we would not be contrasting them with one another. Further, they are also not connected with one another smoothly by transitional sequences, and to this extent they stand apart. All this is true.

On the other hand, it should be emphasized that there is no way that there could be transitional forms as they have often been envisaged and required, namely, forms that are intermediate in every aspect. A placenta cannot be absent and present simultaneously; the two circulatory systems leading from the heart cannot be both separate and not separate; there can be no transition between bifurcate and unbifurcate protosepta in corals, and so on. *Intermediate forms* in the true sense cannot be expected in these cases; the most one will find are *composite types,* which combine features of one group with those of another, and composites like these are, in fact, present in abundance.

Further, we have already determined that the various structural designs and organizations that do appear in sequence *are always connected by a large number of common characters and organs* and that the new organizations build upon the old. This fact cannot be disputed, nor can it be based on chance. Then, however, it must indeed be regarded as infinitely more probable that one type, through modification of particular structural features, *evolved* from another than that its enormously complex organization was, in the biblical sense, *created* from the void—a completely new form, one without reference to any other. In view of the abundance of similar, corresponding organs and functions and, in contrast, the much less important, relatively few different structural designs, there is no doubt that the supposition of a bridging between those designs by some kind of evolutionary process presents far fewer intellectual difficulties than does the claim for independent creation of an entire new type.

In addition, we should refer to morphological convergence,[4] the coming to-

4. ["Convergence" as Schindewolf uses the word here describes a phenomenon that he observed in his reconstructions of phylogenetic trees. If one goes back in several collateral lineages, one observes that they become more and more similar morphologically. The common root stock contains "collective types" which he defines in the text. "Convergence" in the modern sense (as originally defined by Haeckel): The secondary acquisition of a character by two taxa, not derived from a common ancestor.—Ed.]

gether of different lineages in a common root stock. There, we encounter the so-called *collective types,* groups of forms that contain in as yet undifferentiated form the features of several lines that will split off later. In the discussion of cephalopod evolution, we referred to such root groups; they exhibit lability and still bear within them all the potence for the later forms. Similarly, certain ancient, Paleocene mammalian forms combine in their structure features of the lines that split off later—the insectivores, ungulates, carnivores, and primates.

Other undifferentiated collective forms are the protoinsects (Palaeodictyoptera, fig. 2.36), with their still very uniform body segmentation; their four identical wings, which extend laterally even at rest, cannot be folded, and have the simplest of veination; their free mouthparts, simple, many-segmented antennae, and so on. Still other collective types are, for example, the stegocephalians (figs. 2.29, 2.40), the most primitive amphibians, with features such that they can be considered the ancestral stock not only of younger amphibians but of the primitive reptiles as well. Can it be denied, then, that the younger types in question evolved from the collective forms that preceded them chronologically and prepared the way morphologically?

However, because of the absence of gradual transitional forms that might otherwise be expected, this cannot have happened through the slow, smooth processes of speciation. We see the way out of this dilemma in the fact that *the types arose by modification at early ontogenetic developmental stages through the abrupt acquisition of the new typal features.* We shall return to this by way of a summary, after first clearing up some preliminary questions. In support of this view, there are a great many actual observations, the generalization of which explains the unmediated appearance of new structural designs and the absence of long series of links while avoiding the absurd notion that the individual types, having no real, intrinsic connection, fell from heaven, so to speak.

The Dovetailing of Some Types

There was a reason for describing in these pages, in rather extensive detail, the transformation of the Paleozoic pterocorals into the younger cyclocorals. Recently, these two types of corals have repeatedly been presented by opponents of the theory of evolutionary descent (E. Dacqé; O. Kuhn) as proof that the individual structural designs are not connected and that the concept of evolution is untenable. A detailed analysis, which takes into account the entire historical unfolding of the coral stock and, especially, the ontogeny of corals, shows the opposite and, in fact, constitutes excellent proof *for* the genetic relationship between the two differently structured types of coral. Let us return for a moment to this example—which is quite representative—to examine it from this point of view.

To arrive at conclusive conceptions, it is not enough simply to pick out the mature morphologies of individual representatives from the two groups of corals and to compare them; this has already been done many times over. Further, it is not correct to conceive of these as rigid designs, to speak of the pterocorals generally as being a bilateral type and of the cyclocorals as a radial type, and then to claim that one could not have derived from the other because of this supposed profound difference.

We have seen, rather, that even in the pterocorals, certain radial features appear and that, on the other hand, the cyclocorals disclose various signs of a former bilateral structure. There is, as we discovered, an evolutionary trend permeating the entire stony coral stock that effects a gradual transformation of the original bilaterality into radially symmetrical forms.

This evolutionary course (fig. 3.46) is composed of a great many different small events:

1. The protosepta, originally arranged bilaterally and appearing sequentially in pairs, are emplaced all at once, cyclically and radially, in younger forms.

2. The degeneration of two of the sectors, which in the older forms marked a distinctly bilateral feature, gradually falls off in the younger representatives of the line, making way for the complete development of all six sectors, resulting in a purely radial symmetry.

3. Hand in hand with their continuous developmental increase in size, there is an increasing tendency for these two sectors to fill with metasepta to the point that they are finally equal to the other sectors in this regard.

4. The initial bilateral-serial emplacement of the metasepta changes to radial-cyclical.

It is of particular importance that the points at which these two separate evolutionary directions diverge do not completely coincide with the division between the two types, the pterocorals and the cyclocorals; rather, some of these smaller events, those listed as numbers 1 and 2, extend beyond the type boundaries to form a bridge between the two structural designs. With regard to point number 3, there is indeed a difference between the two types in that in the pterocorals, metasepta are completely lacking in the two sectors in question, but in the cyclocorals they are present. However, only in the cyclocorals do all the metasepta ultimately grow out to their fullest; this development is unmistakably connected to the situation present in the pterocorals.

Finally, the radial-cyclical insertion of metasepta typical of the cyclocorals has already been thoroughly prepared for morphologically in the youngest pterocorals through a corresponding size differentiation of the septa, so that there, too, an evolutionary connection is impressively exhibited, regardless of the discontinuity that in the end brings forth the new mode of insertion. *Thus, the two types appear to dovetail in multiple ways.* There is a step-by-step invasion, a replacement of the characters of the initial type by those of the new structural design, even though the critical transformational steps, the ones that ultimately characterize the type, are discontinuous.

To continue, we should, of course, not overlook the fact that the observations made thus far relate to the distinguishing characters of the organization of the two structural designs; or, apart from these features, the designs are alike in the overwhelming majority of morphological and functional features. The basic structure of the soft parts and the skeleton, their general development, arrangement, and very complex structure are in both instances the same; thus, the cyclocorals take over to a large extent the structural elements of the pterocorals and transform them further. In the previous discussion, many individual features went unmentioned, and we cannot go into them in more detail here, either. Nonetheless, we would like to raise the interesting point that in the cyclocorals, the microstructure of the septa refers directly to the developmental state in *the very group of conservative pterocorals* that must be considered the root stock of the cyclocorals for other reasons as well.

Pterocorals are thus in every respect the foundational form upon which the cyclocoral organization builds. In fact, cyclocorals presuppose as their starting point a structure such as the one found in pterocorals, for otherwise their peculiarities would be completely incomprehensible. If, as many authors believe, the cyclocorals represent an independent, newly arising coral type, why do they not simply and directly emplace radial hexametry right from the start? Why do the older cyclocorals detour instead through stages in which two sectors must first gradually diminish in size, a process that is only comprehensible as the result of the original morphology in the pterocorals?

And how are we to interpret the other numerous bilateral features that appear in the cyclocorals, primarily in the structure and emplacement of the soft parts? Further, how are we to understand that in the cyclocorals, the fine structure of the septa develops in such a way that it follows closely that of the pterocorals? Why do they not exhibit any more primitive or completely different kind of structure, which one would expect if the bearer of the features was an independently created line secreting a calcareous skeleton for the first time?

The cyclocorals, therefore, do not make their first appearance with a self-contained, profoundly new structural design but continue the transformational processes already partially introduced by the pterocorals and, with the most varied of references to this group, carry them to their logical conclusion. In view of these objective circumstances, which cannot possibly be based purely on chance, there is only one reasonable interpretation: The type of the cyclocorals does not owe its origin to some mystical process of creation, nor is it simply the manifestation of purely metaphysical connections; rather, it arose *through the natural processes of real, evolution by descent from the pterocorals,* upon whose foundations it continues to build.

In this way, the detailed investigation of individual examples leads to the mandatory conclusion *that the theory of evolutionary descent is not simply an intel-*

lectual possibility but a compelling intellectual necessity. Only by accepting this theory can we explain the existing evidence, which, with its countless individual features, would otherwise be completely inexplicable.

Further Evidence

There are still a great many other pieces of evidence in support of the concept of descent, which, it is worth noting, have grown out of the most diverse branches of biology and modes of observation and yet are in agreement, leading to the same results.

From *living organisms,* we should cite the testimony of *ontogenetic development* in the sense of the biogenetic law, and further, the clues provided by *vestigial organs,* which do not function and are without structural significance for the organism possessing them. These can only be understood as a legacy bequeathed by ancestors in which they were still functional. We do not want to go into more detail on these phenomena here; they are treated exhaustively in all works dealing with the theory of descent.

Compared to *paleontological* evidence, the biological kind just referred to is of less importance, since morphological data from *Recent* plants and animals provide no historical documentation and therefore can never be used to demonstrate the historical course of evolution. This is provable only by *documentation from the past,* the dates and chronology of which must be known. Thus, interpretation of the biogenetic law and of vestigial organs is only of consequence when premised on the course of evolution, which has been deduced in other ways. This evidence is a welcome supplement but can neither establish the history of organisms nor clarify the course it actually followed, for conclusions of this nature cannot be drawn only from the present. (It is another matter when the appearances of the species in question can be observed in *fossilized,* chronological series, as is the case with our example, the corals, which we have already analyzed. Then, insofar as forms with the deduced characters can be directly substantiated, they have, of course, total value as proof. We shall learn of other examples later, in other contexts.)

From the array of specific contributions provided by paleontology, reference should also be made to the important conclusions associated with the *disjunct geographic distribution* of modern organisms. For example, the *family Tapiridae* appears now in two widely disparate areas: the Malaysian Archipelago (Malacca, Sumatra) and in Central and South America. The genus *Tapirus* (in the broad sense) is known as a fossil from the Upper Miocene and Pliocene of China, Japan, and Europe but not from America. In contrast, in the North American Oligocene and in the European as well, the genus *Protapirus,* more primitive in many respects, has been found.

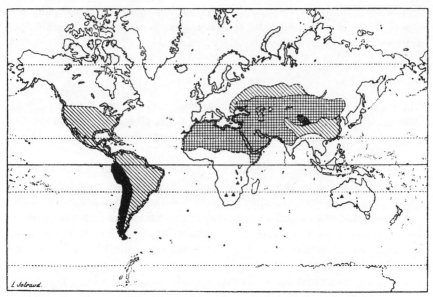

Fig. 3.57. Former and contemporary geographic distribution of the camelids. Solid black areas show the modern disjunctive distribution of camels (Asia) and llamas (South America) in the wild. Cross-hatching indicates the spread of Recent domesticated camels (Mongolia, Arabia, the Sahara Desert). The black triangles indicate regions where camels have naturalized more recently. The diagonal lines show the extended, once coherent ranges of fossil camelids during the Tertiary and Quaternary (North America, the Ukraine, Romania, Algeria, Asia). (After L. Joleaud 1939.)

This very striking modern distribution of the tapirs is easily explained when we assume descent of the genus *Tapirus* from the *Protapirus* group, which had a coherent range throughout the northern hemisphere. In the late Tertiary, there issued from it, as must necessarily be deduced, the genus *Tapirus,* which, in the Western Hemisphere, then migrated toward tropical South America and, in the Eastern Hemisphere, toward the Far East and the adjacent islands. In these two regions, tapirs have persisted until the present, whereas in the areas between they have become extinct.

Similarly, the family of the *camelids* shows today a conspicuously disjunct distribution: *Camelus* in central Asia and the genus *Lama* (guanaco, alpaca, vicuna) in western South America (fig. 3.57). The matter is easily explained when the fossil forms are taken into consideration. Then it can be established that camels evolved during the Tertiary in North America. From there, toward the end of the Tertiary, the ancestors of today's camels migrated, some to South America and others, in the Old World, to Asia, southeastern Europe, and North Africa, where they were widely distributed during the Pliocene. During the Diluvian [Pleistocene], however, they died out in North America and in the regions connecting the modern ranges.

Many similar instances of disjunct modern distributional ranges—other familiar examples are the Indian and the African *elephants,* and the diprotodont *marsupials* (with degenerate dentition) of Australia and South America (fig. 3.150)—can only be understood when the concept of *evolution from a common stock* at an originally undivided point of origin is assumed.

Evolution as the Logical Postulate

In conclusion, we recount an interesting attempt, undertaken by H. Dingler, to prove the evolution of organisms logically as a historical reality.

The conceptual boundary between living organisms and nonliving organic substances, he begins, is, for the time being, not sharply defined. Practically speaking, however, a division can be established in the progressive complication of structure that leads from the inanimate to the animate (protein molecule—macromolecule—viruses—protozoan), and on both sides of this boundary line, a certain zone of transition can be assumed. That which lies below the zone is surely, then, inanimate, and that lying above can certainly be termed animate. Within the undisputed zone of the animate, there can be no direct creation out of inanimate matter; the only possibility for origin is through reproduction. This realization must stand as an unequivocal historical causative conclusion.

Now, in the postulated zone of transition, life either arose on earth by itself or arrived here from some other planet. The transport of living matter from outer space is, however, only possible at the outside for certain spores of the most primitive organisms (bacteria, infusoria, and so on); for all more highly organized forms, the possibility of such transport is excluded. But because such forms do exist on earth, they can only have developed on their own.

Further, a time can be given *before* which they could not have existed on the earth, and Dingler summarizes his observation as follows: "All organisms that surpass in organization the spores of the most primitive organisms arose on their own during a limited, finite period of time. Since we have seen that this can only occur through reproduction, it follows that all the organisms of that period of time could only have arisen from those most primitive organisms by reproduction."

The methodology of the argument presented here is based on the same principles applied in determining historical facts: reconstruction based on existing documents or remains through the application of logic, with the intent of arriving at unambiguous conclusions, all of which establishes a chain of cause and effect in reverse. According to Dingler, this yields the proof for the central point of the theory of descent, that it is not just a systematizing concept but a *historical fact,* with the same degree of certainty and the same validity as all other historical facts.

The Irreversibility of Evolution

Concept, Logical Nature, Validity

The fundamental principle of all evolutionary processes, deduced first by the Belgian paleonotologist L. Dollo purely as a finding based on fossil material, is stated as follows: *Evolution is irreversible.*

Since the proposal of this principle, an entire body of literature has arisen on the subject—whether, to what extent, and with what possible limitations the principle of irreversibility, called "Dollo's law" by Othenio Abel, is valid. Is it really a law or only a rule permitting exceptions? In its most general reading, the "law" holds that an organism can never revert to an earlier state, one already realized once before in its ancestral line. Opposed to that, the science of genetics has repeatedly pointed out that there are *back mutations* and that, therefore, genetic mutations can be reversed, and previously existing characters can be reestablished. This fundamentally contradicts the irreversibility of evolution, depriving it of its character of a law.

We, in contrast, are of the opinion that these kinds of objections are not sound and that the irreversibility of the course of evolution is entirely *unassailable.* This is not some special law or simply an empirical rule drawn from fossil material; the grounds for irreversibility lie much deeper. They are ultimately based on a fundamental property of the organic body as compared with an inorganic one, namely, *mutability, the progressive changes in form and function of organisms over time,* which contrasts with the immutability of inorganic matter in its habit, its chemical and physical behavior. For inorganic matter, therefore, the concept of time is meaningless.

A calcite crystal that formed during the Devonian does not differ crystallographically and chemically from one that forms today before our very eyes. It all depends only on external conditions. When the circumstances are the same, the development of the form, the crystallization, takes place always, everywhere, and at any time, in the same invariable way, and the chemical reactions follow the same regular course. If, on the other hand, external conditions reverse, the reactions are reversible; the inorganic body reverts to an earlier state of equilibrium. The synthesis and reduction of the forms and the chemical compounds follow one another in motley succession; these processes are completely nonindividual and reversible; the passage of time leaves not the slightest trace or alteration.

Not so for organisms. The theory of evolutionary descent teaches us that organisms are subject to continuous change, that they are modified incessantly, sometimes rapidly, sometimes more slowly. Each and every form of plant and animal is thus an individual, temporal-historical structure. It bears the legacy of its past and the aspect of a very particular time, and it is marked by the transformational stages it has undergone and the particular evolutionary level it has attained.

An organism that lived during the Tertiary under particular environmental conditions was no longer the same as, say, a Jurassic ancestor exposed to the same life conditions—just the opposite of what obtains in the inorganic realm. During the interval, the organism changed considerably; all of the evolutionary events that occurred in its ancestral lineage have left their mark, continue to affect it, and simply cannot be made to reverse themselves. Over the course of time the genotype has changed, and even when the same external environmental and selectional factors recur during different phases of evolution, the "reactions" take place in specific, completely different ways.

If we look closely at this truly historical character of evolution and the uniqueness of its individual stages, irreversibility appears simply as a "self-evident" truth. It is not a special law of organic evolution but rather, as Karl Beurlen correctly recognized, the *universal, fundamental categorical, an unconditionally valid concept, independent of experience, of all that is historical,* of "world history" [*Weltgeschichte*] as well as of the history of lineages [*Stammesgeschichte*], or the course of evolution.

The wheel of history cannot be rolled backward. Human history is unique and not repeatable; once an event has occurred, it cannot be undone. It is true that sometimes similar sets of circumstances repeat themselves. In the meantime, however, those who make history have changed, peoples have changed—in their national structures, technology, civilization, culture, and world view, so that similar circumstances are met with completely different preconditions, setting off different effects. In the realm of organisms, all is incessant change; as Heraclitus said, you cannot step twice into the same river.

Even the individual development of the organism teaches us that organic evolution proceeds in a straight line, is not reversible, and never reverts to the point from which it began. Just as an adult human can never revert to being a child and finally, a single cell, evolutionary reversion to a stage once passed through is impossible. Old people may indeed lose their teeth and hair, and their mental faculties may diminish, causing them to resemble their childhood state in many respects; they will never recapture their youth, however, least of all through the symptoms of senility that accompany the involution of aging and that only superficially resemble the youthful state.

*   *   *

If one believes that the irreversibility of evolution can be disproved by the occurrence of the *back mutations* observed today, which cancel out a previous mutational change, it shows a lack of understanding of history. It is entirely possible, of course, that a minute transformational step of the kind represented by a mutation can be reversed immediately upon its appearance and that the latent, existing older genes (alleles) will be activated once more.

The behavior is exactly the same in the realm of human activity: a deed once done can be reversed if it is *instantly* rescinded; it can never, however, be made not to have happened if it has set off other chains of cause and effect or if, in the meantime, the rest of the circumstances have changed. *What has happened is then irrevocable and the past cannot be restored.*

Mutations with *simple* causes may be reversible within a short period of time; for those whose causes are more *complex,* however, the reappearance of the same mutational pattern in exactly the same form will occur much less frequently, and, indeed, this would only be possible as long as the preceding atavistic mutation has not entailed any further, correlating changes, and the rest of the genetic structure has remained unaltered. However, by the time a first mutation has brought about a new state of equilibrium in the complicated net of morphological and physiological interconnections, development has proceeded further, and *a back mutation would encounter changed conditions and never be able to reestablish the earlier state.*

* * *

When during the course of evolution a particular organ or differentiation degenerates, *vestiges* of these once-functional structures are often retained in the descendants for long periods. They continue to be emplaced during ontogenies but persist in an embryonic state, never achieving full development or becoming functional. The potence latent in these vestiges, carried in the genome, may occasionally, but very seldom, be reactivated if the arresting factors are done away with; the vestiges may then go on to develop.

There is no doubt, however, that such an instance is not actually reversion, since the emplacements of the organ in question are still present, having been only concealed, and do not represent a new structure. This can no more be termed atavistic evolution than can the recurrence of the same morphology in offspring or in the F_2 and F_3 generations of Mendelian strains—to use a comparison from H. H. Karny.

For the rest, however, the redevelopment of a once-degenerate organ takes place in a morphological and physiological context that has in the meantime become thoroughly altered; therefore, it could no longer proceed in exactly the same way or follow the same path as formerly. Above all, since the rest of the organism has changed fundamentally in the meantime, *an exact reproduction of a former ancestral form will never come about.* This is what is meant by the irreversibility of evolution.

Thus, all so-called *atavisms*—just like the aspects of senility in humans mentioned above—are by no means complete throwbacks to distant ancestral states of the organization as a whole, but rather structures that show certain resemblances to ancestors *in one or another character* through the development of existing latent vestiges or through the arresting of development.

A reestablishment of once degenerate, vestigial organs is, moreover, a rare exception. In general, recurring similar functional requirements are satisfied in another way, as we shall see in a graphic example. Completely irretrievable, however, are all organs that have been definitively lost during the course of evolution, as W. Haacke demonstrated concurrently with Dollo. When we encounter functions in a later segment of evolution that correspond to those in certain ancestors, they have been realized in completely new and different ways.

The irreversibility of evolution is a logical function of the unidirectional progress of evolution in time. The concept is a completely reliable guiding principle for phylogenetic research and is likewise the fundamental basis of all chronological-geological evaluation of fossils, a subject we have covered in an earlier chapter. If there were an arbitrary reversibility in evolution, if it were not strictly temporally-historically oriented, then in all periods of the earth's history completely identical types of primary, secondary, and tertiary forms could appear, indistinguishable from one another and, therefore, also not usable as temporal indicators. But it is a fundamental characteristic of organisms that they change incessantly, and, indeed, as we shall see further on, that they evolve *directionally,* ruling out the possibility of an evolutionary repetition of identical forms.

Examples

The classic example of the irreversibility of evolution is the *evolutionary history of modern leatherback turtles* (*Dermochelys*) as demonstrated by Dollo. The initial forms in the lineage leading to these turtles were land-dwellers and had a thick, closed, bony shell like the one we are familiar with in modern land tortoises. In the Jurassic, these forms took up a marine mode of life. In such an environment, their shell, adapted for life on land, was no longer required; on the contrary, its weight and rigidity made it cumbersome ballast, and consequently it was gradually reduced. Openings appeared in both the dorsal and ventral parts of the shell (fig. 3.58) and grew ever larger in the younger, Cretaceous, representatives, until finally only a few thin bony struts and plates remained (figs. 3.59 and 3.60).

At the beginning of the Tertiary, these forms left the open sea and returned to coastal regions. The turtles' laborious mode of locomotion, which left them vulnerable to enemies, and the pounding of waves along the shores once more favored the development of a shell. It did not arise, however, through an atavistic reestablishment of the old, reduced shell, which instead persisted in its degen-

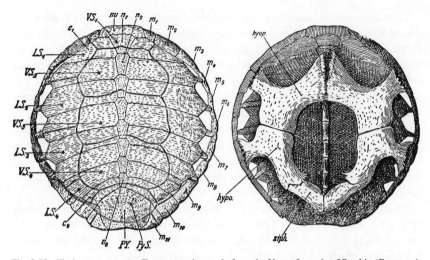

Fig. 3.58. *Thalassemys marina* Fraas, a marine turtle from the Upper Jurassic of Swabia (Germany), showing the beginning of shell degeneration. Much reduced. Left: Dorsal shell from above. Right: Ventral shell from below, with a view of the internal side of the dorsal shell, shown dark. The boundaries between skeletal elements are shown by simple lines, and those between the horny plates that covered the dorsal shell are indicated by double lines. Further, the position of the horny plates is indicated by asterisks. The terms for the individual bony elements are not important here. (After E. Fraas, from O. Abel 1929.)

erate state; rather, a new, secondary shell developed in a completely different way, over the vestiges of the primary one, through the epidermal secretion of thick polygonal bony plates, which connect to form a coherent mosaic.

At the close of the Tertiary, these forms returned to the sea. The new shell, too, degenerated and was transformed into a dermis. In Recent leatherback turtles (fig. 3.61), however, there are still remnants of the once extensive bony plates in the form of little osseous granules in the mosaic of segments of the ventral epidermis. Thus, in this ultimate evolutionary form the remains of two different shells lie atop one another, reminders of the evolutionary fate they have undergone.

This example shows quite impressively that even when the same ecological circumstances are repeated, and the same functions are required and the same preconditions for selection are present, a form does not revert to morphologies present in preceding ancestral lines. Once a structure has been reduced, it will definitely not be reestablished in the way in which it was originally created; once it has been dismantled, the process is not reversed by rebuilding. Instead, *new paths* are followed, and *a substitute of another kind* is created.

But in this process, traces of the previous structural history are retained, making a *reversion to the ancestral state impossible*. This is true for the shell, which,

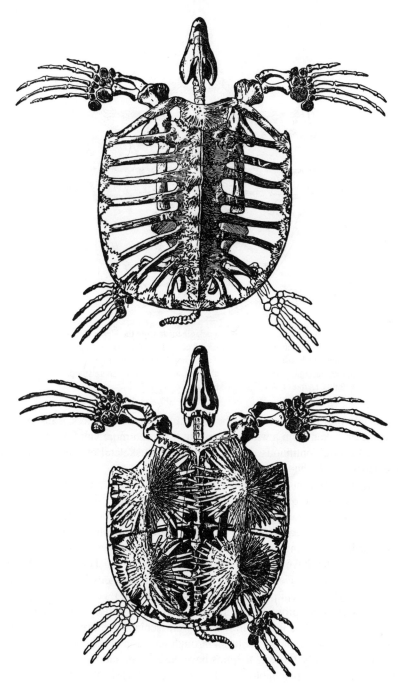

Fig. 3.59. *Archelon ischyros* Wiel., a pelagic turtle from the Upper Cretaceous of Kansas, with much advanced dissolution of the bony shell. About ⅟₅₀ ×. Top: Dorsal view; for greater clarity, the elements of the ventral shell are not shown here. Bottom: Ventral view; the right rear flipper, which the animal evidently lost during its lifetime, is restored here. (After G. R. Wieland, from O. Abel 1929.)

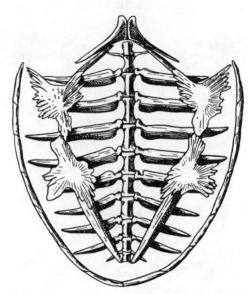

Fig. 3.60. *Protosphargis veronensis* Capell., a pelagic turtle from the Upper Cretaceous of Northern Italy. About ¹⁄₁₆ ×. Ventral and dorsal shell, seen from below. The costal plates (ribs) of the dorsal shell have degenerated into narrow, scythe-shaped bony spans that are connected neither to one another nor to the marginal plates. Likewise, the ventral shell is even more sharply reduced than in *Archelon* (fig. 3.59). (After O. Abel 1929, redrawn.)

with its interplay of reduction and the formation of new structures, is just one set of characters taken from a whole organism; it is even more valid when we look at the manifold, continuous development that the skeletal elements of turtles have undergone in the meantime.

Similar observations can be made in the *beetles.* There, a progressive reduction in the veination of the elytra (wing covers) has taken place. However, the simplified system of veins has no similarity at all to that of the palaeodictyopterans, the primitive lineages of beetles. Thus, regressive evolution did not return to the earlier, ascending path but set out on entirely new paths. In the *realm of plants,* leaves that have once degenerated are not reestablished; rather, the lost leaf surfaces are replaced by other organs, by phylloclades (leafless flattened stems) or phyllodes (leaflike enlargements of the petiole).

Other convincing proof of the irreversibility of evolution, for example, the evolution of the pelvis in certain reptiles (saurischians and ornithischians), the development of a secondary dentition through serration of the edges of jaws and beaks in many birds, the way fish fins have developed, and more, are discussed at length in all larger paleonotological works and in those dealing with the theory of descent, and can be looked up. They are all in accord in teaching that in organisms, the same *functions* appear again and again, but the *structures* that perform the functions do not.

Fig. 3.61. The Recent leatherback turtle *Dermochelys coriacea* L., the ultimate member of our evolutionary series of marine turtles. The former bony shell is here completely degenerate and has been replaced by a thick, leathery skin. Juvenile from the coast of Cameroon. 1 ×. (After O. Jaekel, redrawn.)

Closely related, simple examples are provided by ichthyosaurs and whales, which took up the fish mode of life once more and also acquired certain similarities to fish in outward appearance (fig. 3.110); their anatomical structure, however, remained that of reptiles and mammals respectively and did not in the slightest revert to the organization found in fish. Among other things, they were no longer able to develop gills, and in this respect were like the pulmonate snails, which took up a freshwater mode of life secondarily and have either retained their former atmospheric respiration or use their lungs in the manner of gills.

*   *   *

The evolution of *cephalopods,* already described in detail, also offers several examples of irreversibility. We have seen that the spiral ammonites derive from straight forms, the bactrites and, further back, from the orthocerans (figs. 3.32,

3.34). But several times we find ammonites, especially in the Triassic and Cretaceous, with shells that have uncoiled to become secondarily straight, *Orthoceras*-like forms (fig. 3.41*b, e*).

However, in these instances, too, there is not a total reestablishment of the original situation, no absolute reversion to the initial evolutionary state. Only a portion of the uncoiled ammonite shell—the middle and the end—attains a superficial similarity to the original *Orthoceras* shell. The beginning of the shell, however, has a thoroughly different shape, revealing both its origin from spirally coiled ancestors and the secondary character of the straight form: in these coiled ammonites, the initial chamber (straight in orthocerans and ovate in bactrites) and the first two whorls are coiled spirally, thus preserving unmistakable traces of the spiraling inherited from their ancestors.

Further, these forms show in the design of the septa and sutures the typical features of the evolutionary epoch to which they belong. The straight shapes of the Triassic (*Rhabdoceras*) bear the typical sutures of Triassic ammonoids, and those of the Cretaceous (*Baculites*) have the suture line peculiar to their contemporaries. These features distinguish them fundamentally from one another and from the original orthocerans, with their extremely simple, always unnotched sutures; thus, these forms, similar only superficially and in a very general way, can never be confused with one another.

There is no recoiling of secondarily straight shells; we have already pointed out that these are decadent terminal members of a dying line. In a certain sense, the uncoiling is thus comparable to the features of senility in humans mentioned above. Neither phenomenon ever leads back to the original, initial stages and to the developmental or evolutionary potential inherent in the juvenile state; *uncoiling does not represent a reversal of evolution.*

A similar situation occurs with the secondary regression of the suture, which appears in many Cretaceous ammonites also as a phenomenon of decadence. In these forms, the notching so typical of all younger ammonoids no longer affects the entire suture line as it does in their normally developed predecessors but remains only as a fine frilling at the bases of the lobes (fig. 3.40*b*). In this respect, these forms resemble the Triassic ceratites (fig. 3.40*a*) and are consequently often called Cretaceous ceratites.

Here, too, however, this reduction has in no way rolled back evolution to the early morphological state that once existed. Rather, the regressive suture has retained unmistakable signs of the evolutionary heights from which it has sunk: the saddles, or at least some of them, each show at the apex a deep notch, evidence of their descent from types with fully bipolar notching, that is, from both the lobe and the saddle poles. Thus, it is not an identical form that has returned but only a similar one. The course of evolution subsequent to the Triassic ceratites left behind inextinguishable traces in the so-called Cretaceous ceratites, not only in the suture lines but also in the structure of the siphuncle and other characters.

The situation described earlier in which the same types of ribbing appear at different evolutionary levels of ammonoid stock is different. We recall that Paleozoic goniatites and clymenians, and Triassic and, finally, Jurassic and Cretaceous ammonites all show the same sequence in the evolution of sculpture: originally smooth shells, then simple ribs, and finally simple and then more complex bifurcation of the ribs sometimes superseded by a secondary regression of the sculpture (figs. 3.32, 3.152).

But here, it is not at all a matter of the smooth, terminal forms found in a more recent stratum having issued from those with differentiated bifurcated ribs or their degenerate descendants of the previous cycle and then repeating in their own descendants the evolutionary course already embarked upon once before. Rather, the new cycle of unfolding ensues from the smooth, as yet undifferentiated, conservative forms of the previous periods, whereas the forms with the highest, or secondarily simplified, degree of sculptural development die out without descendants and thus do not carry on into the more recent stratum. What we have here, then, is a sequence of different, independent waves of evolution and not a sort of zigzag course of repeated development and dissolution, not a reversion to former structural processes.

Cephalopods also offer an excellent example of the irretrievability of an organ once it has been lost: The nautiloid line (which included the related ammonoids) has an external shell into which the animal can withdraw for protection. Squids and octopuses are descended from these forms; in these animals, the shell, through extensive reduction, is deposited within the soft tissue, thereby losing its protective function. It finally disappears completely in one line of forms, which later produced at least a partial replacement of the lost shell. In the meantime, however, the mantle in those forms—the normal shell-building organ of cephalopods—had lost its capacity for secreting shell and was no longer able to build a shell like the one its ancestors had.

So the Recent *Argonauta* now produces a shell as a secondary neomorph. It indeed bears a certain superficial similarity to the shell of a nautiloid or ammonoid (pl. 16C) but originates in a completely different way and has a totally different structure: it is produced by a flattened, expanded pair of arms and differs fundamentally from cephalopod shells, which are secreted by the mantle, in that it is unchambered, has no siphuncle, and so on. On the other hand, one modern species of *octopus,* having found itself once more in need of a protective shell, uses—like the hermit crab—a foreign shell, that of a certain clam, for a hiding place.

*   *   *

Finally, a few brief remarks on our *corals.* We observe in the Gothlandian [Silurian] the appearance of stony corals with porous skeletons and then see such forms reappear in the Creataceous or the Tertiary. They are alike in the possession of the one character of porosity, but it is not a matter of a reappearance of identical forms. The older forms are pterocorals [Rugosa] with all their typical features, and the others are definitely cyclocorals [Scleractinia], which means that the similar and probably also ecologically equivalent porosity could not in the least obscure the inner structure of the septal apparatus, which characterizes the specific evolutionary stages of the coral stock.

Even the structures of the corallum—turbinate, discoid, and slipper-shaped (figs. 3.49, 3.142), which appear repeatedly, are always tied in with a particular, different level of organization of the septal structures; the reappearance of similar forms does not mean that an earlier morphology has been reestablished but only that, superficially, there has been a repetition of similar forms, always on new, altered foundations and arising from other roots.

Likewise, the appearance of a columella in many cyclocorals (figs. 3.47, 3.48) does not indicate a throwback to corresponding forms from the pterocoral group but is only an analogous element in a completely altered structural design, as we have already pointed out.

We repeat: *Irreversibility is an unshakable, essential feature of evolution!*

The Periodicity of Evolution

Now, after having discussed some general, preliminary questions in the two foregoing sections, we shall go on to present the specific patterns in the course of evolution, which we deduce from fossil material and from it alone. Our arguments are based primarily on the evolution of the cephalopods and the corals, already described. References to corresponding phenomena in other animalian stocks will demonstrate that the observations made are not special cases but possess fundamental validity.

Different Evolutionary Tempos of Organisms

First, it should be pointed out quite generally that the evolutionary unfolding of individual groups of organisms proceeds *at very different rates.* Thus, there is no one figure for the temporal rate of transformation that wold be the same for all lines of plants and animals. On the one hand, we know of very long-lived forms that transform extremely slowly, and on the other, there are those that change very rapidly and, consequently, are limited to short segments of the geological past. A few figures, which, of course, represent only approximate ranges, will illustrate this.

A known *persistent type* that has remained almost unchanged through numerous geological formations is the genus *Limulus,* the Moluccan, or horseshoe, crab (pl. 4.B; fig. 3.62). Only a few species are found today, in coastal regions of East Asia (the area of the Moluccas) and on the eastern coasts of North and Central America, thus offering another typical example of disjunct distribution, discussed earlier. This crab first appears in the Bunter (Lower Triassic) of the Vosges (perhaps even as early as the Zechstein, or Upper Permian), and from

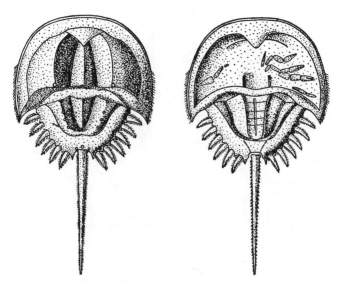

Fig. 3.62. *Limulus walchi* Desm. Upper White Jurassic from Solnhofen (Bavaria). ⅓ ×. Left: Dorsal view. Right: Ventral view. The structure of this species, as far as we can tell based on the hard parts, is identical in all essential features to that of Recent representatives of the genus *Limulus.*

then on, occurs in almost all younger stages: the Muschelkalk (Middle Triassic), the Keuper (Upper Triassic), the Upper Jurassic lithographic slates of Bavaria, the Cretaceous of Lebanon, the Oligocene of the province of Saxony, and so on. Accordingly, the genus has a life span of about two hundred million years. There are also individual genera of corals, ostracodes (figs. 2.57, 2.58), bivalves, snails (fig. 3.15; pl. 30A), and bryozoans (fig. 2.17) that extend from the present back into the Triassic and even further.

The genus *Lingula,* a primitive brachiopod, has an even longer span. This horny-shelled representative has two large, long rectangular valves of almost equal size; a flexible stalk issues from between them to fasten the animal to the ocean bottom or some other foreign substrate (figs. 3.63, 3.64). The several hundred species of *Lingula,* distributed mainly in the Ordovician and Gothlandian, become less frequent later but are nevertheless found in all younger systems; today, the genus lives in tropical seas. Thus, its temporal range is roughly 440 million years.

One might argue that the soft parts, inaccessible to investigation, have perhaps been evolving since the Ordovician and that those ancient forms would not have been completely the same as the living ones. This possibility can be ruled out immediately; large-scale anatomical modifications have definitely not occurred, for the valves of the shell show a complicated system of impressions of the adductor, adjustor, and pedicle muscles, and so on, and these exhibit no changes of any kind. But even when any possible sex difference, concealed from our view,

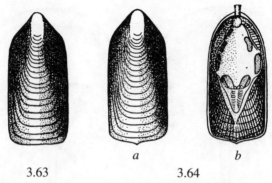

3.63   3.64

Fig. 3.63. *Lingula lewisii* Sow. Gothlandian, from Gothland. 1½ ×.

Fig. 3.64. For comparison: the Recent *Lingula anatina* Brug. 1 ×. *a.* Shell, exterior view. *b.* Ventral valve from the inside, with the impressions of the adductor, adjustor, and pedicle muscles. (After Th. Davidson.)

is taken into account, it would basically not change the fact that here, evolution was almost at a standstill, or at least that it proceeded infinitely slowly.

Genera as tenacious as *Lingula* are also often encountered among the foraminiferans, of which a few Recent genera are known as far back as the Cambrian. For these simple, lower protozoans, however, this is not so surprising, for primitive organizational types are in general distinguished by phylogenetic longevity, whereas higher organization and specialization are bought at the cost of brief life spans. There are exceptions, however, and even among vertebrates there are isolated lines of great tenacity. Among these are the *coelacanthids,* from the order of the crossopterygian fishes (lobe-finned fishes), which extend from the Devonian to the present and have retained their original structure unchanged. The single living representative known thus far, the genus *Latimeria*

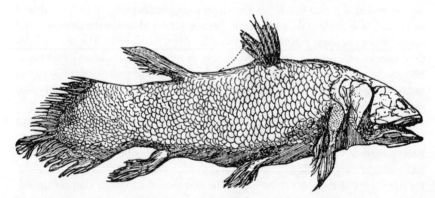

Fig. 3.65. *Latimeria chalumnae* Smith; living representative of the coelacanthids, caught in 1939 off the cape of South Africa near East London. About ⅟₁₃ ×; length about 1½ meters.

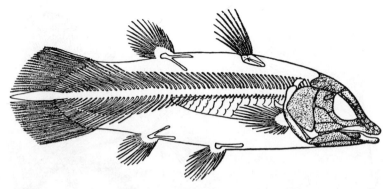

Fig. 3.66. *Macropoma mantelli,* from the Upper Cretaceous of Western Europe; before the discovery of the Recent *Latimeria* (fig. 3.65), the youngest known descendant of the coelacanthids. About ⅕ ×. (From A. Smith Woodward.) The internal skeleton of this primitive bony fish is still partially cartilaginous; in particular, the vertebrae are unossified.

(fig. 3.65), was discovered for the first time a few years ago (1938) in the southwestern Indian Ocean. The find caused quite a stir, for it had been thought that coelacanthids had been extinct since the end of the Cretaceous, the last descendant being *Macropoma* (fig. 3.66). Thus, *Latimeria* is, as it were, a resurrected fossil.

Other relict forms that extend on into the present are the cyclostomes, the fish genera *Polypterus, Lepidosteus, Amia,* and *Neoceratodus* (fig. 3.147), the reptile *Sphenodon* (pl. 31A), and among the mammals, the egg-laying monotremes.

Another very remarkable example is the Recent crustacean genus *Anaspides,* from Tasmania, whose closest relatives are known from the Carboniferous and the Permian.

From the realm of plants, the genus *Ginkgo* (fig. 3.67) should be cited as a "living fossil" in a broader sense—as being the modern remnant of an ancient, long-lived evolutionary line with a geological life span of about 160 million years. Also the genus *Selaginella,* a tropical club moss, represents an ancient persistent type; it has hardly changed since the Upper Carboniferous, that is, in about 250 million years.

But even individual *species* can remain unchanged or show no demonstrable specific changes for long periods of time. For example, the brachiopod *Atrypa reticularis* (fig. 3.68) is found throughout the Gothlandian and the Devonian, a period of 60 million years.

The phyllopod crustacean *Triops cancriformis* yields even higher figures. F. Trusheim found remains of triopsids in the Keuper (Upper Triassic) of Franconia that are distinguished by their outstanding preservational state. Every detail of the structure of the body and its most delicate appendages can be made out—the eyes, the antennae, the mandibles, with their serrated masticatory surfaces, the maxillae, with their rows of fine bristles, the filmy swimmerets

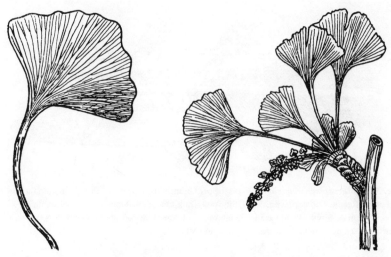

Fig. 3.67. *Ginkgo biloba,* a twig with male flowers; left, a single leaf. The species is the only Recent representative of the gymnosperm class Ginkgoales, which is known from the Upper Carboniferous on and was very important during the Mesophyticum. Even on into the Upper Pliocene, *Gingko* was a member of our central European flora, but it fell victim to the Ice Age, "escaping into the Recent" only in East Asia, a completely isolated relict among contemporary flora. Cycads, taxodians, araucarians, the umbrella pine (*Sciadopitys*), and many other plants are remains of a former abundant flora with wide distribution in the Mesophyticum and Cenophyticum.

(exopodites and endopodites) set with bristles, the brood chamber filled with eggs, and much more (fig. 3.69). As a consequence, a very detailed comparison with the Recent species *Triops cancriformis* could be made, and the author stresses that even the most minor characters are identical. In any event, no qualitative difference whatsoever could be determined that would justify designating the fossil form as a separate species. The only evident difference is quantitative

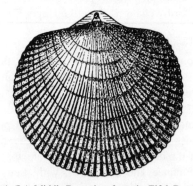

Fig. 3.68. *Atrypa reticularis* (L.). Middle Devonian, from the Eifel. Dorsal view of the shell. 1½ ×.

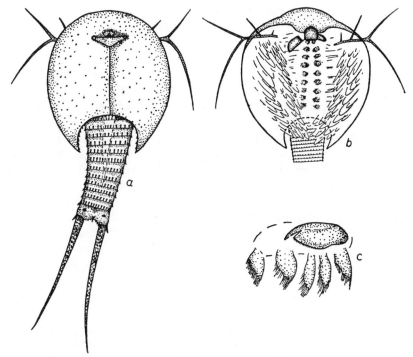

Fig. 3.69. *Triops cancriformis minor* Trush. Middle Keuper of Franconia. *a.* Dorsal view. 3 ×.
b. Ventral view of the anterior section of the body. The somewhat displaced right mandible (to the
left in the illustration), shows the serrate masticatory surface. The legs are only indicated schematically. 3 ×. *c.* Completely preserved extremity. Above, the palette-shaped swimming leg (exopodite);
below, the walking leg (endopodite) with five ciliated appendages. 9 ×. (After F. Trusheim.)

and resides in the smaller size of the fossil representative, which was judged to
be simply a subspecies of the living one. The living species is thereby reckoned
to have a distributional range of about 180 million years.

This figure is surprisingly high but thoroughly consistent with the life span of
triopsids in general. The *genus Triops* itself is known as early as the Bunter
(Lower Triassic) and is very probably present in the Upper Carboniferous, which
would give it a life span of about 250 million years. In general, then, this is an
ancient, tenacious, conservative stock in which evolution is almost stagnant.

Of Recent molluscan fauna, a few long-lived species of clams and snails reach
back to the Oligocene, covering a period of 25 to 30 million years. The same is
true for some species of insects.

*   *   *

On the other side there are the *rapidly transforming* groups of animals, which,
with heightened evolutionary intensity, change their stamp from one rock

stratum to the next and, therefore, like the trilobites and ammonoids, are out-standingly useful as index fossils for relative dating. To give a few rough comparisons, the short-lived genera of the Late Paleozoic index goniatites or the Mesozoic ammonites probably have, on the average, a life span of about a million years, and their species, a span of a few hundred thousand to about a half million.

Similar figures of several hundred thousand years have been calculated for the zones of the older Paleozoic, which are based on the temporal range of rapidly changing species from other groups of animals used in classification. However, even the shortest-lived species of graptolites, which are used to establish the zonation of the Ordovician and the Gothlandian, seem to have had a life span of more than a million years.

Among the mammals, a group of animals also considered short-lived, there is not a single *species* that can be followed back past the Diluvian [Pleistocene], and according to the table constructed by E. Stromer, only a few *genera* extend back into the Middle Tertiary: *Didelphys* (opossums), *Erinaceus* (hedgehogs), *Sciurus* (squirrels), *Mustela* (weasels), *Viverra* (civet cats), *Tapirus* (tapirs), and so on.

In giving other figures for comparison, it should be mentioned that, according to W. Soergel, the famous Heidelberg man is between 450,000 and 500,000 years old, that the Neanderthals from Ehringsdorf lived 130,000 years ago, and that the Recent human, the species *Homo sapiens,* has existed since the late Diluvian, for about 80,000 to 90,000 years. If we set the human generation at 25 years, we have about 3,500 generations behind us. Based on the recent discovery, near Melbourne, of the Keilor skull, it appears, however, that *Homo sapiens* was already living in Australia during the last interglacial, making it a contemporary of the Neanderthal in Europe. The age of *Homo sapiens* would then be estimated at about 150,000 years.

Finally, a few data on the time it takes for mammalian *subspecies* to form. Soergel described a late Middle Diluvian [Pleistocene] cave hyena (*Hyaena spelaea*) from the Lindenthal cave, which, in a comparison with Late Diluvian cave hyenas, is characterized by certain differences in the dentition. The evolutionary time span separating the two forms is about one hundred thousand years, which corresponds to about thirty thousand generations. In spite of this relatively long time span, the demonstrable differences still fall within the framework of the species characters.

Further, W. H. Marshall investigated the mammals of the islands in the Great Salt Lake, in Utah, which have been separated from neighboring shores for about twenty thousand years. Within this period of time, a large number of mammalian subspecies have arisen, which are distinguished from the species on the surrounding land by, among other things, a tendency toward paler coloration, in keeping with the scrubland habitat of the islands. Such differences in pelage, skin coloration, and so on, would never be perceived at all by a paleontologist, meaning that the lowest categories this kind of scientist would set up are generally of a higher rank than the subspecies of zoologists. This circumstance should be taken into consideration when one reflects on the figures given above; nevertheless, only absolute values are affected, not relative ones.

The Phases of Evolution

More important than the general fact already discussed, that the rate of evolution in individual animal and plant groups varies considerably, is the circumstance that also *within one and the same lineage* there are far-reaching differences in the intensity of transformation and that these follow a very particular pattern.

Evolutionary transformation does not flow like a smooth, peaceful river but rather like a stream with many series of waterfalls, rapids, and sharply changing gradients. Evolutionary development is episodic—it proceeds in phases, or in *quantum leaps;* it exhibits an unmistakable *periodicity.* The unfolding of lineages is divided into evolutionary periods or cycles of differing magnitudes, in each of which *three phases of differing evolutionary rates and differing modes of development can be distinguished.*

At the onset of a cycle, there is a brief period of abrupt development of forms. In this phase, a number of different kinds of structural organizations or types are established rapidly, even explosively, in large transformationl steps; during the next phase, these types continue to evolve while retaining their basic nature unchanged. We call this *first phase* the *origin of types,* or *typogenesis.*

This is followed by a *second phase,* one of *type constancy,* or *typostasis,* which entails a progressive elaboration, diversification, and differentiation within the framework of the basic form but does not alter the basic structural design itself. In this phase, evolution is slow, very gradual, and smooth, proceeding in small, individual steps.

This typostatic phase usually lasts much longer than the first, typogenetic period and longer also than the *third phase—typolysis,* or the *dissolution of types,* which brings each evolutionary cycle to a close. This phase is characterized by multiple indications of decline, degeneration, and the loosening of the morphological constraints embodied in the type. Overspecialization and gigantism in the lineages destined for extinction give this period its special mark.

Because this periodicity is an extremely widespread and very general phenomenon, it was recognized early and has been described in various ways. Thus, Ernst Haeckel spoke of *Epacme, Acme,* and *Paracme*—of a rising, a flourishing, and a fading away of lineages; later, Johannes Walther spoke of *anastrophes*—period of profuse, turbulent diversification of lineages alternating with periods of slower, more gradual evolution. Rudolf Wedekind described this set of circumstances in his *Virenz* theory, which holds that from time to time individual faunal lineages enter a climactic periods of expansion (a period of *Virenz*), within which a phase of unstable diversification, a second phase of stable, continuous development, and a final one of excessive morphological development can be distinguished.

Recently, Karl Beurlen, in particular, has elaborated upon the pattern I have just described. He divides the evolutionary cycle into an early phase of explosive development of forms, during which the newly formed structural design breaks up into its various morphological and ecological possibilities; a second period of more gradual, unidirectional (orthogenetic) elaboration of the basic forms created during the first phase; and a final phase characterized by rampant complexity, degeneration, and dissolution of the stable morphology of the preceding period.

Repeated efforts have been made to illustrate these expressions of organic evolution. For example, the evolutionary cycle has been compared to a sort of ornamental column composed of a low, wide base, a long, narrow shaft, and a shallow, slightly flaring capital. Another image used for comparison is that of a skyrocket, which, upon ignition, rains down a great shower of sparks on all sides, then follows a long, straight, ascending course until it finally loses momentum, disintegrates in scattered sparks, and goes out. Such illustrations are always valid only up to a point; nevertheless, the symbol of the skyrocket is, in general, decidedly more appropriate than the old image of the tree with its mighty trunk (in reality not even present!), and orderly branching into limbs of even size.

The three phases of evolutionary cycles described have their counterpart in the developmental stages of the individual cycle—in the youth, maturity, and old age of every single living being, a cycle that exhibits the same characteristics of a brief, rapid morphogenesis, a long period of form constancy, and finally, a decline. Likewise, we know from geology that brief periods of upheaval have alternated with longer periods of gradual evolution, which culminated in a phase of senescence and rigidity in the geological processes.

Furthermore, there is a remarkable parallel to be found in human history and culture, be it in the realm of politics, science, or art. In each, we observe a brief revolutionary, creative age, which breaks radically with tradition, introduces new concepts, and usually determines the course of development for a long time to come. A subsequent period of development, which corresponds to our typostatic period, is based entirely on the achievements of the previous creative period and is concerned solely with elaboration; it proceeds slowly and produces nothing fundamentally new. From these phases there gradually emerges unbalanced exaggeration, obsolescence, cultural degeneration, and inevitable decline, until some new revolution breaks out, heralding a new cycle of development.

According to Karl Jaspers, the same sort of cycle is seen in intellectual history: every new intellectual age begins with the pioneering work of independent, creative individuals, who appear suddenly in large numbers after the collapse of a previous epoch. By resolutely severing their ties to the past, they crack its rigid shell and, in so doing, determine the typical forms intellectual activity will assume in the subsequent phase, when innovation is elaborated upon and diversified, and then preserved as tradition. Finally, the constraints of this epoch, too, break down.

These several analogies indicate strongly that *all temporal-historical development follows the same course* and that, in this respect, phylogenetic evolution fits into a larger, comprehensive framework (all of which, of course, is to say nothing yet about the internal workings of these really only analogous phenomena).

Examples of Major Cycles

The evolution of the *cephalopods* and the *corals* described earlier provided a number of examples of the periodicity of evolution. We saw that, immediately after their first appearance and within the short time span of the Ordovician,[5] the nautiloids brought forth almost all the major structural designs they were ever to produce, discontinuously, without transition, in large, individual steps (fig. 3.21).

The types of the different suborders separated out from small, undifferentiated, highly labile forms. During this typogenetic phase, the new forms, too, were at first extremely labile but soon assumed a stable structure, which then remained unchanged for a long time, usually throughout many geological systems. Evolution in this typostatic phase proceeded considerably more slowly, in smaller steps, and is demonstrable in general by every conceivable kind of form, each reflecting uninterrupted transition. Evolution remained well within the constraints of the original structural design, never once exceeding the established bounds. Only in the final phase, typolysis, did a multiple bursting of the boundaries occur, and the type organization, until then so closely adhered to, began to dissolve.

Further, we refer to the way in which the stony corals diversified (fig. 3.46). The oldest representatives of this group appear in the Ordovician and immediately thereafter set off abruptly in a large number of extremely divergent evolutionary directions, forming superfamilies and families. In this case, too, we see a brief, typogenetic phase at the outset, succeeded by a long typostatic phase.

Within the periodic, overall cycle of lineages, there are also *lesser evolutionary cycles.* From time to time during the extended typostatic evolutionary period, existing types undergo a refashioning and splitting into subtypes of a lesser magnitude. These processes are also cyclical, exhibiting the same, familiar periodicity.

The behavior of the Nautiloidea and Ammonoidea at the boundary between

5. When we speak here of a "brief" period of time, it is of course only relative: we compare the entire course of cephalopod evolution and the abundance of events that crowded the Ordovician with the later paucity of significant transformations.

the Triassic and the Jurassic (fig. 3.32) and of the corals at the transition from the Permian to the Triassic (fig. 3.46), offers proof for this contention. In these cases, a phenomenon characteristic of all typal transformation can be observed with impressive clarity: a dissolution of all the highly differentiated lineages of the previous cycle, which, evidently, were no longer capable of far-reaching evolutionary transformation. Of the original type, only a single lineage of unspecialized forms that had retained their plasticity survived the general mass extinction to become the roots of an altered typal structure and a new evolutionary cycle. The same course of events is then repeated: first, a period of explosive, labile development of forms, which led to the elaboration of new subtypes and soon reestablished the former diversity of lineages; then, a phase of stable, gradual, continuous evolutionary expansion and elaboration of new forms within the established framework; and finally, a period of decline and extinction.

Phenomena of decadence, characteristic of the final evolutionary phase, can be observed, for example, in the clymenians shortly before their extinction in the latest Devonian. At that point, abnormal forms with triangular shells appeared, which we shall discuss in more detail in connection with another matter. Furthermore, we have already mentioned the wild divergence from established form in Triassic and Cretaceous ammonoids, which set in prior to their massive die-off [Triassic] and final extinction [Cretaceous], namely, the dissolution of the typical closed ammonite spiral by uncoiling and transformation into a helical, straight, or even completely irregularly coiled shell; and the formation of abnormal living chambers, the reduction of suture lines, and the like (figs. 3.40*b*, 3.41).

Abnormal shell forms did indeed sometimes appear in ammonites during periods other than the two critical ones mentioned, during the Brown [=Middle] Jurassic, for example. Such appearances do not, however, contradict the concept asserted here. For in these instances, too, the ammonites in question were branches that were dying out—the final stages of lesser cycles; it is prior to their extinction that they exhibit the degeneration of shell characters once so rigidly adhered to.

Furthermore, the final forms of evolutionary lineages that are in the process of becoming extinct are commonly characterized by overspecialization, gigantism of the entire body, or excessive size of individual organs—in other words, by striking incongruities in bodily structure. We have already pointed out such phenomena in the cephalopods and corals; we shall return to this subject in more detail in a later chapter.

*   *   *

The periodicity of evolution just described is by no means a peculiarity of the cephalopods and corals chosen to illustrate the point here but is found throughout the most varied animal and plant lineages.

For example, if we observe the unfolding of the *placental mammals* (fig. 3.70), we also see at the base of the group a brief, typogenetic phase during which almost all the basic structural designs of all the orders were worked out; these designs then continue to develop slowly and gradually during the subsequent long, typostatic evolutionary period in the Tertiary and the Diluvian, but nothing essentially new, nothing that goes beyond the bounds of the individual structural designs, ever appears again. Several of the orders, after having exhibited the characteristic overspecialization, excessive size, and degeneration, died out before reaching the present; but in more recent times, scarcely anything new appeared.

The evolution of the agnathans and the fishes, as well as that of the amphibians and the reptiles (fig. 3.71), presents similar patterns. The *amphibians* show the characteristic contrast between the basal splitting off of the lineages and the long persistence without modification of some of them throughout the entire Mesozoic and Cenozoic.

All the major *reptilian* structural designs were established during the Permian; all the orders were in existence before the end of the Triassic. Later, during the Jurassic, Cretaceous, and Tertiary, the only changes were an increase in size, frequency of occurrence, and individual specializations, but in most lineages, there was also a thinning-out of the diversity of types through extinction. The same splitting of basal types occurred again at lower levels within the vari-

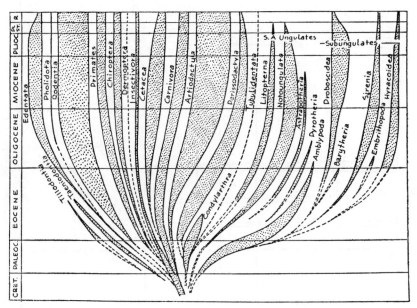

Fig. 3.70. Phylogeny of the placental mammals. (From A. S. Romer 1933.)

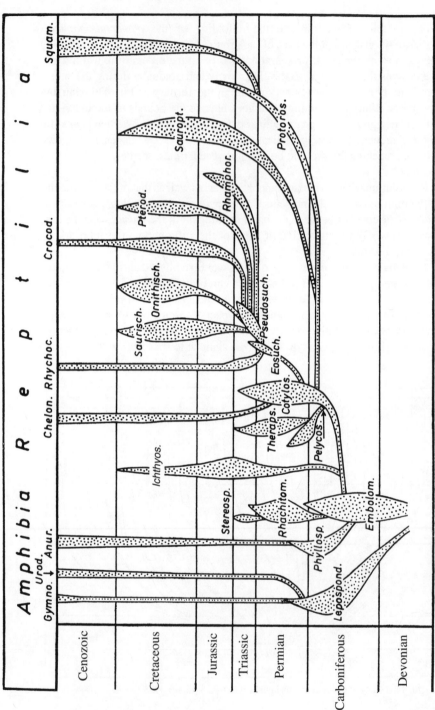

Fig. 3.71. Phylogeny of amphibian and reptilian groups showing the striking contrast between basal splitting (typogenesis) and subsequent lengthier, unaltered persistence of the phyla (typostasis), which then, prior to their extinction, usually exhibit the phenomena of decline, or typolysis. Further, the parallel extinction of many groups of reptiles at the Cretaceous-Tertiary boundary is conspicuous. (Based on A. S. Romer 1933.)

ous orders of reptiles, as, for example, in the Saurischia, Ornithischia, turtles, and so on.

In the plant kingdom, we encounter the *articulates* [Equisetophyta] and their entire, considerable complement of types, at the transition from the Devonian to the Carboniferous; as early as the Upper Carboniferous, these plants were already in a considerable state of decline as giant forms became extinct. No new forms were added later. Only a single, uniform type, the genus *Equisetum,* has persisted to the present. During the later part of the Lower Cretaceous, *angiosperms* and other plants made their explosive appearance, splitting at the very outset into a large number of families that are still living today. It is impossible and also unnecessary to represent here further examples in detail.

Examples of Minor Cycles

Periodicity is characteristic not only of the broader evolutionary cycles of classes, orders, and suborders, of which we have already spoken; it appears to a lesser extent in the lower categories, too, in the *unfolding of families, genera, and species,* which correspond to the lesser and least of evolutionary cycles. The *evolution of the Manticoceratidae* (pl. 22A), a family of goniatites that occurs with a great diversity of forms at the base of the Upper Devonian (Stage I), offers evidence.

The distinguishing character of the family is the appearance in the external lobe of the suture line (M in fig. 3.72) of a medial saddle (itself divided by a medial lobe), an element that appears here for the first time in ammonoid history. At first, in the most primitive family members, it is quite negligible (fig. 3.72a), becoming more pronounced, step by step, in the more advanced forms (fig. 3.72$b–f$). The degree of differentiation in the rest of the suture line varies widely among the individual groups and constitutes the identifying character of the genera. In the most primitive genus (fig. 3.72a), on each side of the shell[6] between the internal lobe (I) and the medial lobe (M), we count only two lobes, or wave troughs, in the suture line, whereas in the most advanced genera (fig. 3.72f), we count sixteen; therefore, the total number of lobal elements in the complete suture line increases from six to thirty-four.

As the illustration shows, this differentiation follows a truly uninterrupted progression; the number of lobal elements increases through a regular, progressive splitting of the internal saddle and its auxiliaries, as we already know. Because the specialized genera repeat every step of this evolutionary path in the ontogeny of their suture lines, the progression can be clearly followed. Consequently, one would expect from this gradual, step-up-step evolution that the individual genera would be well separated from one another in a chronological profile, appearing in a corresponding sequence. This, however, is not the case; the entire complex

6. Because the suture line on one side of the whorl is a mirror image of that of the other, it is customary to use only half of it for observation and illustration.

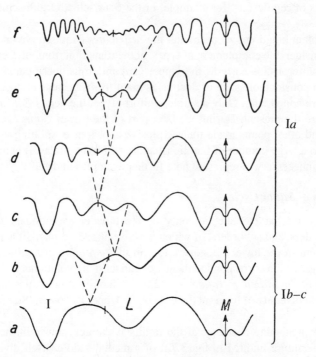

Fig. 3.72. Elaboration of the suture in the Manticoceratidae though an increase in the number of individual elements (formation of umbilical lobes though ventro-alternate splitting of the inner saddle and its derivatives; broken lines) and the more pronounced arching of the medial saddle (M). Each of the stages depicted corresponds to a genus: (*a*) *Ponticeras* Mat.; (*b*) *Manticoceras* Hyatt; (*c*) *Koenenites* Wdkd.; (*d*) *Timanites* Mojs.; (*e*) *Pharciceras* Hyatt; (*f*) *Synpharciceras* Schdwf. The symbols on the right indicate the stratigraphic occurrence of the various genera. As the diagram shows, all of the genera, or all of the different stages of the suture, appeared in the basal zone I*a* of the Upper Devonian. Therefore, the families broke up into their subtypes suddenly, in a burst of evolution, right after their first appearance. During the rest of their existence, in the succeeding zones I*b* and I*c*, there were no further morphological transformations at the genus level (from Schindewolf 1945).

of forms is already present in the early Upper Devonian, which means that all of the forms appeared *suddenly and almost simultaneously.*

It should be added that these observations are not limited to, say, an isolated stratigraphic profile where one would have to take into account disturbances in the original sequence of deposition, or gaps, but can be made everywhere, in every land and on every continent, thereby excluding the possibility that the phenomenon is accidental. In order to evaluate properly this headlong rush of development of forms, this transgression of all normal, gradual evolution, one must realize that our long series of genera, which vary widely in level of differentiation, was produced within a small segment of a single zone, a zone based, as always, on the life span and constancy of a *single species* in which there is no other paleontologically detectable evidence of even the most minute transformation.

In this instance, an *evolutionary cycle at the level of family,* there is also a typogenetic phase of explosive development of forms, which brings forth its entire complement of subtypes—of genera—suddenly, in the briefest of time spans. Remarkably enough, and contrary to expectations, the more advanced of these genera, whose sutures are reproduced in figure 3.72*c–f,* are entirely limited to the basal horizon of the Upper Devonian, zone I*a,* which means that they died out early, whereas the simpler ones (fig. 3.72*a–b*) continued on into the succeeding zones I*b* and I*c,* which means that they existed for a relatively long time.[7] In zones I*b* and I*c* they entered the typostatic phase and remained unchanged as far as the level of specialization of the suture is concerned; the only changes occurred in subordinate species characters, such as the cross section of the whorl, the width of the umbilicus, the shape of the profile, and so on, and these changes occurred slowly.

This periodicity can be followed right down to the *species,* although at that level, only a few imprecise observations are presently available. For example, F. von Huene, in his very careful research on Liassic *ichthyosaurs,* based on extensive material that had been placed stratigraphically with extreme precision, established that, as each species appeared, it exhibited great lability and tendencies toward modification and splitting. Numerous offshoots appeared, which sometimes led to new species. Then came a period of consolidation, a restricting of evolutionary scope and a slowing down of the rate of transformation. F. Zeuner reported similar observations in Diluvian [Pleistocene] *elephants.* A still lower order of periodicity is found in *individuals,* whose episodic ontogeny has already been mentioned briefly.

Conclusions

To summarize briefly, we assert the following: Evolution does not unfold at a slow, even tempo but instead exhibits a pronounced periodicity: a division into three phases based on the rate, potential, and course of evolution; we call these phases typogenesis, typostasis, and typolysis.

The critical period, the one that effectively determines the evolutionary destiny, is that of typogenesis. During this period, accelerated evolutionary change and the sudden increase in the diversity of forms without long series of linking, intermediate forms produce new structural organizations by profoundly transforming the features that determine the type. During this typogenetic period, as

---

7. Moreover, this behavior is not exceptional; on the contrary, it is an often-realized rule that the differentiated forms split off explosively immediately at the base of the cycle in question and are short lived, not surviving the period in which they originated, whereas the simple, unspecialized forms persist.

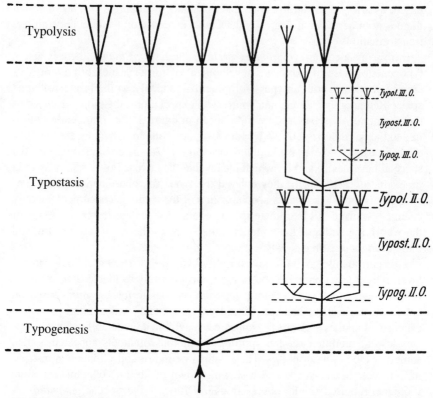

Fig. 3.73. Diagram showing evolutionary periodicity. During a brief typogenetic phase, the stock breaks up suddenly into a number of substocks, or subtypes, whose structural features remain unchanged throughout the lengthy typostatic phase. But in the brief, final typolytic phase these subtypes lose their consistent morphological identity and produce all kinds of degenerative offshoots. During the typostatic phase of the comprehensive type, evolutionary cycles of a lesser rank appear, each with its own typogenesis, typostasis, and typolysis II (of the second order, shown in the diagram only for the subtype on the right). Further, during the period of typostasis II, typogenesis, typostasis, and typolysis III (of the third order) may appear, and so on. (From Schindewolf.)

soon as the class type appears it splits abruptly into the subtypes of its orders, and these, in turn, split to form the suborders, families, and so on.

Therefore, the older notion of a gradual, smooth refashioning of characters and species does not apply to this phase of evolution. *The real problem in evolution thus shifts from the "origin of species," which until now has been in the forefront, to the understanding and explanation of the far-reaching, unmediated differences between types.*

The differences between types are of varying magnitudes, and consequently a continuous, graded series of typal transformations is seen. Differentiation corresponds to taxonomic classes, orders, families, and so on, whose evolution proceeds in individual phylogenetic cycles and exhibits the characteristic periodicity at every stage. During the cycles in which subtypes II, III, and higher unfold there is sometimes a rejuvenation of evolutionary energy: a new period of expansion sets in, and there is once more a rapid, sudden splitting into the subordinate type structures III, IV, and so on. This means that the evolution of a lineage is composed of a *system of ever more comprehensive evolutionary cycles,* which proceed in the same ways and differ only in the degree of magnitude (fig. 3.73).

The Origin of the Types

The Concept of the Type

In the previous chapter, we repeatedly referred interchangeably to "organizational types" [*Organizationstypen*], "structural designs" [*Bauplänen*], or simply "types" [*Typen*] and said that their formation was at the core of evolution. Furthermore, we have already noted that such types can be of different magnitudes. To a certain extent, what was meant should have been relatively clear from the context, without a more precise definition. But before we proceed, it is essential to establish exactly what is meant by the *concept of type.*

In comparative or idealistic morphology, type is based on ideal form, the archetypal form, or the common, basic structure to which a multiplicity of forms that differ in individual characters can be referred. "Type" [*Typus*] and "structural design" [*Bauplan*] are thus conceptual terms for the distinguishing organizational structure of any category in the natural system of organisms—for the set of characteristics common to a broader or narrower group of forms, showing kinship and separating one group of forms from another, related one. We speak, therefore, in descending order, of type (or structural design, or organizational structure) in a class, an order, a suborder, a family, and so on.

We can clarify the morphological concept of type even more by using a specific example. As is generally known, the *insects,* which comprise about three-fourths of all species in the animal kingdom, exhibit an absolutely enormous diversity of forms and abundance of species. Throughout this diversity, uniting all forms, shines a commonality of basic organization, the type of the insect. The type is the immutable basis upon which the immense abundance of forms is built; like a tight ring, it encloses all the countless individual differentiations, joining them to form a whole.

In spite of the most varied forms of individual organs, of divergent adaptive characters, and of the many purely external differences in form, all insects have the same basic structure: They all exhibit a distinct division of the body into a rigid head capsule, a thorax, and an abdomen; the thorax is composed of three segments, each bearing a pair of legs, the abdomen is typically composed of eleven jointed segments, and so on. Within this broad type, there are different subtypes: the Orthoptera (the straight-winged), Neuroptera (the net-winged),

Coleoptera (beetles), Lepidoptera (butterflies), and so on. Each of these narrower groups has, in turn, its own set of common characters, which distinguish it from the rest of the insects.

From this we can infer that the individual characters of an organism are by no means of equal value, nor are they transformable in the same way; rather, there is a *hierarchy* of stability or capacity for transformation, and of the extent of its range of application. The structural characters of the comprehensive type, in this instance the class Insecta, are relatively rigid and constant, whereas the organizational characters of the subordinate types are more mutable; characters limited to an even smaller circle—those that determine the type at the family and genus level—are even more subject to transformation. The most unstable characters are the purely adaptive ones of race and species, characters that interact directly with the environment. This gradation of sets of characters and their hierarchical arrangement within the existing diversity of forms is expressed in the natural system of organisms.

In nature, we never encounter the typal organizations in their pure, unaltered form; there is no individual that would exhibit only the general structural characters of its class. To be viable, each organism needs special adaptations and structures in addition to the general characters of its class. Each individual virtualizes the layered or intertwining character complex of the entire hierarchy of types to which it belongs. Thus, it embodies at once the type of its species, genus, family, order, and so on, and it is a matter of comparative, abstracting observation to single out and separate these type characters, which differ in order of magnitude and extent. We shall address these questions in more detail in the final chapter, which is dedicated to taxonomy.

The Reality of the Type

Because logical abstractions are necessarily involved in establishing and illustrating types, they were often said to be artificial—purely intellectual, idealized human constructs. It is, nevertheless, incontestable that morphological type characters are based on absolutely real facts and causes. When all the individuals of a species exhibit common, identical characters of species, genus, family, and so on, and when their descendants also incorporate the same species, genus, and family types, it is because the same genotype is operating in all these individuals. Therefore, the type complexes of the various taxonomic units have their *objective substrate in the genotype,* in particular combinations of genes, and if the type characters change, it is obviously because the genotype has changed.

Furthermore, we saw in the previous section *that each type has its own evolutionary cycle.* A new form of the type appears at a particular point in a specific typogenetic phase, and its appearance is followed by a very precise course of evolution. The examples cited of the transformation of type from the Triassic ceratites to the Jurassic ammonites or from the pterocorals to the cyclocorals show particularly well that each appearance of a new type signifies a *radical break in the course of evolution.*

When we observe in these examples that at a particular time the specialized representatives and lineages of the older type become extinct almost without exception and that immediately after the transformation to the new type, which takes place in a single surviving lineage, the line begins to flourish and expand, we can no longer say that the morphologically distinguishable types are meaningless abstractions. *They are, rather, the result and the expression of a very concrete, extremely real evolutionary event.*

Nor are the *type characters* we have deduced an arbitrary figment of the imagination; they are, rather, the very distinguishing characters lacking in the types undergoing extinction but developing in the new, flourishing type—characters that clearly must be held directly responsible for both the extinction and the rise. If all the lineages of the previous type that were incapable of transformation into the new type are condemned to extinction, whereas the one lineage still capable of transformation—the one that managed to make the decisive leap—introduces a new, vital evolutionary cycle, it can only be attributable to the acquisition of a new type structure, one that set up the prerequisites for continued evolutionary development.

The type characters as revealed are by no means merely ideal properties; they are not imposed on the organism from without, but are, rather, *the basics upon which all else depends; they are of considerable real, life-determining significance for the affected organism.*

To summarize briefly, we regard types as not just abstract patterns of form or merely idealized archetypes of taxonomic categories but, rather, as the *products of concrete, evolutionary processes,* as *particular evolutionary entities.* And when we go on to treat the *origin of types* (not of the concept but of the types as they have materialized!), it is not a fictive scientific issue, as many who have completely misunderstood the matter maintained. What we are concerned with, rather, is the extremely real *formation of the structural peculiarities that separate a new type from its predecessors* and the elaboration of these peculiarities, which constitutes the decisive process of evolution. Only these *distinguishing* sets of characters will be discussed; the vast majority of organizational features—those that appear in the common supertype—are inherited unchanged from the ancestral type, as has already been pointed out several times.

The Origin of Major Types

We have already seen simple but graphic examples of the origin of new types in the evolution of the stony corals and the cephalopods. We shall refer here primarily to the evolution of the peculiar suborder of the Heterocorallia from the Pterocorallia [Rugosa] (fig. 3.74).

As we recall, these two groups of Paleozoic corals differ from the younger cyclocorals [Scleractinia] in that in the older forms only four quadrants develop and are provided with metasepta. However, in these two groups, these quadrants are not at all homologous, and the developmental paths they follow as they mature are fundamentally different (fig. 3.74*e* and *h*). In the pterocorals, six basic septa, or protosepta, are emplaced, whereas the heterocorals have only four; in the pterocorals, the septa remain simple and unbranched, whereas in the heterocorals, each of the four protosepta splits into two branches toward the periphery.

Hexaphyllia (fig. 3.74*f*), the most primitive representative of the Heterocorallia, at first shows only a splitting of one lateral pair of protosepta (II) and holds to this very simple developmental stage throughout its life span. The genus *Heterophyllia,* which follows it (fig. 3.74*h*), takes another large evolutionary step in the direction already established: the vertical pair of protosepta (I) divides in the same way that the horizontal pair (II) in *Hexaphyllia* did, and in the space that formed between the two branched septal ends, that is, *within* the paired mesenteries of the protosepta (shaded section in fig. 3.74*g*), metasepta formed *cyclically* (fig. 3.74*h*). Thus arise two more profound differences from the pterocorals, in which the metasepta appear not *within* the primary protomesenterial pockets but *external* to them, in the spaces between them (fig. 74*d,* shaded area), and furthermore have not been emplaced cyclically, as a complete circlet, but *serially,* one after the other (fig. 3.74*e*).

As we determined earlier, these *fundamental, qualitative character differences* between the two structural designs make it appear to be completely impossible for the septal apparatus of the heterocorals to have arisen by gradual transformation from the differentiated septal apparatus of the pterocorals. From the very beginning, their structural paths went in different directions. Therefore, the breaking away of the heterocorals can only have come about during a *developmental stage very early in the ontogeny* of the pterocorals, and we were able to pinpoint this stage precisely. It occurred in the early juvenile stage after the emplacement of protosepta pair II and before the appearance of protoseptal pair III (fig. 3.74*b*). At that point, immediately after the larva attached itself to the substrate and began to secrete its skeleton, the decisive switch to the new evolutionary direction occurred, causing both the suppression of the heterocoral protoseptal pair III and the bifurcation of protoseptal pair II.

Gradual, smooth transitions between these two different developmental types are unknown and scarcely even imaginable. It could be conceivable, of course, that the reduction of protoseptal pair III took place gradually; but with respect to either simple or bifurcate protoseptal pair II, there is a *fundamental, dichotomous difference,* which cannot be bridged by a graduated transformation. The two characters are clearly correlated—perhaps owing to compensation of skeletal material; they are part of a coherent type-complex, which first appeared in or was produced by an early juvenile developmental stage *of the ancestral type,* suddenly, without any transition, in a single transformational step.

After the basic type characters of the heterocorals—the formation of only four

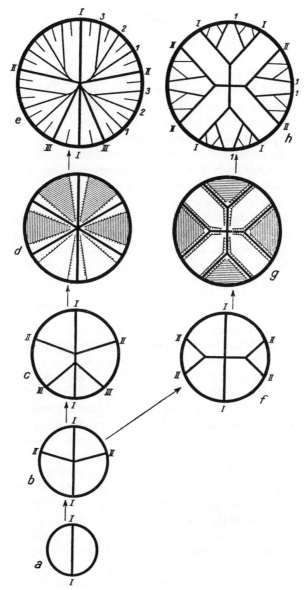

Fig. 3.74. The sudden production of the structural design of the septa in the heterocorals (*f–h*) in an early ontogenetic developmental stage of the pterocorals (*a–e*), schematic, in cross section, greatly magnified. In *d* and *g*, the mesenterial pockets lying between the protosepta (indicated by heavy lines) are reconstructed. Uncertain, but unimportant for our model, is the length of the proto-mesenteries as well as the construction and direction of the musculature, indicated by serration. It is possible that in the pterocorals (*d*) some of the mesenteries formed as micromesenteries and, further, that, analogous to recent Madreporaria, the muscle fibers of the directive mesenteries faced away from one another. (From Schindewolf.)

protosepta and the principle of their bifurcation—were inherited suddenly from the first representative, *Hexaphyllia,* they developed further and were reinforced in the descendants. In *Heterophyllia,* the other protoseptal pairs also split, a development consistent with the new evolutionary direction just introduced. The result was a completely new basis for the structuring of the septal apparatus and the radial division of the corallite, which made it so that when the metasepta appeared, the formerly well-defined area allotted for their development was gone, and they were emplaced in completely different areas than in the pterocorals, namely, in the pocket created by the primary protomesenteries. This resulted in the contrasts between the two types being considerably sharpened.

Moreover, these differing modes of metaseptal insertion also came about by dichotomous developmental processes that cannot be linked by gradual transformation; from the very outset, either one or the other direction had to have been followed. Consequently, we must exclude the possibility that these multiple differences between types occurred in small, individual steps by means of selection and speciation. Furthermore, as we pointed out earlier, the structural characters of the two designs are on an entirely different level from the specific characters that developed within the framework of the two designs.

*   *   *

Another, very simple example of type transformation, one that was given earlier in the book and that can be illustrated with a few simple lines (fig. 3.75), is the very different ways in which the *suture lines of ammonites* develop. We saw that an increase in the number of lobes was usually brought about by the splitting of the inner saddle: as this saddle and its secondary divisions are split again and again, one lobe after another is established (fig. 3.75a). In opposition to this *saddle* splitting is the fundamentally different process of *lobe* splitting, which constitutes (fig. 3.75b) the type character of the superfamily *Popanoceratida.*

The most primitive representative of this type (fig. 3.75b_1) has in its suture line, between the internal and the lateral lobes, two umbilical lobes (U_1, U_2), which arose through the splitting of the primary internal saddle; further, it exhibits a tripartite division of the originally uniform lateral lobe (L). With one stroke, this longitudinal division of the lobe created three new elements, with the result that the two kinds of lobal development not only differ in their mode of origin and in the emplacement sites of the new lobes but also differ considerably in the result achieved.

The conclusions that can be drawn from this example are in every respect completely analogous to those obtained from our previous analysis of the example provided by the corals. Here, too, it can be established first that the two different modes of lobal development are completely and distinctly separate, discontinuous, and in opposition, and can be neither connected in actuality by intermediate, smooth transitions nor bridged in theory.

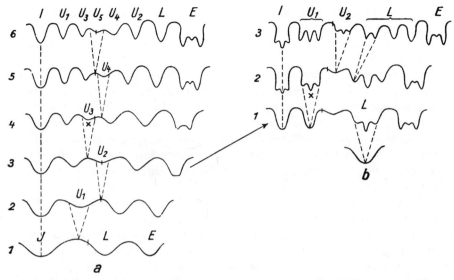

Fig. 3.75. Sudden origination of the suture of the popanoceratid type (*b*), which is characterized by *lobal* splitting, from an early ontogenetic developmental stage of forms with progressive, alternating *saddle* splitting (*a*). In both instances, ontogeny follows the same course up to juvenile stage a_3 of the suture; from then on, the paths diverge (arrow). Note the fundamentally different lobal forms of the last developmental stages shown (a_6 and b_3), which reflect the divergent modes of origin. (From Schindewolf.)

In addition, in this example, too, in the ancestral type, the developmental, or ontogenetic, stage at which the new structural design separates is unequivocal. It is the developmental phase with two umbilical lobes seen in figure $3.75a_3$. This is the point at which the two developmental paths diverge: in the original type, the already established principle of saddle splitting through the development of a third, a fourth, and a fifth umbilical lobe continues; but in the deviant type, the development of the third and subsequent umbilical lobes is suppressed, with the tripartite division of the lateral lobe occurring instead.

Just as with the corals we discussed earlier, this new type structure is *complex;* its characters come into being partly through the suppression of developments in the ancestral type and partly through the introduction of fundamentally new, *qualitatively* different kinds of structural principles. However, the developmental stage of the suture at which the two different paths diverge represents a very early stage of ontogenetic development, since, in the Carboniferous predecessors of the popanoceratids, the first two umbilical lobes formed very early, when the organism was minute, right after the conclusion of the larval period. Therefore, the transformation from one type to another happened *discontinuously, during an early, juvenile developmental phase* and produced mature stages that are fundamentally different from those of the ancestral type and appear suddenly in direct opposition to them.

Here, too, the situation is such that, at first, the oldest representative of the new type only establishes the basic nature of the structural principle that characterizes it, and subsequent forms build on this new principle, following the path already laid out. In the descendants, the new mode of lobal division rapidly takes hold in other lobes too, operating in large individual steps—affecting first one umbilical lobe and then another—and in addition, there is a secondary splitting of the various subelements. Thus, the comprehensive type Popanoceratida divides into its subordinate types of families and subfamilies, within each of which the species differentiate based on lesser deviations in shell shape, whorl section, umbilical width, sculpture, and so on.

The process just described seems to be a general phenomenon operating in the formation of all types. First, a new path opens suddenly and discontinuously and is then developed step by step within the framework of the newly created type until once again a harmonious balance is established and maintained as the unchanging foundation of the type. Thus, the *mammals* exhibit the new typal character of *live birth* and the suckling of young—which trait is only very imperfectly realized in the aplacentals. In the latter, because of the narrowness of the pelvis, young are born in a primitive, undeveloped state; internal gestation is continued externally, so to speak, in the mother's pouch.

Further advances in this direction are the development of the placenta and the widening of the pelvic opening, which made possible a longer period of fetal development within the womb. Clearly, however, this evolutionary trend developed gradually, for even among placentals, the Paleocene creodonts (primitive carnivores) still had a very narrow pelvis and consequently probably also gave birth to incompletely developed young.

*   *   *

We have likewise seen in the example we studied in detail earlier—the transformation of the Paleozoic pterocorals into the younger *cyclocorals* (fig. 3.46)— that the transformation of type occurred in stages and that, since various participating processes go on side by side and work in combination, the critical events in each do not occur contemporaneously. Thus, in many respects, the younger pterocorals already show morphological similarities to the cyclocorals deriving from them, and the older representatives of the latter still have many features reminiscent of the typical characteristics of their ancestors. We have interpreted this as important evidence for the assumption that these coral groups are related by descent.

The actual, final separation of the two types is complete when all six primary sectors of the cyclocorals are provided with metasepta, which, furthermore, are emplaced radially-cyclically. The decisive rupture in evolution at that divide shows that these characters were not chosen arbitrarily by the observer who considered them to be of special significance but are, rather, the marks of a natural separation between types.

These new type-determining characters of the cyclocorals are themselves alternatives, profoundly different qualitatively from those of the pterocorals. They form discontinuously in early ontogenetic stages, at just that point where it must be decided whether or not metasepta will be emplaced in the two as yet undeveloped sectors and, further, whether metasepta I will arise in the middle of the individual radial pockets and the others will be secreted cyclically on either side of metasepta I; or whether the first metasepta will be emplaced marginally, the later ones appearing to one side of it and serially. There is no link or transition between the two possibilities; only an abrupt discontinuity at an extremely early developmental stage creates this kind of transformation.

*   *   *

The general conclusion emerging from these examples is that *all decisive transformations of the basic structures of the higher-ranking types are brought about in large, individual steps with far-reaching consequences, without links or transitions, in early ontogenetic stages.* Their appearance early in ontogeny is consistent with the fact that only those stages still possess the necessary plasticity to meld the newly acquired complex with the old inherited structure and its potential for development in a way that will result in one harmonious form. Moreover, the discontinuous nature of the transformation seems to be a logical necessity, especially for all *alternative changes in important functional organs,* those that *must* be present in either one form or the other if the organism is to be viable.

The *vertebrate lower jaw,* for example, is composed *either* of several separate parts and joined to the skull by the articular, as in reptiles (fig. 3.76*a*) *or*—as in mammals—consists of a single bony element, the dentary, which takes on the function of articulation with the skull (fig. 3.76*b*). Slow, smooth transitions be-

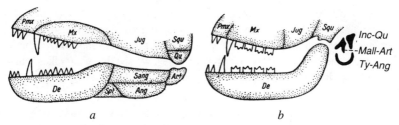

a   *b*

Fig. 3.76. Diagram showing how the reptilian lower jaw (*a*) was refashioned into the mammalian lower jaw (*b*). *De* = dental; *Spl* = splenial; *Ang* = angular; *Sang* = supraangular; *Art* = articular; *Pmx* = premaxillary; *Mx* = maxillary; *Jug* = jugal; *Squ* = squamosum; *Qu* = quadratum; *Ty* = tympanicum; *Mall* = malleus; *Inc* = incus. (From Schindewolf.)

tween these *qualitatively* opposing structures taking place during postembryonic developmental stages, when the jaw mechanism must be able to function, are inconceivable.

To be sure, we recognize in the reptilian lineages that lead to mammals a gradual, *quantitative* reduction of the articular and of the other individual bones of the lower jaw, paving the way for the transformation and bringing the two types closer together. However, the fundamentally decisive, final step—the complete disappearance of these bones or their transformation into elements of the auditory area—must have taken place *discontinuously, suddenly, between one individual and the next, during an embryonic developmental stage.*

Without a doubt, the formation of the typal features characteristic of the main lineages of the animal kingdom took place during the earliest embryonic stages. For example, the differences between the *protostomians* and the *deuterostomians,* those two large phylogenetic units of metazoans, which include, respectively, the annelids, arthropods, and mollusks, and the echinoderms, tunicates, and vertebrates, appear right at the beginning of embryonic development (fig. 3.154). In these two cases, the blastopore of the gastrula was destined to evolve in different directions; in the protostomians the blastopore became the definitive mouth, but in the deuterostomians it became the anal aperture, the ultimate mouth being a neomorph. In addition, there are other alternative differences between the two groups in the way the mesoderm and the coelom develop, in the structure of the gut, and so on, which also could only have occurred suddenly during an early embryonic developmental stage; transitional, intermediate stages and gradual transformation during later ontogeny are inconceivable.[8]

Even in the different types of cleavage of the germ cell certain phyletic peculiarities make their mark, which, like the different types of larvae, obviously cannot arise at a later stage along the path toward gradual speciation, be selected, and then advanced to early embryonic developmental stages.

The Origin of Minor Types

In contrast, the transformation of types of lesser rank takes place in an entirely different way. Here, the characters in question are usually of a *quantitative,* or *meristic,* nature and thus relate to extent, size, and number in the formation of characters. For example, if we look at the *differentiation in the sutures* that occurs when a saddle splits (figs. $3.37a_1 - a_6$; 3.39), we see that what takes place is the simple addition of new lobal elements, and that this does not happen until

---

8. Even when, as recently happened, one postulates for the two lines a hypothetical, neutral ancestral type, one with two openings for the protointestine, *both* of which originally served for the introduction of food as well as for the expulsion of waste matter and only later specialized in one or the other direction, it does not alter the fact that in early embryonic stages two divergent paths were embarked upon, leading to the separation of the two evolutionary lines, which differ profoundly from each other in other respects, too.

toward the end of ontogeny. In this instance, in representatives from one genus the individual development of the suture up to an advanced stage corresponds to the situation found in the preceding genus; it is only just before attaining maturity that the total number of lobes is increased by the appearance of a tiny, new one. In the descendants, this lobe increases in size and is gradually *advanced to an earlier ontogenetic stage;* eventually, another new lobe is acquired in the same way.

The development of a new lobal element is indeed a quantum evolutionary step, but the change it effects on the organism as a whole is relatively minor. And because the novelty does not appear until a late stage of growth, it provokes no aftereffects. Here, continuity is maintained throughout the longest, consistent portion of ontogeny, and the transformation, *gradual and continuous,* is easy to follow, even though it occasionally occurs very rapidly within an extremely short period of time, as we saw in the example of the manticoceratids (fig. 3.72*a–f*).

In contrast, when a completely divergent structural mode intervenes in the formation of the suture line, when lobe splitting instead of saddle splitting appears (figs. 3.37 and 3.75$b_{1–3}$), the paths diverge early in ontogeny. Such an event is a *qualitative leap,* and creates a large discontinuity in that from early developmental stages on, the development of the suture follows a completely different course, one that results in profoundly different, directly opposing mature forms. Here, smooth transitions are not possible; the existing relationships can only be recognized by observing the ontogeny of the suture from its beginnings and making comparisons.

Within lobe splitting itself, however, there is again a *quantitative* progression. Once this new principle has been introduced into one lobe, it also appears later in other lobes. The tripartite division of the affected lobes proceeds gradually; at first, the saddles that form in the base of the lobes are low, but they become more and more pronounced until eventually the subelements themselves take on the form of independent lobes. Further, some of these secondary lobes may also split, an event that does not occur until later ontogenetic stages and causes a change only in the mature form.

We observe exactly the same thing in the *crenulation of the suture,* in the fine frilling of the margins of the lobes and saddles. This phenomenon proceeds slowly and continuously, notch upon notch, the suture of each descendant slightly modifying that of its predecessor. The new individual elements develop late in ontogeny and then in the descendants are gradually advanced to ever earlier developmental stages.

It is another matter, however, when the principles of unipolar lobe frilling dominant in Triassic ceratites is replaced by bipolar frilling in Jurassic ammonites—a radical change of course (figs. 3.32, 3.152, 3.162–64). In this instance, it must be decided early in ontogeny, right when the frilling appears, whether only *one* frilling pole will be established or *two;* for in the ancestral forms of bipolar Jurassic ammonites, frilling begins in very early stages at the bases of the lobes and goes on from there to affect the flanks of the saddles. Thus, the new saddle-pole cannot have arisen toward the end of onotogeny and altered the mode of frilling regressively, so to speak. It is, rather, a qualitatively new kind of type character, a sudden, discontinuous acquisition by the juvenile stages, and as such, it influences the further course of frilling right from the outset.

Conclusions

Only the formation of types of lesser rank just discussed, involving insignificant quantitative differences, corresponds to the usual notion of smooth, gradual transformation, which until now has been commonly thought applicable to evolutionary processes as a whole. According to Darwin's theory, evolution takes place exclusively by way of slow, continuous formation and modification of species: the progressive addition of ever newer differences at the species level results in increasing divergence and leads to the formation of genera, families, and higher taxonomic and phylogenetic units.

Our experience, gained from the observation of fossil material, directly contradicts this interpretation. We have found that the organizing structure of a family or an order did not arise as the result of continuous modification in a long chain of species, but rather by means of a *sudden, discontinuous direct refashioning of the type complex from family to family, from order to order, from class to class.* The characters that account for the distinctions among species are completely different from those that distinguish one type from another. Organizing structure lies at a deeper level than the more external, superficial species characters that are constructed upon it and added to it, one might say, only as arabesques. Consequently, organizing structure cannot be gradually, synthetically built up through the modification of species characters.

The crucial, overall process of evolution does not then consist of the *modification of species* but of the *modification of type,* the direct, transitionless reshaping of type organizations of every grade corresponding to the different categories of the natural system. As we have seen, we have before us a graded series of typal transformations that differ in scope and reflect greater or lesser discontinuities. At one end of this continuum is the relatively negligible modification of species, that is, the reshaping of the most minor type, that of the species, into another specific type of equal rank. In this regard, speciation represents only one, and indeed the least fruitful, *particular instance of the general principle of the modification of types* and is subordinate to it.

At the other end of the continuum lies the formation of the far-reaching, alternative, qualitative differences in the comprehensive, higher-ranked types (class, order), which can only be accomplished through a sudden leap, during a very early juvenile stage. Transformations of this kind, setting in as they do early in ontogeny, entail further correlative modifications as ontogeny progresses, and thus, *with one stroke, a new, complex typal organization arises,* one completely different from and directly in opposition to the ancestral type.

To express this concept concisely, I earlier used the term *typostrophe* to designate the typal modification that is at the heart of our concept, and the term *typostrophism* for the concept itself, which I oppose to Darwinism, for the latter, inadmissably, in my opinion, has simply applied the mechanisms responsible for the formation for race and species to the entire evolutionary process. A later chapter will treat objections to our interpretation and refute them.

In summary: *The course of evolution proceeds through the modification of types in transformational steps of differing magnitudes, each of which sets in motion its own episodic cycle of evolutionary unfolding. In particular, the more sudden the appearance of the evolutionary reshapings of the typal organizations (the typostrophes), the greater the discontinuity, and the earlier the ontogenetic stage at which it appears, the more profound the effect, the greater the qualitative scope of structural difference, and the higher the taxonomic rank of the group of organisms embodying the new type.*

Proterogenetic Evolution

The Nature of Proterogenesis

As we have observed, the redesigning of the basic structure early in ontogeny steers subsequent ontogeny onto a completely new course, bringing forth abruptly and without transition mature forms that differ fundamentally from those of the antecedent groups of types or ancestral cycle. These mature forms often bear scarcely any signs at all of their origin, and it is only possible to determine their lines of descent with the help of the earliest embryonic or larval portion of ontogeny, the only one that is the same as the ancestral ontogeny and that establishes the connecting relationships.

There is one particular instance of juvenile typal modification that illustrates with great clarity the morphological relationship between the transformed descendant and its ancestral type, thereby providing a valuable corroboration for this general evolutionary principle. It is a situation in which a new, diagnostic set of characters is emplaced at a more or less early ontogenetic stage but is not retained until the terminal stage, being replaced instead by a character of the ancestral adult state. In subsequent members of such an evolutionary line, the new type complex *gradually spreads from the juvenile stage ever further into the stages of maturity and full age,* step by step reducing the segment of ontogeny in which the ancestral remnant prevails and finally causing it to disappear entirely.

Contrary to the well-known and more widespread behavior described in Haeckel's "biogenetic law," in this instance it is not the juvenile stages but the *mature stages that repeat the morphology of the predecessors,* whereas the juvenile stages, one might say, point toward the future and anticipate characters that will only appear pure and also be dominant in the mature stages of descendants. In a certain sense, what we have here, as has sometimes been said, is a reversal of the "biogenetic law."

This way of putting it, however, is not beyond question. *Technically and logically,* the objection could be made that there can, of course, be no reversal of a law; in that case, it would no longer be a law. *On material grounds,* it should be pointed out that here, too, consistent with the general processes of early ontogenetic typal modification, the *very earliest* stages of ontogeny at first directly follow the previous state and repeat the corresponding portion of the parental ontogeny. Those, however, are only the *most general* morphological features of the supraordinate types, such as class, order, and suborder, to which the particular type in question belongs but not the special characters that distinguish the family, genus, and species of the ancestral form.

This earliest portion of individual development is followed by the phase of transformation, by the veering off in a new direction, which in turn is replaced by a mature stage that reverts to the special morphology of the ancestral form and repeats it. *In such an instance, then, we cannot use juvenile and intermediate developmental stages to determine the ascendants but must rely on the adult form.*

In contrast to the normal repetitive process responsible for smooth speciation, in this instance the new characters are not emplaced toward the end of ontogenetic development and then in descendants advanced step by step to ever earlier stages; rather, they arise without any transition in a juvenile developmental stage and spread out from there, advancing to the terminal portion of ontogeny. In contrast to the recapitulation of the past, or palingenesis, there is here a *looking forward, an anticipation of the future.*

I have called this ontogenetic-phylogenetic developmental process *proterogenesis.* It is part of the general principle of early ontogenetic type remolding in that here, too, the new characters appear discontinuously, suddenly, in juvenile stages.

Examples

We already know of one instance of the proterogenetic unfolding of characters—in the evolution of the cephalopods: the spiral coiling of the *nautiloid* shell proceeds in this way. We saw, in an evolutionary series beginning with straight-shelled orthocerans and leading to coiled forms, that first the tip of the shell curves slightly and that, further on, the first part of the shell forms a small spiral, which gradually increases in extent until finally the entire shell is coiled (figs. 3.28–31).

Thus, the newly acquired character of curvature or spiraling appears for the first time in early juvenile stages, immediately at the onset of shell secretion, and at first is limited to this stage, whereas the rest of the shell remains straight. Consequently, it is the mature shell that resembles the ancestral form, while the coiled juvenile stages exhibit a character that is different from the ancestral morphology and introduce a new development, one that subsequently extends further and further into the mature stages and finally overwhelms the ancestral remnant.

Further, in the *ammonites* the refashioning of the shape of the shell and its sculpture often takes place by means of proterogenesis. For example, the Brown Jurassic genus *Macrocephalites* has a thick, bulbous shell, its low-apertured coils broadly kidney-shaped in section with nodeless ribs running right across the outer side; from it issue several lineages of high-apertured ammonites with a distinctive elaboration of the external sides and of the sculpture.

One of these evolutionary series is shown in figures 3.77–79, or cross sections. We see in figure 3.77 the already characteristic initial form of *Macrocephalites* and in figure 3.79 the terminal member with high apertural cross section, slightly rounded outer sides, and a row of little nodes at the margins of the whorls. Between these two types, which differ greatly in shell shape and sculpture, there are a number of composite forms, which, in the juvenile state, exhibit the mark of the younger genus but in adult stage show the features of the ancestral form (fig. 3.78).

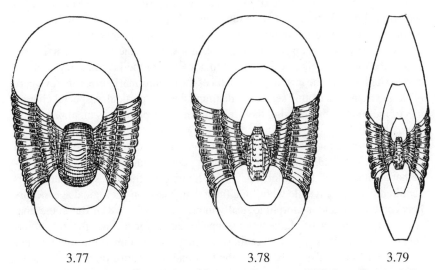

3.77   3.78   3.79

Figs. 3.77–79. Proterogenetic evolution of the genus *Cosmoceras* (3.79) from *Macrocephalites* (3.77) through the intermediate form *Kepplerites* (3.78). All the genera come from the Upper Brown [= Middle] Jurassic but are distinctly separate stratigraphically in the evolutionary sequence shown. Here, as in figures 3.80–82, which follow, the outer whorls of the ammonites are shown in cross section, whereas the juvenile shell is shown three-dimensionally, standing out from the plane of cross section. (From Schindewolf 1936.)

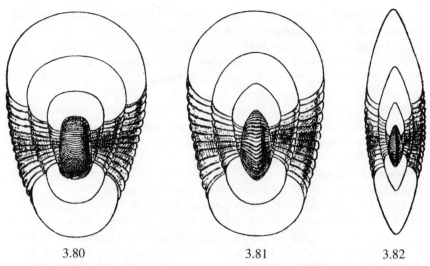

3.80   3.81   3.82

Figs. 3.80–82. Proterogenetic evolution of the genus *Quenstedticeras* (3.82) from *Macrocephalites* (3.80) through the intermediate link *Cadoceras* (3.81) in the Upper Brown [= Middle] Jurassic. (From Schindewolf 1936).

The mature stages of these forms are in direct opposition to those of the genus shown in figure 3.79, and if only these are taken into consideration, it would probably not even occur to one to suspect an evolutionary relationship between the two forms. But when the shell is broken apart and the juvenile whorls exposed, the morphological relationships that lead from the juvenile stages of the composite forms to the terminal members of the evolutionary series are revealed. (That the members of this series are distinguished by particular, special characters of the suture, thus proving their relationship indisputably, is mentioned only in passing.)

Another series that descends from *Macrocephalites* culminates in forms with a high, heart-shaped whorl section, a sharp-edged venter, and ribs that run together at an angle at the venter (fig. 3.82). Here, again, the new shell character is acquired from the juvenile whorl of a form that, based on geological age, occurs between the two and is morphologically intermediate (fig. 3.81). But in the final whorls, the new set of characters disappears again for a time; there, a whorl section and type of sculpture like those its ancestors exhibited throughout their life spans is reestablished. Not until the terminal member of the series does the new mark, beginning at the juvenile whorls, extend completely to the mature shell; through further narrowing of the whorled tube and sharpening of the venter, the "prophesied" high, heart-shaped whorl section finds its ultimate, pure realization.

*   *   *

The cephalopods, the unfolding of whose shell is easy to observe and follow ontogenetically and phylogenetically, are rich in further examples of the peculiar course of proterogenetic evolution; the elaboration of the ribbing (figs. 3.83–85) and of the structuring of the keel and grooves on the outer side follows this mode extremely often. We will take up only one particularly striking example, one that is also remarkable in other respects.

It has to do with the development in certain late Upper Devonian *clymenians* of abnormal triangular, three-lobed shells, the result of deep constrictions (figs. 2.91, 3.86); these clymenians are among the most eccentric of any ammonitic forms. They are the degenerate terminal members of dying lineages and a good example of the typolytic closing phase preceding the extinction of an evolutionary phylogenetic cycle.

This deviation from the normally shaped clymenians, with their spiral coils and wide umbilicus, is significant in that at first it was difficult to discern an incontestable connection between the two forms. Not until I had amassed a considerable collection from the strata lying beneath these aberrant forms was I able to shed some light on the evolutionary course that had led to them. It turned out that three parallel evolutionary series were present; they differed sharply in the structure of the suture and partially also in the sculpture, but with regard to the transformation of the shell, they were alike.

The most completely documented is series III of figure 3.87, which we would like to discuss briefly, leaving out all the details. At the base, there is a form (fig. 3.87*h*) with a broad umbilicus and normally spiraled whorls in the intermediate and mature growth stages. In contrast, the juvenile whorls establish a triangular shape and, further, show the first appearance of three shallow constric-

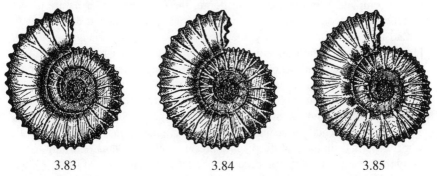

3.83	3.84	3.85

Figs. 3.83–85. Proterogenetic evolution of three-branched ribs with nodes at the point of splitting, which is on the umbilicus, from bifurcate ribs, in which the point of splitting is not nodular and lies at the center of the flanks; in ammonites from the White [= Upper] Jurassic.
3.83. *Orthospinctes.*
3.84. *Prorasenia.*
3.85. *Rasenia.*
(From Schindewolf 1936.)

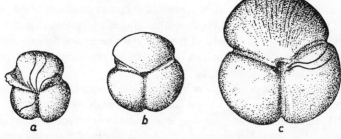

Fig. 3.86. *Parawocklumeria paradoxa* (Wdkd.), one of the last descendants of the clymenian lineage immediately before its extinction. Latest Upper Devonian (*Wocklumeria* stage) of the Sauerland (Rhenish Mountains). The three-lobed, rounded triangular shape of the shell deviates from all others known to the extent that formerly, at a time when only poorly preserved specimens without suture lines were known, the species was not recognized as a cephalopod, let alone a clymenian. Identification was made even more difficult by the fact that the shell is usually broken off at one of the deep furrows, leaving no distinct aperture visible. Without an understanding of the suture line, therefore, it is often impossible to determine which part of the shell is anterior and which posterior. *a, b.* Two juvenile forms. 2 ×. *c.* Adult specimen. 1½ ×.

tions cutting into the whorls. Both characters, however, soon disappear; they leave their mark only on a brief early growth phase before giving way to a normally rounded coiling of the kind found in the basic clymenian type (figs. 2.22, 2.89, 2.90), the ancestral form of the three series.

In the subsequent members of our evolutionary series, the new character complex, first emplaced in juvenile stages, now spreads step by step from the inner whorls out to later ones and finally supplants the remains of the original, normal design completely. In the species in figure 3.87*i,* the triangular nature of the whorls and the constrictions, here more prominently developed, have already spread further; even in maturity, the shell shows a slightly triangular outline. Figures 3.87*k* and *l* show how the triangular shape goes on to become ever more pronounced on the adult whorls and the constrictions gradually extend their effect, already forming a partial but distinct groove on the final whorl.

Further, in these forms (*i–l*), also beginning in juvenile stages, the umbilicus becomes narrower and narrower. In the shape in figure 3.87*m,* this narrowing is even more advanced, yet at first only in the juvenile form, whereas the adult whorl departs from the umbilical spiral present until then, thus returning to the developmental grade of the predecessor (*l*). Then, in the terminal form (*n*), the umbilicus is almost closed even in the adult. The shell, which has become considerably thicker and more compressed than the thin-disked shell of the initial species, henceforth takes on a typical triangular form, the three lobes created by deep constrictions.

In the terminal member of our evolutionary series II (fig. 3.87*g*), the shape of the shell is to a large extent the same as that in series III, and the evolutionary

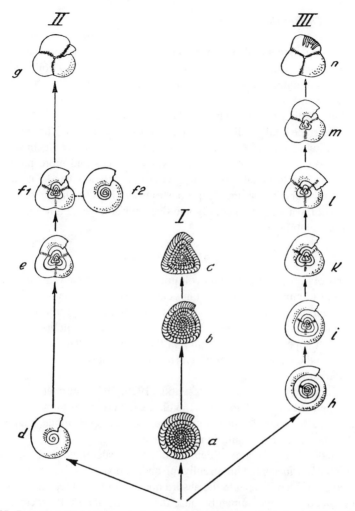

Fig. 3.87. Proterogenetic development of a triangular shell, its three lobes formed by constrictions and its umbilicus narrow, from a clymenian with a normal spiral, wide umbilicus, and no constrictions, in three parallel evolutionary series (I–III). The forms that are on the same stage with regard to shell development are shown at the same level in the illustration. *a. Soliclymenia solarioides* (v. B.). *b. S. semiparadoxa* Schdwf. *c. S. paradoxa* (Mstr.). *d. Pachyclymenia abeli* Schdw. *e. Wocklumeria aperta* Schdwf., juvenile shell. *f. W. sphaeroides* (Richt.): f_1 = juvenile shell; f_2 = adult shell. *g. Epiwocklumeria applanata* (Wdkd.). *h. Kamptoclymenia endogona* Schdwf. *i. K. trigona* Schdwf. *k. K. trivaricata* Schdwf. *l. Triaclymenia triangularis* Schwdf. *m. Parawocklumeria distorta* (Tietze). *n. P. paradoxa* (Wdkd.). All from the late Upper Devonian (*Wocklumeria* stage) of Germany (Rhenish Mountains or Lower Silesia), with a stratigraphic sequence of forms that, together with morphological data, yielded the evolutionary series shown here.

The individual illustrations are all shown the same size for the sake of consistency; some are enlargements and some reductions. f_1 is an enlarged juvenile specimen, and f_2 is a reduced adult specimen. *e* shows the enlarged juvenile whorls of *Wocklumeria aperta;* the adult shell, except for the somewhat wider umbilicus, is the same as the adult shell of *Pachyclymenia abeli* shown in *d.* (From Schindewolf 1936.)

process leading up to it is also fundamentally the same but does not have as many individual stages. Here, the starting point is a normal spirally coiled form (*d*) with a narrower umbilicus than that of the species in figure 3.87*h*. In a succeeding species, *e,* the juvenile whorls (Note: only these are shown, enlarged, in the illustration) show a typical triangular bending and the characteristic constrictions, whereas the adult shell has the shape of the initial form. The same is true for the following species (*f*), whose juvenile whorls (shown again in *f₁*, enlarged), however, have acquired a narrower umbilicus and more pronounced constrictions. The terminal form *g* also shows in the adult stage a narrow umbilicus, triangular whorls, and deep constrictions. Finally, in our series I, the development breaks off prematurely and does not go beyond a triangular stage with a broad umbilicus.

*   *   *

Further proofs for proterogenetic evolution can also be found in abundance in other groups of animals (foraminiferans, corals, graptolites, mammals). Further, the investigation of some Recent animals, the ixodids, a group of mites, has turned up the same phemonena (P. Schulze). Perhaps the evolution of the notochord, which appears for the first time in the embryonic stage of tunicates but disappears in the mature stage, proceeds by proterogenesis in the line leading up to the chordates.

As I attempted to show in earlier works (1929, 1936), ultimately the origin and development of the shape of the skull characteristic of humans and other *anthropoids* also proceeded in a proterogenetic evolutionary mode, beginning in juvenile stages and progressing from that point.

This mode of development, demonstrated by paleontological material, is therefore fairly widespread but usually misunderstood and misinterpreted. Only unawareness of it could lead to the absurd notion, professed repeatedly, that the ape descended from the human because the juvenile ape skull "recapitulated" features of the human skull. In reality, the proportions of the human skull, as shown by a comparison with the skulls of prosimians and other mammals, represent a novelty acquired early in the ontogeny of anthropoids, a novelty which, in the line leading to humans, underwent a progressive evolution beginning in juvenile stages and extending to mature stages but in the ape lineage experienced a gradual regression.

Furthermore, the interpretation that the characteristic human skull arose through fetalization, that it remained at an early juvenile developmental level whereas that of the great apes did not, is incorrect. Rather, what we see in the human is a true proterogenetic process, a phylogenetic advance of juvenile proportions to the adult skull.

*   *   *

Objections have been repeatedly raised against my evolutionary principle of proterogenesis, particularly against the examples used to substantiate it. Ernst
Stromer maintains, as he recently (1944) remarked, that proterogenesis has not
been demonstrated with certainty because the geological and chronological sequence of the individual evolutionary stages of my examples are not indisputably established in every instance.

This objection can be decisively refuted at least in the case of the examples I
have presented from the cephalopods. In the ammonites I have mentioned, the
age correlations are completely unimpeachable, nor have they ever been con-
tested. The arrangement of the clymenian series, however, is based on my own
investigation of about 1,700 specimens collected systematically from a single
profile, the stratigraphy of which has been worked out in detail; the positions of
the specimens within the profile are known with great precision!

Stromer says that I have misinterpreted the long-known fact that in primates,
the development of the brain case accelerates during ontogeny. It has nothing to
do with evolution, he maintains. The notion seems incomprehensible to me.

When we see that, in the various evolutionary stages of the human, in *Pith-
ecanthropus,* in the Neanderthals, and finally in Recent humans, the characteristics of the juvenile skull become gradually more pronounced as they extend to
later ontogenetic stages and determine how those stages will appear, it means
that *the phenomenon is truly evolutionary,* as, moreover, is generally admitted.
J. A. Kappers has recently analyzed this evolutionary process in more detail and
shown that what holds generally true for proterogenesis is orthogenetically directed (see p. 272).

The Significance of Proterogenesis

The theoretical significance of proterogenesis lies partly in its elucidation of the
principle of ontogenetic modification in early juvenile stages and partly in its
clear demonstration of the existing evolutionary relationships through the reversion to ancestral features in adult stages. If such reversions do not appear, if the
newly acquired juvenile character complex continues to develop progressively
in linear fashion during subsequent ontogeny, as is the case in early juvenile
morphological transformation, then the adult stages of the descendants are completely different from those of the ascendants.

For example, in our clymenian series, a mature stage such as that shown in
figure 3.87*l* or *m* would issue directly from the juvenile stage of the form shown
in figure 3.87*h,* and the phylogenetic relationships of such a completely transformed, isolated form would be difficult to document convincingly. One would
look for transitional forms but not find them, for in fact there never would have
been any; and once again, conforming to previous practice, one would hold the
"gaps in the fossil record" responsible.

With regard to the *biogenetic law,* the observations described here show that it is subject to certain limitations and really represents *only a widespread rule.*

I note in passing that Haeckel's formulation, that phylogeny determines ontogeny, is illogical. It is true, of course, that the events of the antecedent phylogeny have their effect on an ontogeny; on the other hand, the reverse is true—that *future* phylogeny, regarded from any point of reference at all, is determined by ontogeny and its modifications.

Phylogeny, or evolution, is not an autonomous, concrete process independent of ontogeny. The only reality is the chain of individual ontogenies, each connected to the previous one, from which phylogeny, as a summary condensing the most important transformational steps produced during the ontogenies, is abstracted. As an abstract concept, consequently, it cannot induce or determine anything. In reality, therefore, ontogeny does not recapitulate phylogeny, which is usually conceived of as the sequence of adult ancestral forms; instead, *it only repeats ontogenies,* the individual developmental courses leading to the adult forms of lineages.

As we have seen, however, ontogenetic modifications of ancestors can appear at very different developmental stages. The earlier they are, the more incomplete is the repetition of the older morphology and the more strongly it is suppressed. Further, the peculiarity of proterogenetic processes requires us to view things in reverse. If one takes the recapitulation theory as a basis, clymenians such as those shown in figure 3.87*h* or *f* would issue from triangular ancestral forms like those in figure 3.87*m, n,* or *g,* whereas in reality the relationship is the reverse, as demonstrated by the clearly evident evolutionary course and the geological-chronological occurrence.

When working out phylogenies, one must always consider the possibility of proterogenetic evolution if false interpretations are to be avoided. The danger of such is reduced, however, by the fact that proterogenesis apparently never affects the entire organism at once but only certain individual organs or sets of characters, as, very generally, in the transformation of types early in ontogeny. In the case of our clymenians, for example, the suture line undergoes a normal progressive, linear development. But because proterogenetic evolution often affects particularly conspicuous morphological features, the discernment of actual phylogenetic relationships based on them can be very difficult, to say the least.

The Pros and Cons of Typostrophism

The theory of typostrophism presented here, of an abrupt, discontinuous, complete transformation of type at various levels, is founded on paleontological evidence. Developed essentially by Karl Beurlen and the author, it is based on data obtained from fossil organisms, the true bearers of the evidence for evolution, on observations of the actual evolutionary processes they exhibit, and on general considerations based on those observations. The theory of typostrophism contrasts sharply with the Darwinian interpretation of evolutionary processes.

Taking experiments in breeding as his point of departure, Darwin sought to explore the origin of *species*. He knew only of the insignificant variability and mutability of Recent species and believed that he could account for the totality of evolutionary processes as being the product of nondirected mutability, which is dependent on chance, and of the regulating principle of natural selection. It would follow, then, that all evolutionary transformations would occur very gradually and infinitely slowly by way of the continuous modification of species; the structural designs and organic configurations of the larger typal groups would be the result of the sum total of the smallest, accidental mutational steps, which would have been selected and added on.

This concept is also shared by many contemporary geneticists, who consider themselves justified in simply extrapolating their observations on *microevolution,* on the morphological changes of the lower taxonomic categories (race and possibly also species), which can be analyzed experimentally, to the evolutionary transformations of the higher systematic units, or *macroevolution.*[9] According to this view, microevolution constitutes the preliminary and intermediate stages of macroevolution. Consequently, all kinds of arguments have issued from this quarter, but have also been advanced by some paleontologists, against the group of concepts presented here; these objections must be examined with great care.

Six Objections and Their Refutations

OBJECTION ONE

Far-reaching early ontogenetic transformations of type are impossible and can be rejected because hereditary transmission works toward bringing about a continuous repetition of the same ontogenies.

This objection is clearly on very shaky ground and is easily refuted. It is true that hereditary transmission is in fact the conservative, regulating principle, directed toward preservation of what already exists; on the other hand, we know that there is a countertendency toward changeability, toward variability and mutability, operating in conjunction with it. If this were not the case, there would be no morphological change and no evolution, but rather only a continuous repetition of unchanging, invariable forms. However, since we have an evolution of lineages, and we see novelties arise, changes must have appeared in the genetic structure that caused modifications in previously existing morphological processes.

9. Recently, B. Rensch (1947) has used the terms "intraspecific evolution" instead of "microevolution" and "transspecific evolution" instead of "macroevolution."

But where would these genetic changes express themselves and play themselves out if not in the ontogeny of individual organisms, which, as we have just determined, alone form the real substrate of evolution? Therefore, if one admits the fundamental principle of the variability of genes—and this is an inescapable necessity—then the supposition that these genetic changes take effect at different stages, early or late, of individual development presents no difficulty.

This is all the more understandable since every single mutation is already present from the very outset of ontogeny, from the fertilized egg cell on, even though it usually does not manifest itself until later, more advanced developmental stages. It should also be remembered that the genotype does not determine the features of the finished organism directly, but only the reaction norm of the cells that construct the emerging organism. Specifically, in the area of human pathology, at least, we know of plenty of mutational effects that develop at very different stages of life.

The upshot of our earlier statements is that there is no *absolute* difference at all between early and late ontogenetic morphological transformation. What we see is more of a graded series of transformations taking place at any of various ontogenetic stages, which causes the transformations to differ in magnitude or exhibit lesser or greater discontinuity in comparison with what went on in the ascendants. To that extent, the formation of the comprehensive structural design type during an early embryonic stage, in one large developmental leap, represents only one extreme of a continuous series of stepped stages [*Stufenreihe*].[10]

* * *

The relationships existing under these circumstances can be illustrated in a simple diagram, one already published by the author on another occasion and reproduced here in figure 3.88. The real links between individual organisms, or in the chain of successive ontogenies, is the continuous germ line, the uninterrupted sequence of cells that progresses from the egg cell to the germ cells of the next generation. This path is represented in the illustration by the abscissas. Upon these are superimposed the individual ontogenies, each of which comes to a dead end.

Each ontogenetic development advances in increasing differentiation from the general to the particular, causing the type characters of the phylum, class, order,

10. [*Stufenreihe* and *Ahnenreihe* as defined by Othenio Abel are two different series of taxa that from a phylogenetic point of view belong together. A *Stufenreihe* is formed by taxa of one phylogenetic tree that follow each other in time but have no direct ancestor-descendant relationship. (They could be side branches of that tree.) An *Ahnenreihe* in contrast consists of taxa that follow each other as ancestors and descendants.

In Abel's terminology the opposite of the evolutionary series (*Stufenreihe* and *Ahnenreihe*) is an *Anpassungsreihe* (adaptational series). This is a series of taxa (irrespective of their relationship) in which an organ is more and more specialized for a certain function (Ulrich Lehmann, *Paläontologisches Wörterbuch*, 3d ed. [Stuttgart: Ferdinand Enke-Verlag, 1985]).—Ed.]

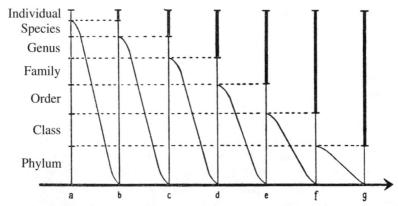

Fig. 3.88. Diagram of the origination of step-by-step increasingly pronounced morphological trans-formations or, rather, of the formation of types progressively earlier in ontogeny, in an idealized chain of ontogenies. (From Schindewolf.)

family, and so on, of the individual in question to form in sequence (fig. 3.88*a*). This law, already discovered empirically by that past master of developmental research, Karl Ernst von Baer, and named for him, is not only valid in a purely morphological sense but has also been supported by more recent experimental investigations into developmental physiology, which substantiate it inductively in the best possible way.

The ontogeny of a descendant, which belongs to the same species as its parents (fig. 3.88*b*), repeats the parents' developmental course down to the last, terminal stage, which brings with it the formation of the *characters of the individual.* This segment, emphasized in the illustration by a heavy line, deviates from the corresponding one in the previous ontogeny; it exhibits novelties, albeit on only a negligible scale.

But when a daughter individual exceeds the framework of the initial species by bringing forth new *specific characters,* its ontogeny is identical to that of its parents only up until the stage at which the formation of the specific attribute in question diverges. From this point on, development follows a different course (fig. 3.88*c*).

Further, when a new genus originates in an individual, that is, in its ontogeny, development deviates at a correspondingly earlier point, taking another path and replacing the old *generic characters* with new ones (fig. 3.88*d*). The same phenomenon appears progressively in all ontogenies that burst the bounds of former families, orders, and classes (fig. 3.88*e–g*).

We saw, then, that the typal differences in the higher taxonomic-phyletic units lie in qualitatively different, alternative structures; they arise, therefore, not through a gradual, continuous development of their respective antecedent type structures, not by building upon those structures, but by a *rebuilding from the ground up.* The set of characters of the ancestral type is directly replaced by the new set of the type that issues from it. Correspondingly, the ontogenetic transposition takes place at a more generalized, which is to say earlier, ontogenetic stage.

Quite generally, in the qualitative recasting of types, the process is such that *the particular organizational features of the corresponding ancestral category of equal rank are not repeated* in ontogenetic development. The deviating ontogeny sets out on its new path *before those features form,* thereby linking up with a stage of the ancestral ontogeny that embodies the general organizational state of the category ranked above it. Thus, characters typical of families, by circumventing the characters of the phylogenetically preceding family type, develop upon the general morphological basis of the order; characters typical of orders develop upon the base of the higher-ranking classes, and so on.[11]

Our illustration expresses this situation diagrammatically, it is true, since in reality, there no such rigid, sharp distinction between the individual classes of characters and the processes by which they are formed, but the diagram is basically correct. Slanting lines joining the individual ontogenies show which portion of the previous ontogeny is recapitulated unchanged—the portion of each development that leads up to the basic structure of its respective higher-ranking category—and from which stage the deviation occurs, establishing a new direction.

From this, we see that in the extreme, the *recasting of the major comprehensive typal structural designs takes place almost at the beginning of ontogeny,* which is consistent with the fact mentioned above, that certain major lineages of the animal kingdom already exhibit their particular differences in early embryonic stages. A new class type issues directly, discontinuously, from the egg of a member of the phylogenetically preceding class type. As Walter Garstang said, "The first metazoan was not produced by a metazoan."

* * *

Thus, the indisputable fact exists that in ontogeny, the particular builds developmentally, physiologically and morphologically, upon the general. Correspondingly, it must be the same in evolution, *that the organic structure of the comprehensive types is formed before that of the types subordinate to them.*

This is also shown to be true by the following consideration: Every phylogenetic-taxonomic category of higher rank shows a *broader* geographic range, and its representatives occupy a correspondingly greater diversity of habitats

---

11. We establish a parallel here—there is no other way to characterize it—between the different degrees of commonality in the developmental levels of ontogeny and the categories of the natural system. This may appear unusual to some, and also doubtless involves a certain schematism, but it is objectively justified, for the morphological abstractions do correspond with one another.

than is the case with the lower categories contained within it. A class may be cosmopolitan and include marine and continental, tropical and arctic, swimming, walking, and flying animals. One of its orders, however, contains only marine, another only continental terrestrial and limnetic forms, and so on. In the individual families and genera within that order, the geographic ranges and ecological limits for their members are ever more restricted.

These differentiations within the framework of a class or an order with regard to particular life conditions have undoubtedly led to a growing *divergence of characters;* it would be impossible, however, given the extreme variety of environmental factors, for selection to result in the formation of *common structural characters,* for the genera and families to merge later, secondarily, to form the typal unit of the order, and the orders to merge to form the structural design of the class. Since they all have the same *basic type,* it must have been laid down *before the splitting into subtypes.*

From all this, only one logical conclusion can be drawn, and it supports our concept: the higher categories do not form additively and gradually; they are not *assembled* from lower categories. Rather, the reverse is true—there is a *disassembly, a descending subdivision of the first-built type of higher rank into the lower-ranking types it encompasses.* After a fundamentally new organic structure has been formed typostrophically through a far-reaching transformation setting in early in ontogeny, there occurs in the new evolutionary cycle thus introduced a descending breaking up of this type *in this way:* the descendants take on a progressively longer initial segment of the modified ontogeny, and the characters typifying the particular subordinate type begin to form at correspondingly later stages.

The actual historical process therefore runs in the opposite direction of the way it is shown, for didactic reasons, in our diagram; the true sequence of ontogenetic-phylogenetic transformational steps is illustrated in the diagram in figure 3.89, the reverse of the one in figure 3.88. The concept expressed there has already been professed by a number of other authors (K. Beurlen, B. Dürken, A. Hennigsen, P. Schulze, J. C. Willis, and others), but has not yet taken hold because it is diametrically opposed to the prevailing way of thinking. Although this latter appears untenable, it has, as we shall presently see, recently found another defender.

OBJECTION TWO

The diagram shown to support our concept of the ontogenetic processes in phylogenetic transformation does not correspond to the actual situation; it should be replaced by another illustration, one that reflects a gradual, cumulative assembly of the more comprehensive typal structures.

G. Heberer (1943) raised this objection and countered our diagram with another (fig. 3.90), which is said to express the traditional Darwinian concept of a cumulative realization of the differences between the higher categories. From a

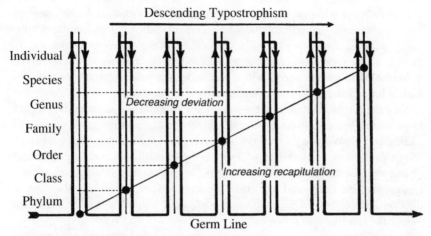

Fig. 3.89. Compressed diagram of a sequence of ontogenies in a typostrophic cycle in the correct order (i.e., the sequence is the reverse of that shown in figure 3.88). There is a descending formation of types, a progressive breakup of the structural designs established first in the higher and then in the subordinate ranks. At the outset, the organizational characters of the class are laid down, followed by those of the order, the family, and so on. Correspondingly, there is increasing recapitulation and reciprocally decreasing deviation of ontogenies, thus an extension of the early segment that corresponds with the respective ancestral ontogenies (thin lines) and, on the other hand, a progressive truncation of the formational segment that has gone off in a new direction (thick lines). The black circles indicate the stages at which the morphological change from the ancestral ontogenetic form appears for the first time. We can say nothing about time, about *when* these changes in the germ line that leads from one ontogeny to another (double line) were *caused*. The only judgment we can make relates to the visible manifestations of morphological changes in real ontogenies. (Thus, when we speak of an early ontogenetic *origination* of the major types, it is in a certain sense metaphorical, referring only to the finished product of processes that in themselves are not accessible to analysis.)

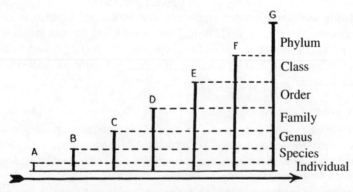

Fig. 3.90. Diagram showing the additive view of the origin of types in an idealized series of ontogenies. (After G. Heberer 1943.)

given species A arises a new species B, and through further mutations, from B to C, from C to D, and so on, a progressive increase in morphological differences at the level of genus, family, and so on, results.

In the light of the facts already described and the conclusions to be drawn from them, there is no possibility that this diagram can be recognized as being correct. Are we to believe, according to this diagram, that the organizational characters of the higher categories were in a certain sense grafted onto those of their respective lower categories? Are, for example, the type characters of the order simply added on to the distinctive features of the family of the predecessor, in which those features constituted the terminal stage of the evolutionary process? Is ontogeny simply prolonged?

This is clearly an outright impossibility and is also not accepted by Heberer himself, for he emphatically says that one should naturally not understand it in the sense of a simple linear addition, merely a cumulative add-on at the end of a particular ontogeny. The truth is, he goes on, that the matter is more complicated, that the changes always affect segments of ontogeny of varying lengths, longer or shorter. Not really an addition, then, but rather a more or less early transformation of the entire ontogeny?

However, if this is what Heberer really means, and the heavy lines of his diagram are meant to show the length of the modified portion of the ontogeny, we arrive at the same conception of the matter as the one we have professed. All one has to do is put the categories into the correct, that is, reversed, order, so that their morphological characters are formed sequentially during ontogeny, and the result is a picture similar to the one we have shown.

*   *   *

Let us study in detail once again which purely theoretical concepts are concerned in an additive formation of a new structural design. If, through progressive transformation of species characters, the boundaries of any particular structural design were exceeded *gradually*, a transitional zone between the old organization and the new one could be assumed, a zone filled with a chain of neutral species that would elaborate the structural characteristics of the higher-ranking categories in small, cumulative, individual steps. They would be "neutral," or intertypal, in that they would no longer belong to the old structural design but, on the other hand, would not yet be part of the new one still under construction.

This is clearly impossible. Every single species not only exhibits specific characters but also incorporates concurrently the typal characters of the genus, family, order, and class to which it belongs. Therefore, it can always be only the member of either one or another order or class, and so on, virtualizing only one or another organizational structure. Therefore, the dividing line between two categories of any given magnitude must necessarily lie directly between two species, between an older one, which exhibits the type of the preceding order, and so on, for the last time, and a younger one, which produces the final basic features of the new organizational design for the first time.

Ruled out, therefore, is a long series of, in a manner of speaking, nonexistent forms, forms that are nothing but species intervening between two orders, which species only then acquire the characters of genus, family, and finally, order sequentially, through the progressive accumulation of extremely small transformational steps. If a species at the boundary line between two such types breaks rank with its former order, the fact is that it no longer exhibits the organizational features of the old order but rather those of the new one.

Correspondingly, the structure of a new typal group must already be developing within the framework of the old, phylogenetically antecedent type; after exceeding a certain threshold in the continuous process of speciation, the new organizational structure would then appear. This development could take place either polyphyletically, which means that all or many specific lines would be transformed into the new type and change over to it, or the transformation would take place in only a single line.

The first instance, again, is of course ruled out. We have already shown that, in view of the very different circumstances of living, adaptation, and selection to which the individual series of species are subject, the formation of common structural characters is out of the question.

There is still the last possibility set out above, that the realization of the new structural design is added on to only a single lineage. This view would at least be consistent with the facts insofar as it can be proven that the move from one type or lineage cycle to another is accomplished by only a single evolutionary line, whereas the remaining lineages of the previous type become extinct.

However, opposing the notion of a cumulative assembly of the new basic structure through individual steps of speciation is the fact that the modification of species affects characters completely different from those of the basic structure, *that the direction of speciation is completely different from and independent of the formational processes that lead to the new basic organization.* Consequently, the new basic organizations cannot be a part of continued speciation, and the formation of the general structural foundation cannot follow ontogenetically upon the development of particular specific characters. Rather, the transformation of the general structural type takes place early in ontogeny; according to von Baer's law, the general is laid down *before* the particular, which then builds upon the general morphological foundation and develops further.

Moreover, we have seen in our examples that these transformations can only proceed discontinuously because of the qualitatively different, alternative nature of features of the structural design. As a consequence of these early appearing transformations—for example, the structure of one order changing into that of a new one—the entire subsequent ontogeny is, of course, modified, so that the characters of family, genus, and species emplaced later in ontogeny are also changed; in one stroke, then, a completely new kind of organism appears. *The structural type peculiar to a large group of forms, therefore, is not the product of a gradual synthesis, not the sum total of a long sequence of individual characters produced one after another, but rather the result of a profoundly new model, of rebuilding, or typostrophism, which sets off a new evolutionary cycle each time it occurs.*

As we have already pointed out, further evolution within the framework of a new organizational type or of a cycle within a lineage then proceeds in the direction of a descending breakup of the more comprehensive type into its subtypes; thus, here, too, there is no addition of higher, more general units but rather a disassembly to lower, more specialized ones. This specialization, however, comes about through reproduction and diversification of individual characters, through the addition of special features. From this, we see that *there is no way that something more general, a more comprehensive structural design, can arise through the addition of individual, special characters to form a prevailing set of characters; on the contrary, a narrowing of the scope and degree of generality of the existing combination of characters sets in.* This is probably the most fundamental objection to the assumption of an additive origin for general structures.

* * *

A simple diagram might serve to illustrate our view concisely. At a particular time the tail end of an evolutionary lineage that belongs to order A will be present in some one species *a*. From species *a* arises then a species *b*, which, compared with *a*, exhibits certain pronounced structural differences (fig. 3.91*a*). To it are joined other species, *c* and *d*, again with different transformations, which carry on the development initiated by *b* and build upon it, at the same time forming certain particular specializations within the framework of the newly acquired general structure (fig. 3.91*b*). As more species are added on, a larger

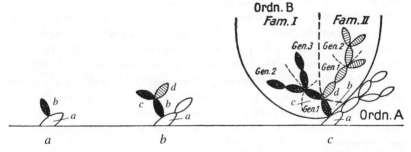

Fig. 3.91. Diagram showing the origin of a new type at the level of the order (Ordn. B) from the species *a* of an order A (Ordn. A) that precedes it phylogenetically. The dividing line between the two types of organizations runs between species *a* and *b*. The general structural design of the new order, the basic features of which came from [species] *b*, breaks up abruptly into its lower-ranking subtypes of families (Fam.) and genera (Gen.). (From Schindewolf.)

group of forms comes into being, which, in its basic organization, is so different from order A that we judge it to be a new, independent taxonomic order, B (fig. 3.91*c*).

Now the question is, At which point does the new order begin? Are the forms *b, c,* and *d* at first, for all intents and purposes, just species, which as yet exhibit no characters of a more comprehensive type at all, meaning also that they do not belong to any higher category? Have their specific characters only gradually intensified to the point of being generic characters, through the addition of extremely small transformational steps, the generic characters, in turn, becoming features of families, and so on, so that finally, several rows of individual chains of species would finally coalesce, so to speak, to form the order and to be its constituents?

As we have seen, this is certainly not the case. Rather, the step that forms order B should be shifted to the point between species *a* and *b,* the latter being the one that virtualizes for the first time the set of characters typical of order B. This general, basic organization is common to all other later members of the order and overlies their special characters. Therefore, the formation of the basic organization must precede that of the special differentiations; it cannot appear only at the end, emerging synthetically and polyphyletically from all the individual, particular forms.

One might, perhaps, counter this by saying that only the later fate of species *b* or of its descendants decides whether one should assign it or them to a new order. If the species had left no descendants behind, one would regard it simply as an extremely aberrant variant, a monstrosity, or such. This, in itself, is correct, but does not alter the fact that through species *b,* a large-scale transformation is introduced, one that exceeds the organizational structure of order A and falls outside of that framework. In later descendants, if there are any, the differences between the new evolutionary type and the characters of the ancestral type usually become more pronounced (as we saw in our concrete examples of the heterocorals and of the lobe splitting in the popanocerates); *the critical basic features of the modified evolutionary direction, however, are already present in their first constituents, which is why they must be assigned to a new order.*

Just as inappropriate is the possible objection that systematicists often have differences of opinion on the categorization of any particular set of characters. The structural change, as such, is still present, independent of whether one author bases an order on it, another a suborder, or a third, perhaps, only a family. Since the general character of typal modification at all of these levels is the same, these kinds of differences in evaluation are not of fundamental importance for our interpretation.

The formation of the types of lower magnitude is also indicated in our rough diagram. As we have described with regard to the character complex of the order, the structure of the family—at least in its general, basic form, which may later be refined—must also be laid down *before* the particular forms of the subordinate genera, for these have only specialized later upon the foundations common to all of them. The same is true for the typical features of the genus in relation to the species included within it.

The individuals composing our species *b* do not, therefore, simply embody only one species; rather, they are at the same time the first representatives and bearers of a new genus, family, and order. Technically, they indeed belong to a species in that they also have specific characters; but *as the ancestral form of a new family and order,* such a species, because of its morphological stores and potential for development, is different from species that form as final, marginal elements within the framework of those categories. These bring forth *only* new specific characters but, for the rest, remain within the framework of the category, whether genus, family, or order, from which they issued and which exhibit the corresponding characters.

Each of the typal sets of characters of these higher units, however, has formed abruptly and discontinuously in a single step of greater or lesser scope, all of a piece, independently of speciation. *Correspondingly, there is a deep gulf between the first representative of a new type and its parents,* which belong to the ancestral type and, therefore, perhaps to another order. Consequently, Garstang's radical way of putting it, namely, that the first bird was hatched from a (modified) reptile's egg,[12] which I have often quoted in the past, is correct. To say, *Natura facit saltus* is absolutely valid; nature does indeed take leaps!

Furthermore, it has been pointed out correctly that in the latest Cretaceous and earliest Tertiary, when the mammalian orders suddenly formed (fig. 3.70), there were at the start only as many species as we have orders today. Naturally, this should not be taken completely literally, for the breakup into the various orders did not take place absolutely simultaneously, but it is correct in the sense that each order was first embodied in a single species, which brought forth discontinuously a new organizational structure and produced the germ cells of the multitude of forms that went on to constitute the order.

OBJECTION THREE

The complexity of the differences between types rules out the possibility that they arose suddenly through a single, discontinuous step; they could only have been formed slowly, in a long chain of extremely small mutational steps.

This argument refers to the circumstance that, in reality, the lowest taxonomic units, races and species, differ from one another not just in a single character but in a whole array of characters, and that in the construction of each, several genes

12. Similar ideas were also professed by Arnold Schopenhauer in his *Generatio in utero heterogeneo.* Accordingly, "there would issue from the uterus, really from the egg, of a particularly favored pair . . . once, as an exception, not its own likeness but a directly related higher form."

participated simultaneously, and numerous cumulative small mutations were involved. This would be all the more true in the structural organizations of the higher types.

In our examples of profound reorganization in the development of the suture line in ammonites or of the septal apparatus in corals, we have shown that in fact the formation of new typal characters is complex and always affects an entire group of characters. The process consists partly of the suppression or transformation of previously existing structures and partly of the introduction of new organs. What is significant, however, is the fact that these type-determining individual characters relate to a *coherent complex,* to the suture line, the septal apparatus, and so on. It may be assumed, then, that there are *correlations between the individual transformations,* and we have already indicated that perhaps the development of deficient and exaggerated structures is related to the compensation of organic material.

This raises the question of whether mutations that influence not only individual characters but an entire, independent set of characters are currently demonstrable experimentally. As a matter of fact, they are. Here, the investigations of K. Kühne on the vertebral columns of humans and rats are especially informative. The results showed that a single gene with pleiotropic or polyphenic (multiple) effects controls not only the typal variation of the vertebral column and the ribs (number of vertebrae and ribs in the various regions, displacement of the boundaries of the regions toward the head or the tail), but also the particular construction of the related dorsal musculature, the mode of development of the neural plexus of the arm and leg, certain blood vessels, the location of the diaphragm, and more. Here, then, the related bones, muscles, nerves, and blood vessels form a unit, and if, in such an instance, a profound, extensive transformation were to be introduced in one stroke, a new, harmoniously integrated pattern for the structural design would arise.

Further, we refer to the complex mutations that H. Stubbe and F. von Wettstein achieved in the snapdragon *Antirrhinum majus.* There, a single mutational step caused, among other things, a transformation of the bilaterally symmetrical *Antirrhinum* flower into a radial form. In this instance, too, it is a matter of an entire set of characters: not just the corolla, but also the ovaries, the calyx, and the number of stamens were affected by the transformation to radial symmetry.

These data, corroborated experimentally, make it appear a certainty that, in keeping with our concept, *the type-determining sets of characters could very well have changed in a single step, as an integrated mutational pattern.* In this connection, it should not be overlooked that we are only talking about the fundamental differences between types, those differences that denote the actual boundary between two structural designs and are decisive for the radical phylogenetic change appearing at this boundary. And these, even in types of a higher magnitude, are by no means as enormous as one usually imagines; sometimes they are of no greater extent than the mutational step in *Antirrhinum,* mentioned above.

All the rest is incidental and has nothing directly to do with the characters that actually determine type. If, in the formation of a new, higher type, a more complex mutational step appears at an early ontogenetic phase, it follows necessarily that the profound transformation of the organic structure unleashes corresponding effects in the developmental physiology of subsequent ontogeny and also influences other characters. In other words, if the typical structure of the order is changed from the ground up, it is not to be expected that the specific and generic characters of the ancestral form will remain completely unchanged.

On the other hand, we know from our previous explanations that the first representative of any type at first expresses only the basic features of the new, decisive typal complex or the modified structural principle, and that the rudimentary complex, once emplaced, undergoes further development in the descendants. Thus, one should not compare the most highly evolved terminal members of two typal complexes; rather, one should go back to the point at which the two evolutionary paths diverged. What is critical, then, is only *the differences between the corresponding ontogenetic stages of the old and the new types,* for it is these that bring about the decisive turning point in the shaping of the type.

*   *   *

In our examples of the transformation of the pterocorals into the cyclocorals, we have seen particularly clearly that different structural processes work together to form the differences existing between the Recent stony corals and the oldest representatives of the pterocorals. Some of the features that, in addition to the actual typal characters, are also typical for cyclocorals have already been initiated in the youngest pterocorals and represent there a morphological approximation of the younger type; other features will be added later, after the boundaries of the type have been exceeded. Similarly, in the evolutionary series that lead from the reptiles to the birds and from the reptiles to the mammals, many accessory characters of the more recent type are already formed within the framework of the ancestral type, and others do not form until after the type has been transformed.

To illustrate this, here is another simple example. The lines of the *even-toed ungulates* and the *odd-toed ungulates* can certainly be distinguished by a great number of characters. The structure of the extremities in their Recent representatives, the camel (fig. 3.92*e*) and the horse (fig. 3.93*e*), for example, shows profound differences in the number of toes, the manner in which the toes have degenerated, the structure of the tarsus and the metatarsus, and so on. And yet, these differences ultimately go back to one *small-scale transformational step,* which nevertheless influenced all of subsequent evolution in a decisive way.

In the even-toed ungulates, we know that the weight of the body is borne by

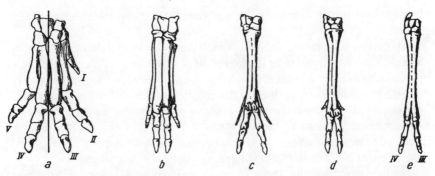

Fig. 3.92. Stepped stages [*Stufenreihe*] in the evolution of the forefoot of the even-toed ungulate. *a. Oreodon.* Oligocene. The thick line represents the axis of the foot, which in these paraxonic forms runs between digits III and IV; compare with figure 3.93*a. b. Leptomeryx.* Oligocene. *c. Blastomeryx.* Lower Miocene. *d. Merycodus.* Middle Miocene. *e. Camelus.* Pliocene-Recent. (After Matthew, Romer, and Scott, from Schindewolf.)

two toes, or digits, III and IV, whereas the rest of the digits, I, II, and V, of the original, complete mammalian foot, have degenerated. In contrast, in certain odd-toed ungulates, the Recent equids, only the middle toe, III, remains functional; the rest have all but disappeared, leaving only vestiges.

At the roots of these different evolutionary directions are forms with the full complement of five digits on each foot, which, in both instances, is structured in a very similar way, showing only insignificant deviations in the position of the digits. The foot of the ancestral form of the even-toed ungulates (fig. 3.92*a*) has a *paraxonic* structure, which means that digits III and IV are located at the center

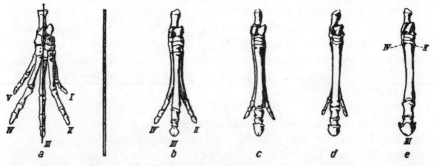

Fig. 3.93. Evolution of the foot of the horse (hind foot, *b–e*) as an example of one-toed ungulates, linking up with the complete, five-toed foot of the protoungulates (*a*). *a. Tetraclaenodon.* Palaeocene. The axis of the foot, seen in the illustration, shows the mesaxonic structure of this form, in contrast to that in figure 3.91*a. b. Eohippus.* Lower Eocene. *c. Miohippus.* Oligocene. *d. Merychippus.* Late Miocene. *e. Equus.* Upper Pliocene–Recent. (After Matthew, Cope, Osborn, and Romer, from Schindewolf.)

of the foot; the axis of the foot runs between these two digits, which together bear most of the body weight. In contrast, in the *mesaxonic* ancestor of the odd-toed ungulates (fig. 3.93*a*), the axis and pressure line of the foot passes through digit III, which stands by itself in the middle of the foot.

The differences there may unhesitatingly be conceived of as the result of *a single mutational step*. After this transformational step, the typal differences of paraxonism and mesaxonism, previously not very prominent, were intensified by consistent, orthogenetic development; the result, brought about through different modes of digit reduction, the transformation of the carpal and tarsal bones in correlation with the different way of distributing pressure, and so on, was two very different structural forms.

Thus, the objection under discussion here does not hold up; on the contrary, there are experimental results from Recent animals and plants that justify our conceptions, provided that the distinguishing typal complexes are evaluated properly.

OBJECTION FOUR
The assumption of large developmental leaps is superfluous; geologic time is sufficient for an understanding of evolution based on small mutations and natural selection, the path of gradual speciation.

This claim, made by many authors, has indeed not gone unchallenged, and it seems by no means so certain that the expanses of time available are really sufficient, particularly with regard to recessive [inferior] mutational characters that appear infrequently at first or that have a selective advantage of little value. Some information we have already presented in connection [with the phases of evolution] shows how extraordinarily slowly, expressed in absolute values, subspecies formed even in the mammals, which, geologically speaking, evolved relatively quickly. Here are a few more figures for other groups of animals.

According to fairly certain determinations, the period of time in which certain endemic races of Scandinavian birds developed was between ten and fifteen thousand years. On the other hand, G. Kramer and R. Mertens have calculated that races of lizards found on the islands of western Istria [Croatia] are at the most nine thousand years old. Further clues can be found in marine fauna of the Atlantic and Pacific oceans found on either side of the Panamanian land bridge. Near there, in the area of Nicaragua, there was once a link between the two oceans; it disappeared about two or three million years ago. The faunal isolation that ensued has thus far had only slight influence on the animal world; in spite of the lengthy period of time, many species of fishes and crabs on both sides of the isthmus of Panama are still identical. For example, according to S. Finnegan, eleven species of brachyures (crabs) are absolutely identical, whereas thirteen others have split into vicariant pairs of forms since that time. Other examples: According to the survey made by Bernhard Rensch, a *racial* age of between three hundred thousand and five hundred thousand years can be assumed for many Central European birds, the hedgehog, toads, and so on; and according to F. E. Zeuner, in the terrestrial forms he discussed (mammals and insects), the length of time it took *species* to form was between one half and one million years.

If such long periods of time are required in every instance just for the formation of minor races, there is no doubt, as Richard Goldschmidt and other workers have already concluded, that the geologic eons, in spite of their vastness, would *not be long enough* to account for the origination of the differences among contemporary structural designs in the plant and animal kingdoms through raciation and speciation. In this regard, it should be noted that the racial differences of the forms named lie below the threshold of paleontological observability, whereas in the evolutionary transformational phases for which we have offered proof, there is, comparatively, an enormous amount of material. Furthermore, we should take into consideration that many of the examples we have presented here involved small populations, in which evolution proceeds more rapidly than in large, widely distributed ones, and the pace of speciation was much reduced owing to the prevalence of panmixia. Thus, the data cited above must be considered to reflect particularly favorable circumstances. But we do not want to assign too much weight to these figures, and we concede for the moment that the magnitude of the mutational and selective pressure known to us is, or was, sufficient to allow for all evolutionary processes to have proceeded in extremely small transformational steps within the available period of time.

But the calculations that are usually drawn up as evidence are based on an erroneous presumption, completely ignoring the phasic or quantum nature of the course of evolution. They focus on only the starting and ending points of the unfolding and, ignoring actual evolutionary processes completely, divide up the intervening time evenly and fill it with hypothetical small modifications of form; this is inadmissible.

Walter Zimmermann (1934) has provided a typical example of such an incomplete calculation. He relates the pasqueflower (*Pulsatilla,* fig. 3.95), a modern flowering plant, to the oldest type of terrestrial plants, the still algaelike Middle Devonian genus *Rhynia* (fig. 3.94). This assumed ancestor of the flowering plants is separated from *Pulsatilla* by a period of about three hundred million years, which might correspond to about sixty million generations. If you figure on one rather important, constructive mutation every thousand generations, this means that the chain of species leading from *Rhynia* to *Pulsatilla* had sixty thousand genetic changes available to it (Zimmerman wrote "600,000," either a typographical or an arithmetical error). Now, if you think that the morphological distance between the two compared forms could be resolved in six hundred thousand individual steps (actually only sixty thousand according to our reckoning!), then each of these modifying steps would be roughly of the magnitude of experimentally observable mutations.

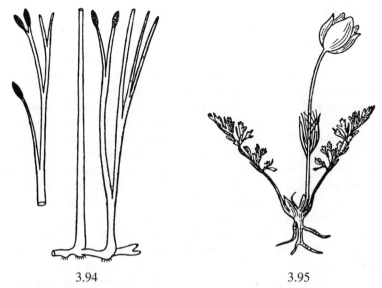

3.94   3.95

Fig. 3.94. *Rhynia maior* Kidst. and Lang. Middle Devonian of Scotland. Reduced; the species grew to ½ meter. (Reconstruction after Kidston and Lang.)

Fig. 3.95. *Anemone pulsatilla* (pasqueflower). Reduced. Recent, Central Europe. (After W. Zimmermann 1934.)

Of course, this is not conclusive proof that evolution did in fact proceed in this way, in countless small, individual steps. One calculation by itself can never provide actual proof; all it shows is that technically and arithmetically, the existing temporal framework was potentially sufficient for such a process.

However, no consideration at all has been given to the fact that in actuality, the course of evolution is not evenly distributed over time; rather, periods of extreme transformational intensity and rapid evolutionary rate alternate with very long phases in which the pace of unfolding is sharply reduced and only negligible morphological alterations are brought forth. In the plant world, one such period of frenzied, explosive unfolding took place at the boundary between the Devonian and the Carboniferous. During a period of time generously estimated at sixty million years, the simplest vascular plants gave rise to meter-high horsetails, treelike club mosses, giant ferns, and what is more, also to the oldest seed plants (pteridosperms)—all completely new types. On the other hand, since the beginning of the Tertiary, a period also of sixty million years, there has been no appreciable evolutionary advance whatsoever in the plant world.

In agreement with traditional notions, we have assumed that in the portions of phylogeny where the pace of evolution is slow, our typostatic periods, it proceeds essentially as a gradual, continuous transformation of species. In view of the paucity of events in these extended phases, during which only a correspondingly small number of transformational steps occurs, evolution must have proceeded more slowly than the rough calculations assume. On the other hand, for the brief typogenetic periods, with their frenzied remodeling of forms, one would have to assume that within an extremely short period of time, there occurred a quantity of individual steps so enormous as to exceed the capacity of the mechanism [slow, gradual evolution] presumed for them, particularly since, at the beginning of a new evolutionary direction, when the biological advantage of the new mutants is extremely slight or a selective value is still lacking entirely, natural selection works especially slowly, completely in contrast to the rate of evolution, which is so sharply accelerated at exactly this point.

In the example presented by Zimmermann, consideration should be given, among other things, to the fact that already in the Upper Devonian, and therefore very shortly after the appearance of *Rhynia,* not only the higher vascular cryptogams developed but also even the pteridosperms, and that, further, the angiosperms, their organization complete and, from their very beginning, already broken up into the modern classes and subclasses, appear in the upper portion of the Lower Cretaceous suddenly and without a long series of transitional forms showing gradual change. *The periods of time in which these profound transformations took place are not long enough for gradual, slow modifications in small mutational steps*—that type of change exists only in times of stagnant evolution. In opposition to the view expressed by Zimmermann elsewhere, that the period of time that has elapsed since the appearance of modern angiosperm families in the Lower Cretaceous is sufficient for the formation of family differences through the accumulation of numerous small mutations, it should be emphasized that these families, just as the entire angiosperm type with its subtypes of the monocots and the dicots, appear suddenly and without transition in the Lower Cretaceous and, therefore, must have arisen in a comparatively very short period of time through a few large transformational steps.

* * *

The unfolding of the goniatite family Manticoceratidae, described earlier, has shown that the base of the Upper Devonian, within the brief span of one zone, a rush of evolution brought forth almost simultaneously six successive genera possessing quite significant differences in the degree of differentiation of the suture line (see pl. 22A and fig. 3.72). If we estimate the duration of the entire zone to be, at the most, a few hundred thousand years, where, then, is the time for the "hundreds of thousands or millions of generations" that are presumed by many authors for the bridging of *one, single* significant morphological transformation?

The same thing holds for the certainly considerable morphological difference between the corals *Hexaphyllia* and *Heterophyllia,* which always occur side by side in the very same layer, without connecting transitional links and unseparated by any temporal interval; they must have issued from their root stock, the ptero-corals, abruptly and, for all practical purposes, almost simultaneously.

Another example: Earlier, we touched briefly on the evolution of the odd-toed ungulates, in particular, on that of the horse. The unfolding begins in the Lower Eocene with *Eohippus* (fig. 3.93*b*),[13] whose hind foot has three toes, and pro-ceeds through a long, consistent series of progressive reductions of the lateral toes (II and IV) to Recent horses with functionally one-toed feet, but still with vestiges of digits II and IV (fig. 3.93*c–e*). *Eohippus* itself derives from the late Paleozoic mesaxial protoungulate genus *Tetraclaenodon* (fig. 3.93*a*), which was still in possession of a complete, five-toed hind foot.

A comparison of the illustrations cited allows us to recognize that the magni-tude of the transformation from *Tetraclaenodon* to *Eohippus* is greater than the difference in foot structure between *Eohippus* and *Equus*, which required the rest of the Tertiary from the Lower Eocene on, a period of about fifty million years. In contrast, *Tetraclaenodon* and *Eohippus* are only slightly different in age and separated in time by only about five million years. The reduction of toes I and V, that is, the shaping of the typical equid foot, proceeded with extreme rapidity, a great deal more rapidly than the degeneration of toes II and IV, which is incomplete even today.

If that change took place, as is likely, through gradual evolution in small in-dividual steps, the large transformational effect in the brief early portion of the phyletic cycle cannot have been brought forth through the same slow transfor-mational mode. Further, since no connecting transitions of any kind between *Tetraclaenodon* and *Eohippus* are known and since, as we shall presently see, nothing can be attributed to gaps in the fossil record, it must be assumed *that the evolution of the foot of the horse was introduced discontinuously in a single large or a few important transformational steps;* that here we have a distinctly typogenetic phase produced by mechanisms other than those of smooth speciation.

Thus, in view of the actually very long period of geologic time, it should not be overlooked that the actual, decisive formational phases are only *brief epi-sodes. But the brevity of these episodes is not in the least sufficient for us to postulate a gradual, smooth transformation in a long chain of species* that could have linked the individual typal structures with one another.

OBJECTION FIVE
The links between the individual types, which are required by the presumed continuous, smooth transformation of species but have not been found, were originally present; their lack is accounted for by gaps in the fossil record or by

13. The genus should probably be combined with *Hyracotherium* and more properly presented under that name.

the migration of the lineage in question from some distant, as yet unstudied, area to the place where it was found.

We confronted this issue earlier in a general way; the objection, raised repeatedly and usually uncritically, that the fossil material is incomplete and inaccessible, seeks to discredit paleontological data or, preferably, when such data are a nuisance to the customary views, to exclude them from the discussion entirely.[14] Now, after we have become acquainted with certain patterns of the evolutionary process, we are ready to take a definitive position on the matter.

To carry the analysis to its conclusion, let us first refer to the last example given. From the family of the equids, that is, from the evolutionary series leading from *Eohippus* to *Equus,* about thirty genera are known; these are connected by documented gradual, uninterrupted transitions. Of these, the genus *Eohippus* contains about ten species; in each of the genera *Mesohippus, Miohippus, Parahippus,* and *Pliohippus,* seventeen to nineteen species have been distinguished; *Hipparion* includes twenty-seven, and *Merychippus* as many as thirty-six species. And these species are represented in some cases by many thousands of remains.

In view of the richness of the fossil record, is it just by accident that between *Tetraclaenodon* and *Eohippus,* which is to say, between the protoungulates and the equid type, not a single link has come to light, even though, because of the significant morphological difference, at least the same abundance of forms as between *Eohippus* and *Equus* would have to be expected if a similar, gradual transformation had taken place?

Such an accident of preservation cannot be taken seriously, leaving, therefore, the only reasonable conclusion *that such intermediate forms never lived, that the initial development of the horse type must have proceeded not only very rapidly but also discontinuously, without transition,* whereas subsequent development within this type went on slowly, gradually, with every conceivable transitional stage, in a large number of species that merged imperceptibly with one another.

The concept of a migration of *Eohippus* out of some other part of the earth can also be rejected, since *Tetraclaenodon* and *Eohippus* both come from the North American Tertiary and their evolution was completed in that area. Moreover, the extensive investigation of the North American Tertiary appears to rule out the possibility that essential features of the view we presently hold will change. And on the other hand, the Tertiary of almost all other countries is so

---

14. As proof, in this context we refer once more to the excerpt already cited from the pertinent principle enunciated by the *zoologist* J. Schaxel (1922, p. 49): "Phylogeny [here, the group of phylogenetic concepts inferred from Recent organisms] violates paleontology by not allowing the historical evidence to speak for itself, requiring, rather, that it verify the evidence of morphology and furnish proof of the phylogenetic stages that, according to the biogenetic law, are repeated in ontogeny."

well known that we cannot expect a long, hitherto concealed series connecting the ancestor of the horse with the protoungulates (Condylarthra) to be found.

* * *

Similarly, we can most assuredly exclude the possibility that the manticoceratid genera, which appeared discontinuously at the base of the Rhenish Upper Devonian (pl. 22A and fig. 3.72), migrated to the place where they were found from some other area, which a slow, gradual transformation had already taken place. This is to say nothing of the fact that these genera appear suddenly and simultaneously not only in the Devonian of the Rhenish Schiefergebirge but also in all countries and parts of the earth where the strata in question have thus far been studied. One might still object, saying that perhaps there may yet be a place in the world we do not know about where the situation is different.

Therefore, it is more important to be able to demonstrate that the temporal separation of such a mixture of forms through immigration is out of the question on purely theoretical as well as actual grounds. We saw earlier that the distribution of marine forms through active or passive migration proceeds quite rapidly, much more rapidly, at least, than does the transformation of species at the evolutionary center of any series of forms whatsoever. Based on this, the zone fossils in the most different, widely disparate areas always appear in the same sequence; they are never arbitrarily mixed up as a result of uneven rates of distribution. Therefore, the migration times of individual index species are insignificant when compared with the duration of the zones based on them.

Thus, if the manticoceratid genera had developed somewhere in extended sequences, step by step, species by species, genus by genus, they would have to appear everywhere they are found at the same intervals in which the species originated. A mixing of *genera,* which correspond to a considerably longer series of chronological stages, must be considered completely out of the question. Let us recall once more that this abundance of manticoceratid genera appears discontinuously in the basal portion of zone I*a* of the partial member of the Upper Devonian stage I, which includes several of such zones, and that one of these genera, *Manticoceras,* is used to characterize that stage.

Thus, the conclusion is unavoidable—what we see here is an *explosive, precipitous proliferation of forms* at a rate many times greater than that involved in normal speciation, which takes place very rapidly in these index goniatites anyway.

* * *

As mentioned earlier (Chap. 2), in the later Upper Devonian we also encounter repeatedly the discontinuous, unheralded appearance of genera and families of cephalopods, which show no continuous connection with forms from their respective preceding horizons. Here, the entire stratigraphy is based on one such sequence of disconnected types of cephalopods layered one atop another. In these cases, it can also be ruled out with certainty that the fossil record is incomplete, that there were changes in the facies, or that there was migration to the site from somewhere else—in short, that layers are missing from the profile. *Therefore, we must consider that large typal discontinuities exist; no connecting links of any kind ever lived anywhere.*

On the other hand, there are naturally also examples of the appearance of forms from elsewhere. One we know of is the sudden appearance of genera and species of ammonites in the Central European Jurassic. There, however, it is easy to demonstrate that the animals in question immigrated, and in this instance, we can also determine with precision the area from which they came. These ammonites were native to Tethys, the once extensive "mediterranean" sea, where they evolved gradually, without interruption. These are things that are well known to paleontologists, who are careful to take them into account.

With regard to the peculiar group of the heterocorals, we have seen that no connecting, intermediate forms are known either between *Hexaphyllia* and the pterocorals or between *Hexaphyllia* and *Heterophyllia*. If such had lived, there is no reason that they would not have been found. Thanks to their calcareous skeleton, they would have been just as susceptible to preservation as were all other stony corals and the representatives of the two named genera that are present in massive quantities. Nor can gaps in the stratigraphy or other geological events be held responsible for the lack of connecting links and the discontinuous appearance of the types in question; for, at the same time and in the same profiles, the individual lineages of the pterocorals show continuous evolution, which can be followed step by step.

Any transitional forms whatsoever between the pterocorals and *Hexaphyllia* were, therefore, *not present originally* and in fact should not even be anticipated theoretically, since the development of the features typical of *Hexaphyllia* can only have taken place discontinuously at a larval stage. Similarly, the lack of connecting links between *Hexaphyllia* and *Heterophyllia* must be considered as being primary, the actual state of affairs. It would indeed be much too remarkable a coincidence if, within one enormous distributional range, representatives were found, and these by the thousands, of only two finished, morphologically very distinct genera out of the long chain of connecting species that we would have to presuppose if the development of forms had proceeded smoothly but from which not a single bit of proof has been preserved!

* * *

In connection with this, the widespread but erroneous opinion that paleontology is based only on isolated fossils and accidental discoveries must be countered. The truth is that the science has at its disposal, at least for most groups with skeletal structures susceptible to preservation, an absolutely *overwhelming amount of material.* And this is to say nothing about invertebrates, for example, brachiopods, echinoderms,[15] corals, bivalves, snails, and ammonites, which have been collected by the millions upon millions and used as the basis for investigation. This abundance of material is also available for many vertebrates. About six hundred specimens representing some few species of the enormous pterosaurs of the genus *Pteranodon* (pl. 27B) have been recovered from the Upper Cretaceous of Kansas. There is also much evidence of other species and groups of flying reptiles. Nevertheless, there is no evidence of gradual, continuous development of the flying reptile type; it appears before us discontinuously, its organization complete.

Evidence of the early Ice Age saber-toothed tiger (*Smilodon,* pl. 23; fig. 3.121), from the asphalt swamps at Rancho la Brea, in California, is present in the remains of about three thousand individuals. Considerably more have eluded collectors through decomposition or are still embedded in the ozokerite at the locality. Many thousands of specimens of Diluvian [Pleistocene] cave bears are known, and on the Great Plains and in the Badlands of North America, there are inconceivably massive accumulations of mammals. A single locality in the Nebraska Miocene contains the remains of an estimated sixteen thousand individuals of a small rhinoceros (*Diceratherium*), making it completely impossible to collect all of them and place them in museums. But in any event, that is not the goal of the paleontologist, no more so than it is the task of the zoologist to catch all living animals and preserve them.

And yet charges of this kind are sometimes leveled against paleontology. It has been calculated that about twenty thousand individuals of the Miocene horses of the genus *Anchitherium* (fig. 3.96) must have lived in southern Germany, and the comparison has been made that from that enormous number only the remains of about seventy animals have found their way into our collections. This is said to constitute vivid proof for the frightful incompleteness of the fossil record.

In rebuttal, we must ask the question: Are these seventy specimens not sufficient for us to understand the forms in question thoroughly? What essential advantage would there be if thousands more specimens were piled up in museums? In practice, how many species form the basis of investigations into morphology, phylogeny, taxonomy, and so on that are carried out by zoological and botanical morphologists? How many examples of rare Recent species of plants and animals are found in museums, zoos, and botanical gardens?

* * *

It is completely understandable that of all the countless generations of fossil plants and animals that have ever lived only a comparatively very few would fall

15. As an example, many species of echinoderms from the Permian of Timor [Indonesia] are represented by more than 100,000 individuals and were studied more exhaustively than many Recent species with regard to ontogeny, variability, deformities, and so on.

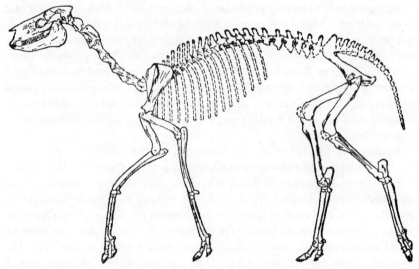

Fig. 3.96. *Anchitherium agatense* (Osb.), an American representative of the genus *Anchitherium,* which is indigenous to North America and Europe. Lower Miocene of Wyoming. Reconstruction of the skeleton; the parts outlined with the broken line have been added. ¹⁄₁₂ ×. (After A. S. Romer, from O. Abel 1929.)

into our hands, but these are completely adequate for the purpose. They are sufficient to document uninterrupted transformational series in places where they are actually present.[16] Since we are not practicing the genealogy of closely related individual organisms, and the family tree is only one portion of the total reality, it is not a matter of coming to know the complete sequences of generations and chains of individuals. For phylogenetic purposes, rather, it is quite enough *to be able to grasp the more distinct individual steps of transformation; and with our countless closed evolutionary series, we are in a position to do this.*

With great regularity, however, these continuous series break off at their bases, always when we come to the point where the lineages must have been merged, where we have to look for the emergence of the closed typal complex virtualized by each of these lineages from the likewise closed and continuously documented typal complex of the ancestral stock. Indeed, we know of morphological approximations among the individual types of structural design in space

16. Opponents of the view put forth here have repeatedly striven to obtain evidence that such uninterrupted evolutionary series do exist, which no one doubts. We paleontologists even know of far more of them than the zoologists and botanists who raise this objection seem to think. There is, however, an inadmissable generalization in the supposition: if continuous transformation is observed in some instances, then, according to the preconception, it would also have to exist in *all* others. In contrast, our efforts must be directed toward considering the facts of the matter objectively and explaining them as they present themselves to unbiased consideration.

and time, but not of any truly smooth transitions in the transformation of the actual typal features, which, as we must emphasize again and again, we do not simply *regard* as decisive—they *are* so in reality. There is always a crucial dividing line that allows us to say that a particular phylogenetic unit still belongs to the older type but that the next closest one is already part of the younger type, and *this dividing line is never bridged by a long chain of species with fluctuating morphological modifications.*

Similarly, the secants and tangents of a circle are typologically different. This difference is not only conceptual but exists in nature, in being, in the "function" of the two elements. They may approximate one another and be joined by transitions, but the profound, qualitative typal difference is not thereby obscured.

It has been charged that it would be a pure fluke if one were to find the *exact* ancestral form of any one lineage or *one* particular connecting link. This is certainly true. On the other hand, the issue is not at all one of finding a *particular individual form.*

The differences between two immediately successive types of organization are quite radical in nature and essentially more significant than the relatively trivial specific deviations formed entirely in the individual, continuous evolutionary series. If, in the process, the typal structure is to be bridged by gradual transitions, as is supposed, there would have to have been *a very long chain of closely related species* separated by an enormous number of generations for the smooth transformation to be demonstrated.

Of them, however, *many* of the individual elements would perfectly fulfill the requirements for an ancestral form or a transitional form. Accordingly, it would be an even greater coincidence if nowhere a single bit of evidence of this anticipated abundance of forms could be found, whereas the succeeding chain of species is present in as uninterrupted a sequence as could be desired.

Also incorrect is the often repeated assertion that the members of the primitive lineages, which are regarded as ancestral forms, were short-lived and, thus, would have left few remains behind. The ancient, primitive lineages and groups, which often represent decidedly collective types and are close to the point at which the branching once took place, are on the contrary distinguished by extreme longevity. Occasionally, they even extend on into the Recent, as illustrated, for example, by the onychophores, enteropneustans, tunicates, and acranians, some of which are already present in the Cambrian. Where such connecting links, in spite of a potential for fossilization (our statements are based only on such groups), are missing, this lack is surely primary and natural.

Nor should one set one's hopes on the chance of existing gaps being filled in the future. Since, for example, in the later Gothlandian the individual families of fishlike animals (agnathans) are present throughout their entire range in their already finished form, independent and clearly set apart, one imagines that their phylogenetic connections will someday turn up in the Ordovician and Cambrian. It must give pause for thought, however, that as yet nothing in the way of transitional forms has been found, although each of the two systems has yielded extremely rich invertebrate faunas—some of considerably more delicate structure than that of the agnathans—among them remains of tunicates, which stand at the base of the chordate line. The future will certainly grant us many informative fossil discoveries; whether much will change with regard to the picture we now have of family isolation is, based on our experience thus far, doubtful.

Since the time of Darwin, such expectations have been repeatedly disappointed. All the more so, as K. Diener (1920, p. 51) once said in resignation, but aptly, in that "the discovery of many new, provisionally isolated groups of forms has increased the number of gaps rather than reduced it." Furthermore, more extensive investigations have shown many times over that the differences between the particular organic structures are more significant than was earlier assumed, and many fossils that were initially greeted with enthusiasm as being transitional forms could not be justified as such upon closer examination. Thus, the discrepancy between unbroken, smooth evolutionary series involving relatively trivial transformations, on one hand, and the large contrasts between basic structural designs, unconnected by transitions, on the other, has not diminished, as might be expected, but increased.

For us today, there is no longer any reason to resign ourselves to some perceived insufficiency of fossil material; rather, we see in the regularly recurring pattern of the fossil record a reflection, incomplete in detail yet on the whole faithful, of the actual situation: *a natural lack of intermediate forms and the existence of real gaps between individual types.* And we also know the reason now that the anticipated connecting links never lived: *the recasting of types was carried out as a far-reaching restructuring during early juvenile stages, with the result that in the mature stages, the new interfaced directly with the old.* Contrary to the usual interpretations, then, there are no gaps in the fossil record impinging on the long series of *connecting links of adult forms;* the only inaccessible evidence is possibly the *embryonic and early juvenile stages* at which the decisive, morphological novelties come about and which, alone, constitute the transitions.

OBJECTION SIX
Changes in type of the kind we postulate here are impossible because corresponding phenomena are unknown in the present.

This objection is untenable in two respects, first with regard to the argument itself, and then, to the conclusions drawn from it. If analogues to the large mutational steps called for based on paleontological material were not, for the time being, observed in Recent material, this would not constitute conclusive proof that there is no such thing and that, therefore, the phenomena could not have appeared during the course of evolution. We also know of events that took place in the geologic past that have no comparable counterpart in either kind or extent in the present. And yet, according to the evidence we have found, there is no doubt that they once occurred.

We should particularly bear in mind the brevity of typogenetic episodes, which we have emphasized again and again, compared with the long duration of the typostatic phases, with their gradual, and only negligible, transformation. The statistical probability that one of the few species of plants or animals selected by geneticists for investigation would be in the midst of a typogenetic period and, consequently, capable of thoroughly refashioning itself is therefore extremely slight.

Furthermore, extensive changes in types, even in the infinitely long evolutionary history of organisms, have been extremely rare events. The number of major structural designs in the animal and plant kingdoms is strikingly small, and this indicates how *extremely seldom* typostrophes of the rank of taxonomic phyla and classes have appeared or have represented vital, successful evolutionary solutions.

Therefore, we should never expect to witness experimentally the origin of new basic types of that magnitude. The only things we can count on observing are *monstrosities,* which, in a certain sense, are unsuccessful attempts (or perhaps also initial, promising ones) to exceed the constraints of the former type, or *typal transformations of limited extent,* which affect the structural system and may serve as *models* for the development of the comprehensive types to which they are connected by intermediate stages of every degree.

*   *   *

Both expectations are fulfilled. *Monstrosities* are known well enough in all groups of the animal and plant kingdoms, making it unnecessary to describe individual examples here. They are understood as teratological phenomena, as deformities, and in by far the majority of instances they are probably indeed such in the sense of being aberrant, pathological, and nonviable. In many instances, however, they may be thoroughly "hopeful monsters." To the extent that they are viable and able to survive, which is to say, to the extent that their *monstrous* features are tied to suitable adaptive characters, they may bear within them the germ of the future evolution of a base that is henceforth profoundly altered; success alone will decide the matter.

Without a doubt, the first Early Tertiary representatives of the pleuronectids, flatfishes such as the turbot, the plaice, the flounder, and so on, with the torsion and asymmetry of the head and the displacement of the two eyes to one side (figs. 3.97–99), were "monstrosities" that were able to survive because, by lying flat on the bottom of the sea, they found a mode of life consistent with the altered form. Then, with the changing colors of the pigmented upper side, a further adaptation was acquired, which guaranteed the persistence of this form and its wide distribution. The origin of the gastropod type through the torsion of

3.98

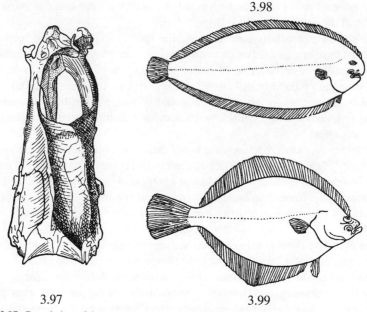

3.97   3.99

Fig. 3.97. Dorsal view of the asymmetrical skull of the plaice (*Pleuronectes platessa* L.), with displacement of the left eye to the right side of the head and torsion of the forward part of the skull. The keel in the center of the brain case indicates the original center line or plane of symmetry of the skull.
Fig. 3.98. Lateral view (upper side) of the sole (*Solea vulgaris* Qu.).
Fig. 3.99. Lateral view (upper side) of the plaice (*Pleuronectes platessa* L.).
(After O. Abel 1912 and 1929, modified.)

the complex of organs within the mantle cavity of about 180 degrees from back to front, which occurred at an embryonic stage, was originally "monstrous" in nature.

Similarly, the new typal characters that we discussed earlier—recall, for example, the separation of the protosepta in the heterocorals, the subdividing of the lobes in the popanoceratids, and the paraxonia in the even-toed ungulates— were, compared with the "normal" features of their ancestral types, "monstrous" mutations at the time of their inception; one would not have been able to tell right off whether they were a substrate capable of evolution—whether they had scored a hit—or not. Since those lineages did produce offspring and introduced new, flourishing evolutionary cycles, the new characters were not lethal monstrosities; in a certain sense, they were legitimized after the fact by their success and, through the currency they acquired, lost their monstrous quality.

In contrast, without that kind of prospect for the future, the abnormal forms appearing in the typolitic phase of decline of the evolutionary cycles are doomed to extinction. Thus, there are two different classes of monstrosities; they can scarcely be demarcated morphologically and their difference becomes obvious only through their phylogenetic position or their ultimate evolutionary fate.

* * *

Most important, however, especially for our problem, is the fact that recently, *complex macromutations* have been described in certain plants; these mutations can be related to our discontinuous transformation of types, thereby shifting our paleontological data into the realm of experimental research and genetic analysis. That such viable macromutations have been demonstrated in plants first is certainly no accident, for such profound transformations occur more readily in plants. And, of course, what applies to plants is basically also applicable to animals.

For example, the investigations carried out by H. Burgeff (1941) on the liverwort *Marchantia* seem to be of great fundamental importance. In this plant, Burgeff observed the discontinuous appearance of morphological characters that appear again in other Marchantiaceae as generic characters and represent such changes as the outstanding botanist K. Goebel expected to find in the phylogeny of the liverworts for morphological reasons. Based on his results, Burgeff distinguished between two forms of genetic mutability (1) *micromutations of limited extent pertaining to specific characters;* and (2) *generically typical macromutations, which overlie interspecific mutability.*

According to Burgeff, it is very likely that evolution proceeded not through an accumulation of micromutations but through *mutational discontinuities of varying magnitudes, which could be classified according to the category they typify—species, genus, family, or phylum.* Thus, a geneticist [Burgeff], basing his observations on Recent material, has expressed as conjecture what we, based on our fossil material, believe must represent an ineluctable necessity.

The simultaneous investigations carried out by H. Stubbe and F. von Wettstein (1941, 1942) on the snapdragon *Antirrhinum majus* led to similar results. There, macromutations could be achieved that led to the development of organizational characters that were diagnostic for other genera of the family in question (Scrophulariaceae). One of these macromutations shows a transformation of the bilaterally symmetrical flower into a radial form and a simultaneous increase in the number of stamens to five. Thus, with a single transformational step, a combination of characters was created that is typical of the genus *Verbascum* (mullein). We have already pointed out in another context the particularly important fact that this is not just a matter of the transformation of a single character or organ but of an *integrated set of characters.*

Other macromutations in the snapdragon led to the elaboration of a spur or to a decrease in the number of stamens to two, which are likewise typical generic characters of other members of the family. All these mutations were observed many times, so that the potential for breeding and capacity for perpetuation exists.

The intense *lability of morphological design* accompanying the macromutations should be pointed out as a significant parallel to our paleontological findings. For example, in the snapdragon, the innovation of the spur varies widely in position, size, and shape and includes modes of development that have their counterparts in several genera of the wild forms. The large mutational step does not yet set the details of the profoundly altered organization firmly in place; not until subsequent, somewhat smaller mutations have occurred influencing the same character or the same set of characters it is likely that, as Stubbe and von Wettstein assume, the original scope of developmental potential will be restricted and the harmonious incorporation and stabilization of the new structure achieved. These phenomena are completely in accord with our paleontological experience, that the large transformational steps of the explosive typogenetic phase exhibit a pronounced lability in design and that only the subsequent typostatic period brings stabilization to the characters and smooth, continuous further development in small transformational steps.

These correspondences can hardly be coincidental, and since we have seen that the picture is repeated in typostrophes of different magnitudes, at all evolutionary levels of the lineage, *we may indeed also link large-scale typal discontinuities to the phenomena observed at the generic level.* There is thus a smooth series of evolutionary discontinuities or of the macromutations ascribed to them, which differ only in their level of magnitude and the category they affect and not necessarily in their nature.

Stubbe and von Wettstein themselves take into account that macromutation intervenes in the formation of the organizational features of the more comprehensive taxa and doubt that the mostly clearly discrete genera and higher units of the system were joined by a continuous series of connecting links and arose through the progressive addition of small mutations, for paths proceeding by way of many intermediate stages would have been too long. Accordingly, the accumulation of micromutations would only be the evolutionary mode for the modification of adaptive characters in geographic races and species. Our analysis of the way evolution proceeds led to exactly the same interpretation.

Since evolution is a unique, historical, and irreversible process, we cannot expect to be able to repeat it experimentally, to be able to resurrect the evolution of, say, an entire family or order artificially, before our very eyes. We may only hope to illustrate the *principle* of typal transformation through individual experiments that model the phenomenon, and *this is what happens in the macromutations in the plants just described.*

Very recently, the *animal kingdom* has also yielded a model, one that can be drawn upon at least as an analogy for the explosive origination of types that we support. We refer to the macromutations lately achieved by W. Goetsch with the help of vitamin T; these mutations accelerate evolution intermittently and bring forth new forms discontinuously, which then interface with the parents directly and without transition. For example, in termites, genetic mutation in combination with the major modifier vitamin T results in the creation of soldiers and workers; one would never figure that these all belonged to the same species, let alone to the same genus and family, if their actual specific affiliation were not known. There are no transitions between them, for according to Goetsch, their early larval development is solely a matter of "either/or," and development sets out in different directions right from the early stages. In addition to these significant morphological transformations, other micromutations are then observed that affect coloration and spotting.

These new observations pull the rug out from under the objection discussed here. They go even further, offering strong justification for our interpretation, and we mark as noteworthy the fact that, in this instance, paleontology anticipated a new explanatory direction, one that, based on corresponding experimental data, could only be taken into account by genetics very recently and permits us to come closer to a physiological understanding of the problem. Therein lies convincing proof of the strength of the evidence yielded by fossil material and of the inferences derived from it.

Further Proof for Typostrophism

The arguments we have just discussed, which allegedly counter our theory of the discontinuous transformation of types of different ranks, have, upon closer examination, turned out to be important arguments *in support of it*. In addition, there are a number of other bits of positive evidence and further considerations that compel us to see the typostrophe as the only possible important evolutionary event and to challenge the notion that the process responsible for speciation—smooth transformation through minor mutations—is of universal validity.

Let us put the significance of selection aside for the moment; we can discuss this matter in detail later, after we have become acquainted with phenomena reflecting other evolutionary patterns and principles. Meanwhile, we shall remain with the question that has occupied us up to this point: gradual, continuous refashioning of species or abrupt, discontinuous transformation of types?

An analysis of the nature of *specializations* speaks unambiguously in favor of the latter view. Such an analysis shows that increased specialization and differentiation—the processes most responsible for the gradual modification of species—are not the agents of evolutionary advance but, on the contrary, constitute an *obstacle to ascent*. To counter this process, which is fundamentally antithetical to evolution, there must be, therefore, another one that offsets the restricting, rigidifying effects of specialization, once again creating generalized, undifferentiated foundations upon which evolution can build. *The agent of this new impulse resides in the far-reaching transformation of types.*

Furthermore, only the presumption of this kind of evolutionary event removes the difficulties for our evolutionary explanations otherwise presented by the extremely common phenomenon of the *cross-specializations* [*Spezialisationskreuzungen*]. But these problems can also be taken up later, after the groundwork has been laid in the next chapter.

Just one point should be emphasized here, that only when we take into account the evolutionary recasting of types of different rank does the way open to a true understanding of the *graded diversity of forms* and the *natural system of organisms.* Taxonomic categories are, accordingly, not arbitrary divisions of an uninterrupted continuum of forms but instead conform to actual evolutionary realities.

The boundaries of the categories are not accidental, the result solely of external factors creating a situation in which here and there longer or shorter segments of the originally continuous chain of species have not been preserved; rather, they reflect the transformational discontinuities, large and small, of the typostrophes. Thus, the higher units or types of the system are by no means abstract morphological concepts imposed upon organisms by the human mind; instead, they, or the divisions between them, are, like the concept of species, based on the *realities of concrete evolutionary processes.*

Further, from this point of view, we understand how it is that the types are telescoping, encompassing one another stepwise; that all the genera of a particular family are held together by the structural design of the family to which they belong; and that the set of characters that distinguishes an order is present in every one of the families encompassed by that order and constitutes the basic structure upon which the particularities of these families are built. Only in this view does the *graded diversity of forms* as expressed in the natural system become *true evidence in support of the theory of evolutionary descent.*

Other Principles of Evolution

The evolutionary interpretation of fossil material has led—and, to some extent, this has been true for a long time—to the determination of two empirical principles, which we shall take up presently. Originally, they stood as separate, isolated statements, but now, under the general, more comprehensive point of view of typostrophism, they come together as obvious conclusions.

Specialization and Unspecialized Descent

One of these principles, formulated by the North American paleontologist E. D. Cope, is the *law of unspecialized descent.* It holds that the types of any geological epoch do not tie in with the most highly specialized lineages of the previous temporal unit but rather derive from the most primitive, least highly specialized representatives of the ancestral type. Thus, contrary to what one might expect, there is no continuous straight line of upward evolution in the sense that the latest forms are an advance upon the highest evolutionary state of precursors, but rather the opposite: evolution builds upon that which has remained simple.

This principle is probably almost without exception and is valid for all ranks of phylogenetic transformation. Mammals do not evolve from the mighty giants among reptiles, not from the higher ranked marine or flying reptiles so extremely well adapted to aquatic or atmospheric life, but from a completely unlikely, little specialized lineage of small forms (ictidosaurs).

Among the mammals, in turn, the high-ranking order of the primates does not appear in a direct evolutionary line following upon one of the next lower orders of mammals, which, with regard to mode of life and feeding habit, however, are as highly differentiated as, say, the carnivores, the whales, and the ungulates. Rather, they go directly back to the lowest group of placental mammals, to the insectivores with primitive, five-digit extremities and a complete, as yet unspecialized, dentition. Further, within the primates, the highest and geologically youngest unit, the family of the hominids, is not rooted in such differentiated specialists as the modern anthropoid apes but instead, together with these, derives from lower, as yet simple representatives of the primates.

Among the nautiloids, all of the lines in which the siphuncle is highly specialized—the Endoceracea, the Actinoceracea, and the Cyrtoceracea, were sterile (figs. 3.21 and 3.32). They no longer possessed the capability for any kind of far-reaching transformation and did not give rise to a single new, flourishing line; instead, after a relatively brief period of proliferation, they died out without descendants. Only the long-lived, conservative line of the Orthoceracea, which had remained primitive and simple, retained the plasticity necessary for further transformation. This line not only formed the starting point for the groups just named but later also furnished the root stock for new types, the Ammonoidea and the squids and octopuses.

At the boundary between the Triassic and the Jurassic, we saw among the Nautiloidea with coiled shells that the highly evolved lineages with differentiated suture lines and sculpture died out, while only one group of small forms with an extremely primitive suture and without any secondary sculpture survived; it introduced a renewed period of proliferation in the Jurassic (fig. 3.32).

At the same time, there was a radical dynastic change among the ammonoids. There, too, all the lines with the most complex sutures, composed of many elements and frilled in the extreme, with the most highly developed sculpture and, in some instances, with enormous shells, became extinct. Evolutionary progress was carried on by a rather insignificant group of smooth forms, simple in every regard, which accomplished a far-reaching transformation of type, thus forming the basis for a new and powerful unfolding.

In the stony corals, it was not those with the most highly developed septal structure, not the differentiated lineages of pterocorals, with their complicated tabulae, columns and such, that brought forth the new type of the cyclocorals. All of those died out without descendants; rather, the transformation to the new type took place in the low-ranking line of the zaphrentoids, which, of all of the pterocorals, had retained their simple structure the longest.

Increasing differentiation and narrow, highly specialized adaptations lead, therefore, not to actual evolutionary progress but to the eventual extinction of the forms subject to them. They either become extinct as soon as a new, rapidly unfolding evolutionary type appears or live side by side with it for a while without ever changing over to it and without ever managing to flourish again. These forms are frozen in their specialization, have been made sterile by it, and are no longer capable of any far-reaching transformation. In this regard, one can say *that the processes of differentiation and adaptation are inimical to evolution.* These processes do indeed represent a certain kind of higher evolution, but *only within the framework of each restricted typal structure,* which is to say, only for a brief stretch of the way.

Counter to Darwin's notions, therefore, progressive differentiation and the selection-driven adaptation of races and species are not the agents of ascending, continuing evolution. The originally beneficial specializations inevitably intensify to become detrimental overspecializations, evolutionary dead ends. Highly specialized adaptations lead to limitation, to the crippling and drying up of the potential for continued evolution and, ultimately, to the extinction of the lineages subject to these forces.

Do Lineages Age?

A few years ago, a French author (H. Decugis) wrote a fascinating book on senescence in the organic world, in which he placed particular emphasis on the degenerative effects of differentiation. According to Decugis, we encounter everywhere, in the present as well as in the geologic past, the primacy of indications of extreme overspecialization and the morphological and metabolic disharmonies, the decline and extinction, that they entail. This author believes that these phenomena argue for a progressive aging of lineages and contradict the belief in unlimited progress held by Darwin and Lamarck. However, Decugis's notion, clearly the product of a decadent, escapist view, is unquestionably one-sided.

The aging and extinction of lineages is only one side of evolution. Paleontological material teaches us not only that the individual lines of plants and animals age and die out but that an inexhaustible abundance of new forms appear continuously to take their places, and it is in the latter aspect of evolution that the truly important events take place. If one made cross sections of the individual epochs, they would in fact show almost consistently a predominance of well-, that is, narrowly, adapted, highly specialized or overspecialized forms, for which no extended life span could be predicted. And yet, life is renewed again and again, as one group of organisms after another embarks upon a new period of proliferation.

Accordingly, we note in passing that we also do not share the view held by the author just named and many others as well that evolution has now reached its conclusion because all the various lineages have attained a high point beyond which they cannot go, and that the modern organic world consists of senile forms incapable of further transformation, only a few of which are persistent types with prospects for continuation. Had there been a thoughtful observer around during the Carboniferous or Cretaceous periods, he would undoubtedly have had the same impression. The giant lepidodendrons (pl. 24A), sigillarians (pl. 24B), and horsetails (pl. 24C), the giant insects (fig. 3.139), and all the other organisms of the Carboniferous that were already completely adapted to their environment, as well as the giant ammonites (fig. 3.141c) and giant dinosaurs (pl. 28) of the Cretaceous, would unquestionably have seemed to be the non plus ultra.

And yet, an abundance of unforeseeable organisms and new evolutionary directions appeared, indeed, from simple, unlikely root stocks. The organic world of any epoch is in a state of harmony and balance, of wondrous adaptation, yet new evolutionary possibilities have kept cropping up, and the continuous unfolding has never come to a standstill. Therefore, it is difficult to see why the present and the future should be different just because there are human beings around to reflect on the matter. In any event, there is no solid evidence for this view.

From all that has just been said, it follows that the processes of specialization, which are, on the whole, negative, must be accompanied by compensating *positive events,* which would offset the freezing of evolutionary potential brought about by specialization and break through to charge the lineage with new evolutionary impulses, or as Karl Beurlen would say, to rejuvenate it. This positive principle, which opens up the potential for new advance, lies in the *far-reaching transformation of type early in ontogeny;* such a transformation bypasses previous extremes of specialization and exceeds the boundaries of the existing type to bring forth once again more generalized morphological types, thereby setting the stage for new trends in specialization and adaptation. *Accordingly, typostrophes, the relentless antagonists of differentiation, are the true vehicles of evolution.* This conclusion, which necessarily follows from the considerations above, is another important piece of evidence for the necessity of accepting them.

A corresponding bipartite division of processes, striking in its similarity, lies in *geologic* evolution. There are the brief, episodic movements of the earth's crust, expressions of endogenous dynamics, which provide the critical impulse for shaping the earth and setting new developmental cycles in motion. These forces correspond to the phenomena of our typogenetic phases and are supplanted by other, exogenous processes such as weathering, erosion, and sedimentation, which are of importance only in elaborating and modeling what is already in place; these forces result in slow change and work toward equilibrium, thereby automatically bringing the evolutionary cycle to a standstill and terminating it; then, suddenly, there is another shifting of crustal blocks, upsetting the equilibrium and creating new impulses.

Since we shall see that typogenesis is based on events that reside within the organism itself, whereas in the phenomena of typostatic specialization environmental factors play a role, the existing parallels go considerably further: In geologic history, too, it is, as we have already indicated, an *endogenous,* inherent course of events that provides the critical, *revolutionary* driving force for geologic change, whereas *exogenous* dynamics only serve an *evolutionary,* or developmental, role of gradual equalization and progressive "adaptation."

Specialization and Structural Design

Going back to our original train of thought, we repeat that the highly specialized, extremely well-adapted lineages of any one structural design never exceed the boundaries of that design and do not change over to the new evolutionary type that issues from it and replaces it. The structure of the new type does not lie in the continuation of any specializations whatsoever of the ancestral type but rests on features of a completely different cast, which are independent of the specializations; the elaboration of the new type takes place within a series of forms of the ancestral type that has to a large extent maintained its original simplicity.

We have already seen that in the change from the reptilian to the mammalian type, all of the adaptations of the former directed toward a special mode of life, toward swimming or toward flying, are not continued—they do not form the starting point for the mammalian type. Rather, an unspecialized line of reptiles, the ictidosaurs, become the creators of the new generalized type, one whose structural features lie in a completely different direction from flying and swimming. One of those features, which, among others, indicates a typal difference between the ictidosaurs and the mammals, we have already become familiar with in the makeup of the lower jaw (fig. 3.76*a, b*). Only on the basis of this newly created, at first undifferentiated general structural design does renewed specialization take place in the most varied of adaptive directions—running, jumping, swimming, and flying.

We encounter the same situation in a comparison with the aplacental mammals. Among the aplacentals, there are specializations in dentition, body shape, and mode of life that have led to the evolution of predators (pl. 25A; fig. 3.100), herbivores, rodents (fig. 3.102), burrowers (fig. 3.104), insectivores, and flying creatures, and these specializations have been repeated at an evolutionarily higher level in the placental mammals in exactly the same ways, sometimes with

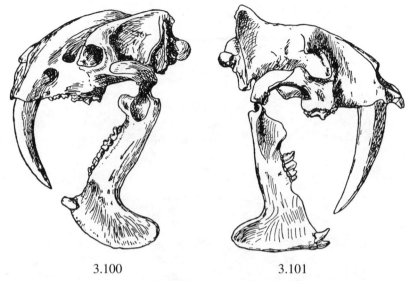

3.100   3.101

Fig. 3.100. Skull of *Thylacosmilus atrox* Riggs (fam. Borhyaenidae, aplacental). Pliocene of Patagonia. Reduced.

Fig. 3.101. Skull of *Eusmilus sicarius* Sincl. and Jeps. (fam. Felidae, placental). Oligocene of South Dakota. Reduced.

The skulls of carnivorous marsupials and of true carnivores show an extremely surprising similarity in overall habitus and, in particular, in the unusual overspecialization of the upper pair of canines. The similarities of form are present even in such details as the structure of the large flange on the lower jaw, designed to guide and protect the upper canines. (After E. S. Riggs and after Sinclair and Jepsen, respectively, from O. Abel.)

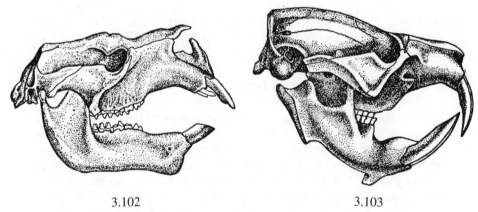

3.102   3.103

Fig. 3.102. Skull of *Diprotodon,* a giant, rodentlike aplacental from the Diluvian [Pleistocene] of Australia. About $\frac{1}{17}$ ×.

Fig. 3.103. Skull of the Lower Miocene rodent *Palaeocastor.* Somewhat reduced.

The structure of the incisors and the anterior portion of the jaw is surprisingly similar in these two analogous forms. (After R. Owen and A. O. Peterson, respectively, from A. S. Romer 1933.)

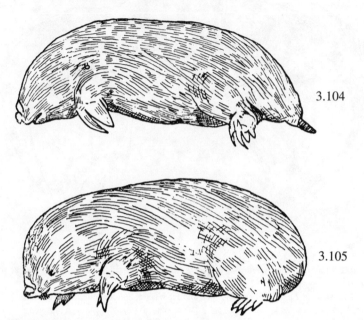

Figs. 3.104–5. Extensive similarities in the shape of the body in an aplacental and a placental mammal with the same mode of life. Both reduced.
3.104. South Australian marsupial mole (*Notoryctes typhlops* Stirl.).
3.105. South African golden mole (*Chrysochloris aurea* Pall.).
(After Heck-Matschie and Brehm, respectively, from E. Dacqué.)

astonishing external similarities (pl. 25B; figs. 3.101, 3.103, 3.105). But here, too, there is no connecting lineage that runs from the marsupial carnivores or rodents to their equivalent lineages among the placental mammals. None of these specializations are taken over by the placental type; all were first cast off, to reappear later on the transformed, generalized base.

However, it goes without saying that the primitive, initial forms of a new evolutionary type and of its first representatives as well are by no means *completely unspecialized*. There is no such thing. Each organism, as a living being, must have the mechanisms necessary for life—adaptations and certain special features—that go beyond the general structural design. Consequently, there is no form or group of forms, however primitive, in which all the characters and organs are equally unspecialized; rather, there is always a mixture of primitive, unspecialized features and those already specialized. An excellent example of such a combination of characters reflecting different levels of specialization is offered by the human, in which, compared with other mammals, the simplest of structure in the hand and foot is paired with the highest of specialization in the cerebrum.

The more primitive lineages, from which a new evolutionary type develops, are therefore only *relatively* unspecialized, that is, *most* of their organs have remained at a simple, undifferentiated evolutionary level. As a consequence, such a lineage is not set in any particular path but rather still has the plasticity and undiminished potential necessary for far-reaching transformation.

Cross-Specializations

Now, it frequently happens that in the ancestral form, the specialization (generally negligible) of one or several organs lies in a different direction or has attained a degree that causes it to contrast with the specializations in the descendant; this is called cross-specialization. For example, in one form, a lower, primitive evolutionary level of dentition is combined with a more advanced developmental stage in the limbs, whereas in a related form, the reverse combination of features is present. An extreme instance of such a crossing of lower and higher degrees of specialization of numerous organs occurs in two families of baleen whales, the balaenids and the balaenopterids, and has been summarized by Othenio Abel in an interesting diagram (fig. 3.106).

These cross-specializations used to present great difficulties for the understanding of evolution in general and for the elucidation of individual pathways of evolution. In the search for the ancestral forms of individual lineages and evolutionary cycles, one always encountered in the forms regarded as root stock certain differentiations that contradicted those in the evolutionary successors; the conclusion was drawn that such evidence ruled out the possibility of a relationship of descent. Although, based on their other structural characters, the species in question fulfilled all the prerequisites for the sought-after ancestral form, they were relegated to collateral lines or only to a position near the point at which the actual branching took place. Efforts to find the "real" ancestors continued without ever, even in the presence of the most abundant material, achieving the goal. Ultimately, there were nothing but "collateral lines"; the actual main stock remained elusive.

These difficulties fall away once we have recognized that *the transformations of the structural design take place abruptly at a more or less early stage of ontogeny and that the particular specializations that did not occur until later in the ontogeny of the ancestor are not recapitulated but rather stripped away.* Only such an individual developmental stage, one *preceding* the formation of the characters of specialization, corresponds completely to the undifferentiated, generalized structure that has always eluded scientists looking for ancestral forms; this early ontogenetic stage is the *objective, actual point of linkage, the point at which every far-reaching change of evolutionary direction takes place.*

The Belgian zoologist S. Frechkop was correct recently when he pointed out

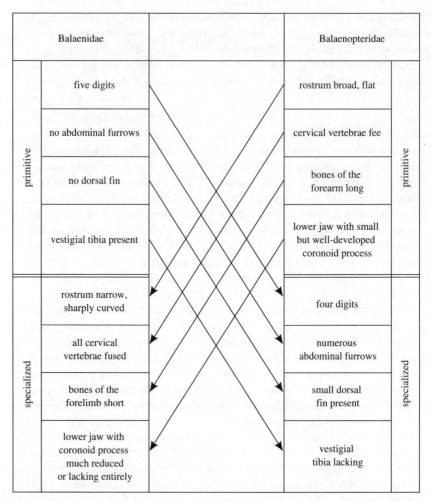

Balaenidae		Balaenopteridae

primitive

five digits	rostrum broad, flat
no abdominal furrows	cervical vertebrae fee
no dorsal fin	bones of the forearm long
vestigial tibia present	lower jaw with small but well-developed coronoid process

specialized

rostrum narrow, sharply curved	four digits
all cervical vertebrae fused	numerous abdominal furrows
bones of the forelimb short	small dorsal fin present
lower jaw with coronoid process much reduced or lacking entirely	vestigial tibia lacking

Fig. 3.106. A typical example of numerous cross-specializations in the Balaenidae and the Balae-nopteridae. Each of these families of whales shows a mixture of more primitive and more advanced characters. The developmental level of the limbs contrasts with that of the cervical vertebrae, that of the dorsal fin with that of the lower jaw, and so on. It is impossible, therefore, to trace one family back to the other, at least based on the adult state. Rather, both families are descended from a common ancestral group or from a juvenile developmental stage in which a pronounced, particular differentiation of the character in question is still lacking. (From O. Abel 1929.)

how important it is for ascending evolution that certain evolutionary stages and states are not assumed by the ontogenies of later representatives but, instead, are *avoided*. Many specializations or unbalanced evolutionary trends are, as he says, *faux pas*, which, if evolution is to be progressive, must be circumvented. One example is how the growth form of the adult tunicate is avoided by the rest of the chordates, which link up with the state represented by the larval ascidian; another is how the reduction in the number of digits in *Anoplotherium* is not taken on by the rest of the even-toed ungulates, which outlived that genus.

However, Frechkop followed the usual practice of consistently relegating forms exhibiting any kind of specialization to collateral lines, which, in many instances, is most certainly unnecessary and unjustified. If the lineage in question is not actually a collateral line in the process of becoming extinct, then, as we now know, typal transformation early in ontogeny bypasses the ancestral characters impeding evolution—an indispensable requirement for evolution, comprehensible only in the context of the ideas presented here.

An example of certain specializations that are passed over and avoided by descendants is provided by the corals. We recall that the root stock of Triassic to Recent cyclocorals [Scleractinia] is assumed to be one particular conservative group of pterocorals [Rugosa]—the zaphrentoids, and specifically, their young-est Carboniferous-Permian branch, the polycoelids, which consist of the poly-coelians (strictly speaking) and the plerophylls (fig. 3.107a, b). However, among these polycoelids there is not a single known representative that answers all the morphological claims that one would generally place upon an actual ancestral form. At late age, the existing genera consistently show a reduction, a retardation of some of the protosepta—of the counter or the cardinal septum or both and of the counter lateral septa as well. These specialized structures represent typical generic and subgeneric peculiarities of the polycoelids but are completely lack-ing in their Triassic descendants, the most primitive cyclocorals.

This state of affairs, however, by no means rules out that the true ancestral form is to be sought among 'the polycoelids, perhaps in *Polycoelia* itself (fig. 3.107a) or in *Plerophyllum* (fig. 3.107b), for the transformational path from the pterocoral to the cyclocoral type does not lead through the fully developed, mature states (a_2 and b_2, respectively) but, avoiding these, links up directly with the early juvenile developmental stages in which the protosepta are still com-pletely developed (a_1 and b_1, respectively), and from that point, takes off in a new direction.

The evidence of cross-specialization, or rather, of the "avoidance" of highly specialized ancestral differentiations during ontogeny thus provides another ar-gument in favor of the abrupt transformation of type early in ontogeny. This is the only way in which such specializations can be circumvented; *continued spe-ciation, which entails ever more specialization and adaptation, would never lead to this result.*

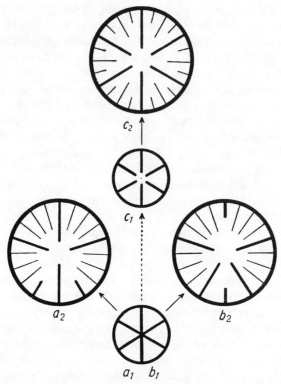

Fig. 3.107. Descent of the cyclocorals [Scleractinia] (c_1, c_2) from the early juvenile developmental stage of a representative of the polycoelids (a_1, b_1). Evolution does not proceed by way of the mature stage of *Polycoelia* (a_2) or of *Plerophyllum* (b_2), the septal apparatuses of which are specialized in particular ways and thus represent dead ends, so to speak, which will be "avoided" by the descendants. In evolution, a form never "derives" from the mature stages of its ancestor. The continuity of evolution is effected only in the germ line, with its developmental potential. When this potential is altered, the ontogenetic paths are the same only up to a certain developmental stage, after which they diverge. It is in this sense that the somewhat exaggerated way of putting it—that the "origin" of a new type of form links up with the juvenile stages of an ancestral type—is to be understood. The juvenile stage in question is the one prior to which ontogeny has run parallel. In the example given here, it is a stage common to the entire polycoelian group: the protoseptal stage (a_1, b_1), with its potential for further development still intact.

The law of unspecialized descent, discussed earlier, is thus closely tied to this general evolutionary principle and in this context acquires an even broader scope: *Not only the relatively unspecialized lineages of individual types but also, within that framework, the as yet unspecialized, or incompletely differentiated, juvenile stages form, through alterations in ontogeny, the base and point of departure for the new structural designs.*

Reduction of Evolutionary Scope

Another law of evolution that should be mentioned here is that formulated by D. Rosa, *the law of progressively reduced variation.* This way of putting it is unfortunate. What is meant is a reduction of evolutionary potential as the course of evolution unfolds, an increasing narrowing of morphological possibilities. It would be better to speak of a *diminished evolutionary capability,* which, however, would only relate to the realm of the individual evolutionary cycles.

In the way it is stated, however, the law tells us nothing new; rather, it seems to be only a consequence of evolutionary typostrophes and describes the situation we are already familiar with: within the individual evolutionary cycles there is a descending step-by-step breakup into narrower subtypes of a lesser order.

We have already said often enough that each new cycle of unfolding begins with relatively unspecialized forms, which, possessing great formative lability, still combine creative potential with opportunities for the most varied of evolutionary directions. During the typogenetic phase of their evolution, they split immediately into different subtypes, which from then on are organized in specific ways; it is these subtypes that provide the rigid framework, never to be exceeded, for the long-lasting lineages of the typostatic period. Within these lineages, typostrophes of a second, third, and so on, degree lead to the step-by-step elaboration of structures of lower rank, the developmental scope of which continues to decrease. These processes, as we have seen, extend right on down to individual species; at first, during their phase of origination, species, too, have broader possibilities for variation, which later decrease.

In general, then, diversity of form is greatest and most extensive at the beginning of each evolutionary cycle. Large numbers of the most varied of what one might call experimental models are constructed, and all existing possibilities are explored. A considerable portion of these experiments, however, do not prove successful in the long run; sooner or later, some lineages die out, ultimately leaving behind only a few types out of the former abundance of forms. For example, in the ancient group of the algae, as Walter Zimmermann pointed out, there were a multitude of structural solutions and many different kinds of assimilation pigments and modes of reproduction that were later given up and are lacking in the younger division of the cormophytes. Similarly, in the vertebrates there is a conspicuous contrast between the many different models of structural designs of the most primitive representatives, the agnathans and the fishes, and the morphologically consistent basic structural design of the tetrapodes, which are a continuation of a single type from among the fishes.

There is no need to discuss further examples here; we have already pointed out this situation frequently, in particular in the evolution of the cephalopods.

Further, all those groups of plants and animals that are represented in the present only by sparse relics of a former period of proliferation, such as the brachiopods, pelmatozoans, cartilaginous fishes, Articulatae [horsetails], and so on, offer significant proof for the phenomenon of a reduction in the breadth of a phylum and a decrease in the original multiplicity of types.

In the final, typolitic phase of each evolutionary cycle, however, considerable variability appears once more, so much so that often even the basic structure of the type in question breaks up. But here, it is only an *appearance* of increasing formative potential; in reality, the phenomenon is destructive, not constructive, and contains no germ of ascending evolution.

This is the reason that we do not accept Rosa's expression "reduced variation" but replace it by the concept of diminished evolutionary scope or decreasing evolutionary capability. These phenomena, however, occasioned by progressively narrower specialization, by degeneration and extinction, are *actually present quite generally* within each evolutionary cycle, until once again, a large evolutionary step toward a new, more general basic type breaks the existing bonds and brings about a marked broadening of evolutionary scope.

Orthogenesis

These last remarks lead to another important question: Is evolution directional or does it proceed at random? Both interpretations have been advocated. In general, paleontologists hold to the former view and regard evolution as a process that is somehow guided according to principles. On the other hand, geneticists place particular emphasis on the mutability of organisms, a nondirectional phenomenon, subject purely to chance, into which a certain orientation is introduced only secondarily by natural selection and progressive adaptation.

Both views are at once, and to a certain degree, correct and incorrect; it depends entirely on the phase of evolution under consideration. *In the initial, typogenetic period of the major evolutionary cycles, freedom reigns in the development of forms.* There, the generalized type splits into subtypes spontaneously, in every direction. These, however, bring with them organs and preconditions already developed for this or that mode of life, and because of this, each of these subtypes is already marked out for a particular kind of development, or expressed in the opposite way, other directions are ruled out right from the beginning. Every descending breakup of type brings about a further narrowing of morphological potential; thus, *in the typostatic phase, the course of unfolding is compulsory,* following a linear, consistent progression, which is called *orthogenesis.*

Many attempts have been made to disavow orthogenesis, but it is a fact, and there is no way around it. Once any principle whatsoever is introduced into a particular development, we see it follow its course as if from inner compulsion, automatically, like clockwork, no matter whether it leads to ascendancy or decline. It continues on its way until an end is reached, a point that cannot be transcended; this usually also means the end of the lineage in question.

Examples of Orthogenesis

In *ammonites,* after the frilling of the suture has been introduced as a fundamentally new process, it continues to develop step by step until the last tiny bit of lobe and saddle margin is broken up into extremely fine teeth and notches. Further, as soon as the principle of the differentiation of the suture line through saddle splitting has been acquired, it is unswervingly pursued, and one by one, one after another, new lobal elements are emplaced. This saddle splitting may affect different saddles, either the inner or the outer ones; at first, the choice was open. But after the decision was made in favor of one site or the other, further development was inevitable, preordained. The same is true for the increase in the number of lobal elements through lobe splitting. Once this mode had been "invented" by a particular form, its descendants carried it on; the mode prevailed, and there was no stopping it, no going back, and no breaking away from the evolutionary direction once it was established.

In the *nautiloids* and the *ammonoids,* the coiling of the shell progressed in an orderly way. In the process, however, a decided difference appeared between the two groups, as we have seen (figs. 3.34 and 3.35): in the ammonoids, the axis of coiling runs through the protoconch, located at the center of the shell; in the nautiloids, however, the protoconch is eccentric, lying *next to* the axis of the shell. Thus, the further course of evolution is dictated in advance by the respective initial forms: as the move toward ever tighter coiling progressed, the protoconch of the ammonoids had to participate in the process and acquired a spiral torsion; in contrast, in the nautiloids, in order to arrive at as tightly closed a spiral as possible, one with no perforation, the protoconch had to become increasingly smaller and assume a flat, cowled form. Once the preconditions were established, no other mechanical possibilities were open to the protoconch, and we then see evolution proceeding in a straight line along the path marked out for it.

The unfolding of the *stony corals* is dominated by a progressive replacement of the original bilateral arrangement of the septal apparatuses by a radial one (fig. 3.46). The direction of this course is determined ahead of time by the decidedly hexamerous stage of the six protosepta, which makes a temporary appearance early in the ontogeny of the pterocorals. Thus, the structural design of the lineage is laid down from the beginning and is executed as a complete, pure realization of this hexamerous emplacement by suppression and progressive dissolution of the bilateral features, which at first dominated the mature stages of the pterocorals. In those mature stages, as we recall, only four quadrants were completely developed, and remarkably, this peculiarity was also passed on to the heterocorals, which issued from the pterocorals, as a general morphological capability, although there, it was carried out in a completely different way (fig. 3.74*d* and *g*).

*   *   *

Excellent examples of orthogenetic courses of events are provided by the progressive reduction of digits on the fore- and hind feet of the *ungulates*. This process is best and most completely known in the evolution leading to the modern *horse* (figs. 3.93, 3.108, 3.109). The ancestral form, *Eohippus* (or, for that matter, *Hyracotherium*), from the Lower Eocene, shows four fingers on the forefoot and three functional toes on the hind foot, next to which are tiny vestiges of digital rays I and V, displaced far to the rear. As was stated earlier, *Eohippus* is in turn descended from a representative of the protoungulates that had five normally developed digits on both the fore- and hind feet and that, through its mes-

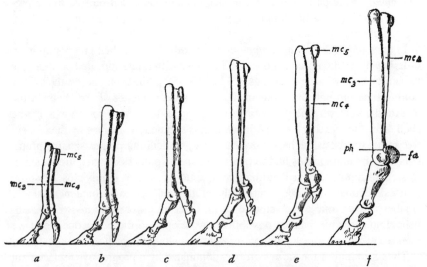

Fig. 3.108. Lateral view of the skeleton of the left front foot of some horses from the North American Tertiary, to illustrate the increasing reduction and decreasing use of the lateral toes. The second toe, not visible in the figure, is reduced, analogous to the fourth. All about ⅓ ×. *a. Mesohippus bairdii* Leidy. Middle Oligocene of Nebraska. *b. Miohippus intermedius* Osb. and Wortm. Upper Oligocene of South Dakota. *c. Parahippus atavus* Osb. Middle Miocene of western Nebraska. *d. Merychippus primus* Osb. Middle Miocene of western Nebraska. *e. Merychippus eohipparion* Osb. Middle Miocene of northeastern Colorado. *f. Pliohippus lullianus* Trox. Lower Pliocene of South Dakota.

mc_3 = third, mc_4 = fourth, mc_5 = fifth metacarpal bones. ph = the fused elements of the fourth digit in *Pliohippus;* they are still separate from the distal end of mc_4. (From O. Abel 1929.)

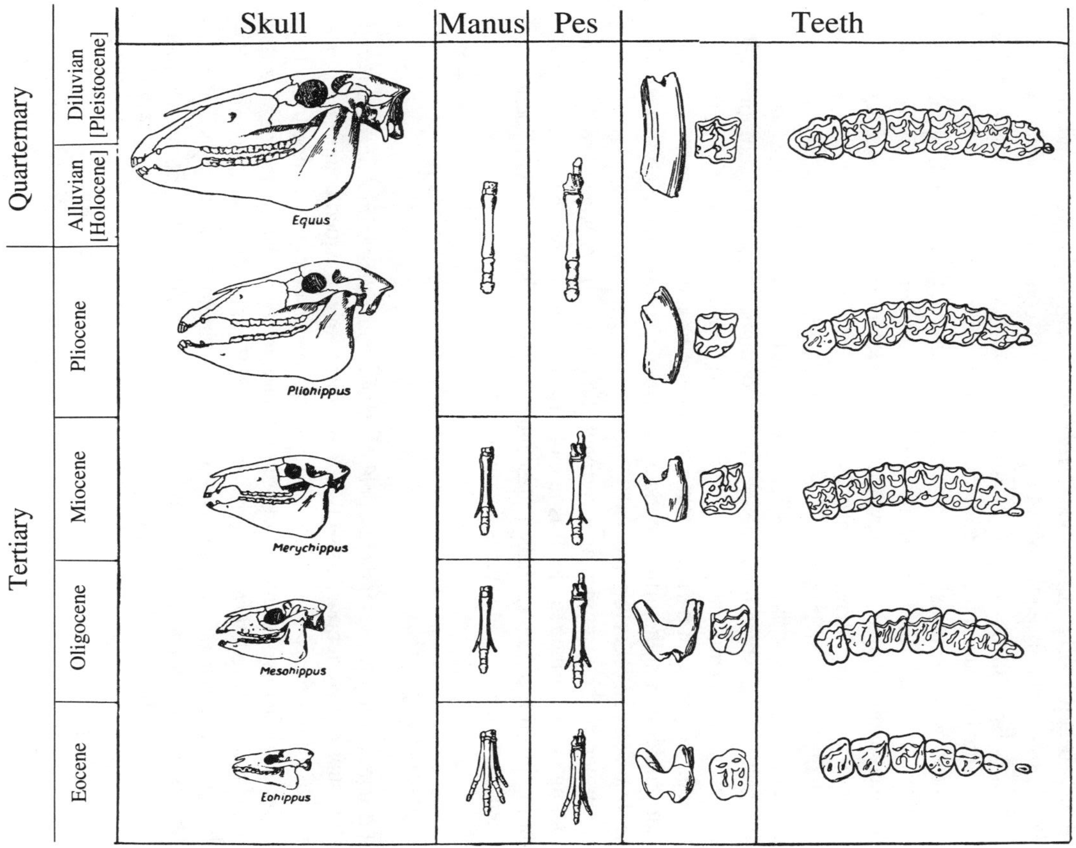

Fig. 3.109. Main stages in the evolution of the horse. The first column illustrates the development of the skull, its increase in size, the lengthening of the face, the displacement of the eyes from the center of the skull to behind the center, the progressive closure of the orbits by a postorbital bar, which is originally short and narrow, later sturdy and connected to the zygomatic arch, and so on. The fore- and hind feet show the evolution of the original four-digit and three-digit forms to the functionally one-digit form in both extremities; this is described in more detail in the text. In the dentition, it is characteristic that the incisors, large in *Eohippus,* become progressively smaller; the originally low-crowned cheek teeth become high crowned and acquire a complex crown structure with cementing, pulp, enamel ridges, and secondary columns at the outer edge; and the premolars of the upper jaw (the four teeth to the far right in the rows of teeth illustrated) become increasingly molarized: originally triangular and different in form from the cheek teeth, they become four-sided, and their development is now exactly the same as that of the molars. (After H. F. Osborn.)

axonia, already portends, as it were, the evolutionary mode to come. A closed evolutionary sequence follows upon *Eohippus,* which, documented by an abundant material, allows us to follow step by step all the individual stages of the further reduction of the toes.

To single out only the main steps from this evolutionary progression, we cite the genus *Mesohippus,* from the Oligocene, which has three matching digits on both the fore- and the hind feet. The lateral toes (II and IV) on either side of the dominant middle toe (III) are here still relatively well developed and in contact with the ground; in addition, in the forefoot, there are still vestiges of digital ray V present. In the Miocene genus *Merychippus,* these have disappeared, and further, the lateral toes have grown even shorter, to the point that they no longer serve to bear the weight of the body. *Hipparion,* from the Pliocene, and the genus *Pliohippus,* probably a member of a collateral line, continue this evolution, until finally, in Recent horses, only the strongly developed middle toe and tiny vestiges of the two lateral toes remain on the splint bone.

Parallel with the transformation of the extremities, the dentition of the horse lineage also undergoes gradual orthogenetic evolution: the originally low-crowned teeth become high crowned; the enamel folds become more complex, the cement coating increases in thickness, and the premolars progressively approach the form of molars (fig. 3.109). This evolutionary cycle, which took place in the major lineage during the Tertiary of North America, engages about ten different genera, which illustrate the main evolutionary steps. (In addition, there are several collateral lines of varying duration that undergo the same transformational sequence in parallel.) But even within the individual genera, a consistent progression in the orthogenetic evolution of the fore- and hind feet and the dentition can be observed in detail.

Further, in the examples of proterogenetic evolution we have described, unfolding is also orthogenetic. It is demonstrated when a new character acquired during juvenile stages spreads in descendants, step by step and with increasing intensity, to later growth stages and transforms them.

Explanation of Orthogenesis

It is unnecessary to present additional evidence from other animal lineages here; all groups of invertebrates as well as vertebrates have provided countless examples of orthogenetic evolution. The unwary observer could easily form the impression that evolution is purposeful, that right from the beginning it is directed toward a predetermined goal and that the path it follows is determined by the goal. Such a finalistic explanation, however, cannot be seriously supported; there is no basis for it in natural science, and the observed facts do not warrant it in the least.

Rather, things are just the opposite, in that it is not the conceptual final point but *the concrete starting point that determines and brings about* the orientation of evolution. Such a view can be based on actual, causative mechanisms and does not have to take refuge in mystical principles of any kind. The explanation lies in the fact that *the set of rudiments in the first representatives of each lineage largely determines later evolution, and that subsequent differentiational steps entail a progressive narrowing of evolutionary creative potential.*

A comparison with ontogeny is obvious and offers further enlightenment: the ontogenetic development of an individual is largely predetermined by the genetic base inherited from its ancestors. Any greater morphological freedom exists only in the plasticity of early morphogenetic stages, during which the rudiments of the individual organs are emplaced. After this has taken place, subsequent ontogeny is to a large extent determined.

Once an organ has been emplaced, it can indeed grow slowly or more rapidly, attaining lesser or greater dimensions—to this extent, there is still a certain latitude at the start—but each and every further developmental step marks the subsequent one with its stamp and sets off correlative consequences. The further ontogeny progresses, the more the existing possibilities for shaping characters are reduced, with the result that toward the end, an obligatory course of development has been established.

Orthogenesis, then, is the characteristic evolutionary mode of the typostatic phase of evolution. It corresponds to the unfolding phase of ontogeny, with its extensively determined periods of growth, whereas the typogenetic phase of evolution can be compared to the morphologically plastic early portion of ontogeny. During the typostatic period, the transformation of form is accomplished by the gradual addition of new characters or subsidiary organs to the ontogeny of ancestors (the incorporation of new lobal elements, further notching of lobes, and so on). These novelties are in turn progressively advanced in the ontogeny of descendants, making room once more for new characters along the same line.

In this manner, the compulsory course of events leading to the final stages of one ontogeny is transmitted through evolution to the following one, which then carries it further. No longer can there be any far-reaching changes of course; all morphological possibilities not included in the evolutionary course originally embarked upon are blocked.

Mutations appearing in such a series may be essentially directionless, that is, they do not in the slightest have to be consistent with the established direction. However, only those that fall within the scope of the predetermined path have the prospect of being retained and of undergoing further evolution, unless, through a large, far-reaching mutational leap in an early ontogenetic stage, a new, successful typal structure arises, opening up new evolutionary possibilities once again.

To this extent, our explanation does not maintain in the least that evolution is prefigured right from the beginning and in every detail. Rather, any typogenesis opens the door once more to unfettered morphological opportunities, with the reservation, of course, that these are within the framework of the creative potential of the evolutionarily higher-ranking type.

Parallel Evolution

Other informative glimpses into the nature and causes of orthogenesis are provided by parallel evolution (the "geitonogenesis" of Kleinschmidt, homologous evolution of Plate), an extremely common, probably even universal phenomenon of the typostatic phase of evolution. The unfolding of the individual stocks does not take place in a single lineage but rather in a varying number of parallel lineages at the same time. These differ from one another in certain characters, which are indicators of their independence, but agree in other, supraordinate characters, those that demonstrate their relatedness; parallel lineages exhibit similar remodeling even though they are related only at their roots and, moreover, evolve independently.

Every group of plants and animals provides multiple examples of parallel evolution. Even the orthogenetic phenomena just discussed—the notching and saddle-splitting of suture lines, the radial elaboration of the septal apparatus in corals, the reduction of the toes in the horse group, and so on—run their courses in whole clusters of parallel lines; in that context, however, we did not examine this aspect in detail.

Here, using the *cephalopods* as an example, we refer to the reversal of the septal vaulting, from concave to convex, to the evolutionary about-face in the siphuncular necks from backward- to forward-facing (fig. 3.36), or to the progressive increase in the number of elements in the primary suture (fig. 3.38). The evolution of all of these characters permeates the entire ammonoid stock, and in many series of forms, no matter what their special taxonomic position, the characters evolve in the same way. Each evolutionary stage in the reshaping of the septa, the siphuncles, the sculpture, and the elaboration of the suture line is realized by representatives of a number of parallel lineages at the same time.

In the *clymenians,* the elaboration of the normally round shell to a triangular one takes place independently, but in exactly the same direction, in three different lineages evolving in parallel from a common ancestor (fig. 3.87). This ancestral form must have acquired the capability of developing a triangular shell and, through whorl constriction, a three-lobed one, and transmitted it to subsequent lineages.

The commonality of morphology seen here cannot just be the result of chance, nor can it be explained simply by a disturbance in the capability of developing a normal shell spiral; if that were the case, shells in one lineage or another would, perhaps, also be four-sided or more, like the four-sided forms actually known in other ammonoids. Therefore, the common ancestor must have transmitted a specific factor for three-sidedness, which, in the three lineages, then evolved along a directed, undeviating course.

The palaeodictyopterans (fig. 2.36), the ancestral forms of all flying *insects,* show, in addition to other primitive features, an ancient situation in which the four identical wings cannot be flexed but rather spread out laterally even when the insect is at rest. These ancestral forms gave rise during the Carboniferous to the protoblattoids, in which the wings were folded back horizontally over the abdomen. Further evolution led to a situation in which the forewings, at first delicate and soft-skinned, became increasingly chitinous, hardened to form elytra, and took on the function of protecting the hind wings and the abdomen. In the Permian, then, there are about five or six lineages, descendants of the Protoblattoidea, among them the beetles, the cockroaches, and so on, that sooner or later independently underwent parallel transformation of the forewings to elytra. Likewise, the ability of this group to fold the hind wings crosswise was acquired independently as the result of similar latent evolutionary potential.

In the *agnathans* and the *fishes,* to single out just a few more, we encounter in the progressive degeneration of the armoring a very general, widespread evolutionary trend carried and passed through by a large number of parallel lineages. Numerous commonalities in the reshaping of the tail fin, the scales, the endocranium, and so on characterize the two lineages of the crossopterygians and the dipnoans, which evolved parallel from a common ancestor.

Within the *actinopterygians,* a subclass of the bony fishes in the broad sense (Osteichthyes), there is a broad evolutionary trend distinguished, in particular, by the transformation of the tail fin from the original heterocercality to homocercality; by the gradual degeneration of the thick, rhomboidal ganoid scales to thin, round cycloid or ctenoid scales; by the restructuring of the endocranium; and by the progressively lighter construction of the cranium, increasing ossification of the vertebral column, reduction of the clavicula, and so on. Based on the different evolutionary stages of these characters, three orders were recognized: the primarily Paleozoic Chondrostei, the Mesozoic Holostei, and the Teleostei, which dominate from the Cretaceous on. We now know, however, that these groups are polyphyletic, which means that they consist of a cluster of separate, independent lineages, which, with regard to all the transformations cited, developed in parallel in the same direction and reached and passed through the various evolutionary stages more or less at the same time.

Among the *amphibians,* the three orders of the labyrinthodonts, the phyllospondyls, and the lepospondyls show a broad parallelism in the transformation of the skull. The commonalities consist of the similar flattening of the originally high skull, a decrease in ossification, the disappearance of the interorbital septum, the displacement of the brain and the cranial nerves, an increasing perforation of the originally solid palate, and so on. Yet another example—the *reptilian* group of the Therocephalia breaks up into different lineages, which also bring forth the same sequences of characters in parallel evolution.

The cheek teeth of *mammals* undergo a similar kind of parallel evolution in several lineages and in widely disparate areas, which is to say, under different kinds of environmental conditions. Here, the original trituberculate molar, which the lineages in question all had in common, limited subsequent evolution (increase in the number of tubercles, formation of the zygomatic arch, and so on) right from the start to a particular direction; it was not simply left to chance. Similarly, we see that the original basic number of forty-four teeth in the oldest placental mammals represents an upper limit, one that is never exceeded. Here, too, morphological freedom is already limited by the first representative of the stock. Only a reduction in the number of teeth can occur, and this happened repeatedly orthogenetically and in parallel in the various phyletic units, so that, in general, younger representatives of the individual orders and families have fewer teeth than older ones.

The rhinocerotids, like the horses already discussed, also show parallel evolution in several lineages with regard to the height of the crowns, the development of the roots, and the ridges in the cheek teeth. Other similar instances in the evolution of the sea cow, the deer, the mastodonts, and so on, have been described; it would be impossible even to list all the known examples.

Naturally, the same kinds of phenomena are also known in the *plant world*. Here, we mention only the evolution from isosporic to heterosporic and, finally, gymnosporic reproduction, which, according to Walter Zimmermann, took place within the cormophyte stock independently in at least three parallel lineages.

As our examples show, parallelism in evolution expresses itself in quite different categories and orders of magnitude. It is found in lineages within genera and families as well as in categories of higher taxonomic and evolutionary rank, where the phenomena are the same as in the smallest unit, the species. And in species, too, there is parallel evolution in numerous separate reproductive lines, as we see in individual clans, races, and subspecies. We know further that in *different* species, similar, parallel races appear, the similarities being occasioned by the portion of the genetic constitution they have in common and by the similar capacity for mutation, both of which factors limit variability.

When we see within genera, families, and other units that evolution is being guided in the same direction in several parallel lines, it is obvious that the reasons are the same for all. We may assume, then, *that the orientation and parallelism of individual lineages is essentially guided by a common genetic base, which reacts the same way in each line.*

The effects of external factors cannot explain parallel morphology for the following reasons:

1. The transformation of individual lineages has taken place many times under extremely different environmental conditions in widely disparate areas, and despite this, the result has been parallelism.

2. On the other hand, the points at which transformation takes place in the individual parallel lineages are by no means always simultaneous, which means that the same external influences cannot be inferred. To take an example from the amphibians mentioned above, the phyllospondyls attain as early as the Permian the evolutionary stage of skull transformation that does not appear in the labyrinthodonts until the Triassic. Further, the parallel reduction in toes in the individual lineages of even- and odd-toed ungulates causes the final stage of two toes or one toe respectively to arise at quite different geologic times.

3. Finally, external factors alone would not explain the circumstance that it is always only a certain cluster of closely related lineages that brings forth these parallel transformations, whereas other series of forms existing at the same time, which are subject to the same external influences, behave in a completely different way and show a different evolution of characters. Thus, the critical, deciding factors are always internal and depend on the potential for evolutionary creativity of the organism itself.

*   *   *

In addition to more or less *simultaneous* parallel evolution there is yet another parallelism that takes place *sequentially,* during the various evolutionary stages of the lineage as they come into play one after another (the "hypogenesis" of Kleinschmidt).

An example of this kind of nonsimultaneous parallelism is the identical course of development of the sculpture in the Triassic and the Jurassic-Cretaceous *ammonites* (figs. 3.32 and 3.152). At a higher evolutionary level, later ammonites repeat the same sequence of, first, smooth shells, then shells with simple ribbing, then with split ribbing, and finally forms with multibranched ribs—the same sequence in which the sculpture was elaborated earlier by Triassic ammonites. The individual sequential stages of sculpture in turn show parallelism in the similar repetition of changes in shell shape, which always takes place in the same way with similar final results.

In the *corals,* we see among the pterocorals and the cyclocorals as well, which is to say, at different evolutionary levels of the stock, the same types always being repeated: forms with solid or porous skeletons, with or without columns, with cylindrical or top-shaped corallites (figs. 3.47–49), and so on. Further, we have already referred to the fact that *placental mammals* take up the same morphological tendencies, such as specializations as predators, rodents, flying creatures, and so on, that were present at the more primitive level of the aplacental mammals (pl. 25A and B; figs. 3.100–105).

These phenomena are called *iterative,* or repetitive, morphology, or also, often, *convergence.* The use of the term "convergence" for such cases seems, however, to be not quite appropriate, for it implies the idea of evolution coming together from separate points, of converging on one point, whereas in reality, the situation is one of a *parallel* evolution of the same, *homologous organs* at different stages of the lineage as a whole. I have therefore recommended that such phenomena be termed *nonsimultaneous (heterochronous) parallel evolution,* and that its resulting morphological similarity be given the neutral designation of *homeomorphy.*

Apart from the *sequential* aspect, what we have here is absolutely the same as in *parallel* evolution, which, moreover, as already stated, also may exhibit certain temporal differences in the appearance of equivalent transformations. One characteristic example of parallel evolution not yet mentioned occurs within the *mammals* and consists of the elaboration of the postorbital arch,[17] which provided a posterior bony partition between the orbit and the temporal fossae, closing them off and thus providing better protection for the eye. This postorbital arch appears in individual mammalian lineages at very different times, resulting in a parallelism that is sometimes simultaneous and sometimes not.

The internal reasons for parallelism are also the same in both instances: *they reside in the matching genotypes linking the lineages in question, which allow only a limited number of possible directions.* The remarkable thing about heterochronic parallel lineages is simply that the potential for evoking them, in other words, the appropriate alleles, has been preserved unchanged in an organic base that has in the meantime become profoundly transformed, often over extremely long periods of time.

Actual *convergences* contrast with these simultaneous or nonsimultaneous homeomorphies. They signify morphological similarities that, through evolution that is literally convergent, are sometimes produced in lineages that are not related. Consequently, they are not concerned with equivalent homologies but only with *analogous organs.* Well-known pairs of examples are the wings of insects and of birds, and the eyes of vertebrates and of squids and octopuses. In addition, the external morphological similarities of animals with equivalent modes of life (fig. 3.110) and of plants that are subject to the same conditions of habitat and climate (pl. 25C) belong in this category. These phenomena have nothing to do with parallel evolution in the sense described above and have a completely different significance.

17. A postorbital arch like this is already found in the reptiles, but there, we should note that it is differently constituted than in the mammals. In this regard, as Ernst Stromer (1944), in particular, has recently pointed out, the mammalian postorbital arch does not repeat some earlier evolutionary stage, but shows instead a typal novelty constructed from different bony elements: the irreversibility of evolution!

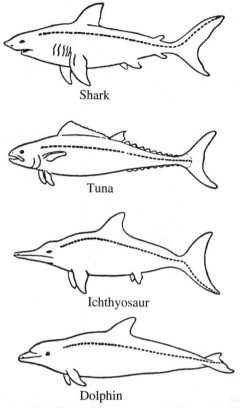

Shark

Tuna

Ichthyosaur

Dolphin

Fig. 3.110. Extensive atypical similarities in the outer shape of the body in marine vertebrates that are adapted for rapid swimming, from the groups of the cartilaginous fishes, the bony fishes, the reptiles, and the mammals. (After Jacobs, from F. Reinöhl 1940.)

Overspecialization

Orthogenesis consists in undeviating, progressive specialization, which to some extent leads to increased adaptation but also, quite often, affects organs and takes off in directions that do not directly have anything to do with adaptational processes. Insofar as an improvement in adaptational relationships is attained, it may be assumed that, in addition to the primary influence of genetic structure, selection also plays an important role in the orienting of lineages.

We frequently observe, however, that orthogenetic development, once embarked upon, continues on its unswerving course long after an advantageous adaptational state has been attained or even exceeded, when continuation quite

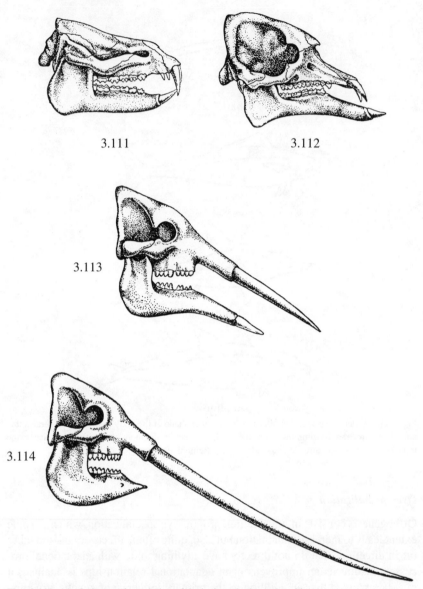

Figs. 3.111–16. Orthogenetic evolution of the incisors in the proboscids.

3.111. *Moeritherium trigonodon* Andrews. Lower Oligocene of Fayum, Egypt.

3.112. *Phiomia wintoni* Andrews. Lower Oligocene of Fayum, Egypt.

3.113. *Tetrabelodon longirostre* (Kaup). Lower Pliocene, Europe.

3.114. *Anancus arvernensis* (Croiz. and Job.). Middle Pliocene and early Upper Pliocene, Europe.

obviously leads to impairment. In other words, orthogenesis is not limited to our second evolutionary phase of typostasy, but extends on into the final, degenerative, typolytic phase. There, specialization becomes overspecialization, the appropriate becomes inappropriate, and harmony of bodily structure yields to pronounced disharmony. The contribution made by selection is not immediately comprehensible; on the contrary, it seems even nonsensical, but the details of this problem do not concern us just now. Rather, we shall remain for a moment with the facts themselves and take a look at a few examples.

The lineage of the *Proboscidea* (elephants) is characterized by a gradual increase in size of the upper (and sometimes also the lower) incisors. Originally, these were short and straight or slightly curved; they aided in feeding, and were used to dig and grub in the ground to expose roots and the like. In the early

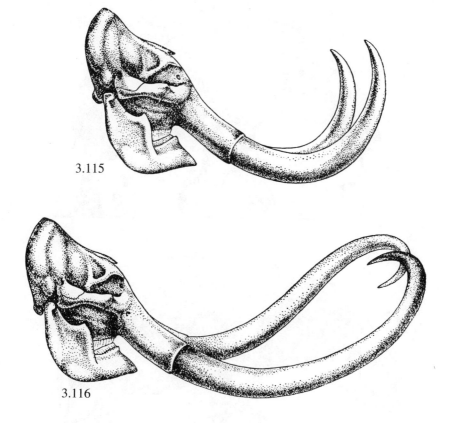

3.115

3.116

3.115. *Elephas* (*Mammonteus*) *primigenius* Blumenb. Diluvian, central Europe.
3.116. *Elephas* (*Mammonteus*) *columbi* Falc. Diluvian, North America.
(After O. Abel, C. W. Andrews, and A. S. Romer.)

3.119

3.118

3.117

forms (figs. 3.111 and 3.112), the only slightly elaborated dentition permitted various applications, and the "tusks" of the lower jaw still served as organs of food gathering. During the course of evolution, the trunk increased in length, as can be seen from the enlargement of the attachment surfaces for its musculature. Conjointly, the lower tusks and the processes of the lower jaw bearing them degenerated while the upper ones grew progressively longer (figs. 3.113, 3.114, 3.117, 3.118) and came to be used as dangerous thrust weapons.

Finally, in certain Diluvian [Pleistocene] elephants that were the terminal members of dying collateral lines, the tusks reached enormous lengths, completely out of proportion to the skull and body, clearly exceeding by far the optimal size relationship. Furthermore, since in the mammoth (fig. 3.115) these teeth curve sharply upward and in the American *Elephas columbi* (fig. 3.116 and pl. 22B) begin to spiral backward and inward, they could no longer serve their original purpose and could fulfill their other function as weapons only incompletely; they had become a heavy burden, an impediment, for these animals. To illustrate their loss of function, O. Abel has compared them with nonfunctional rodent teeth, which also undergo that kind of spiral coiling.

Further, we refer to the overspecialization of the long, saberlike upper canine teeth of the American *saber-toothed tiger* (*Smilodon,* pl. 23; figs. 3.120*d* and 3.121), which arose through excessive growth of the normal canines of predator dentition (fig. 3.120*a* and *b*). The original biting function was thus rendered impossible, especially since the enlargement of the upper canines was connected to a reduction in size of the lower ones. The biting off and intake of food was also undoubtedly made much more difficult by this overspecialization, and even full utilization of the teeth as weapons, ferocious as they looked, was no longer guaranteed. Furthermore, the canines themselves were endangered by their disproportionate length: they frequently broke off, as numerous fossils testify.

3.117–19. *(Opposite)* Reconstruction of three mastodonts from the European Late Tertiary that stand in direct line of descent and show uninterrupted connection through transitional forms.
3.117. *Trilophodon angustidens* (Cuv.). Middle and Upper Miocene.
3.118. *Tetrabelodon longirostre* (Kaup). Lower Pliocene.
3.119. *Anancus arvernensis* (Croiz. and Job.). Middle Pliocene and early Upper Pliocene. (After O. Abel 1929.) The pictures are intended to illustrate the effect of the orthogenetic shortening of the lower jaw and simultaneous lengthening of the upper tusk, shown in figures 3.113 and 3.114, in living animals. Even in the oldest representatives of the mastodont line, the trunk had evidently attained a considerable length and participated with the lower incisors, which came together like a shovel, in looking for food and picking it up. Later, it was the lengthened upper incisors and the trunk that took on this function.

Ultimately, because of their orthogenetically exaggerated increase in length, the upper tusks were also excluded from the function of taking up food, and the task fell exclusively to the lengthened trunk, which had now become free along its entire length and therefore more mobile. The same evolutionary course occurred in the line of the true elephants, which does not derive directly from the forms reproduced here—despite the fact that *Anancus arvernensis* is a pronounced elephant type—but is a branch of the proboscids that evolved parallel to them.

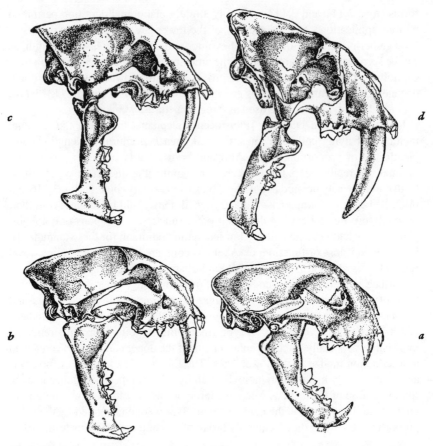

Fig. 3.120. Orthogenetic enlarging and overspecialization of the upper canine in the family Felidae. The forms illustrated are not a genetic lineage but represent evolutionary stages that have only a loose phylogenetic relationship with one another. All much reduced. *a. Metailurus,* a cat from the Pliocene of Europe and Asia. *b. Dinictis,* a cat from the Oligocene of North America. *c. Hoplophoneus,* a saber-toothed tiger from the Oligocene of North America. *d. Smilodon,* a saber-toothed tiger from the Diluvian [Pleistocene] of North and South America. (After W. D. Matthew and O. Zdansky, from A. S. Romer 1933.)

Other examples of curious exaggerations are the enormously long spiny processes of the dorsal vertebrae in the Permian *reptiles Dimetrodon* (fig. 3.122), *Edaphosaurus* (fig. 3.123), and the perhaps related *Ctenosaurus,* from the central German Bunter. These processes formed the support structure for a high crest, the functional significance of which is not apparent. The length of these spiny processes, up to sixty centimeters, is definitely out of proportion to the centra, four centimeters in diameter at the most, of the vertebrae that bear them.

Fig. 3.121. Reconstruction of the head of *Smilodon californicus* Bov. Much reduced. (After J. C. Merriam and C. Stock.)

Another decided overspecialization is the double row of thick bony plates rising from the back of the *Stegosaurus* (fig. 3.124), an animal from the Upper Jurassic of North America, which grew to as much as nine meters in length. Originally smaller and better proportioned, the plates were evidently some kind of protection for the vertebral column, perhaps against attack by larger predatory dinosaurs. During the course of continued orthogenetic evolution, the plates eventually reached a size and weight that were not in the least necessary for this purpose. This protection seems additionally problematic in that the animal's flanks remained quite vulnerable to attack. On the other hand, this dorsal burden compelled our *Stegosaurus* to assume an altered stance and locomotion. Its ancestors were bipedal, moving about only on the hind legs, the forelegs being shortened to half the length of the hind legs. But in *Stegosaurus,* the weight of the heavy dorsal plates forced the forward part of the body down upon the short forelegs and back toward the ground. The front feet would have to have changed back from gripping to walking, the pelvis would have been displaced, and so on; the overspecialization, itself useless, necessitated a series of correlated transformations that were by no means beneficial, thus proving itself unquestionably to be a disadvantage.

The fossil *sawfish Propristis* (fig. 3.125), from the Eocene of Egypt, has on its snout a swordlike process over two meters long and provided with lateral teeth; this structure considerably surpasses the "saw" of Recent sawfishes and is undoubtedly longer than the optimum consistent with its function of churning up the sea floor.

Exaggerated to the point of being fantastic are the antlers of the *huge Diluvian*

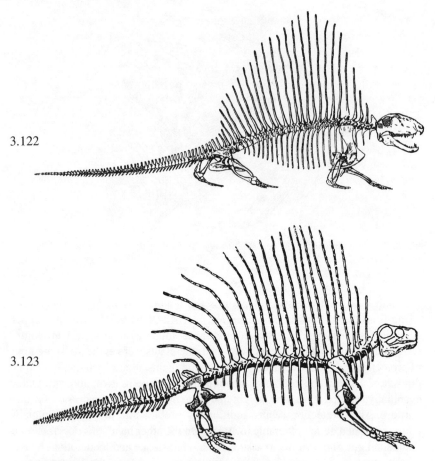

Fig. 3.122. *Dimetrodon,* a pelycosaur from the Lower Permian of Texas, with excessive development of dorsal spines. About ¹/₁₇ ×. (From A. S. Romer 1933.)

Figure 3.123. *Edaphosaurus,* another long-spined pelycosaur, from the Lower Permian of North America. About ¹/₁₇ ×. This genus shows, especially in the dentition, many differences from the *Dimetrodon* shown in figure 3.122 and is assigned to another family. The dorsal spines here must have evolved independently of those of the other genus. (After S. W. Williston, from A. S. Romer 1933.)

Irish elk, which reach a span of three and one-half meters (pl. 26). They undoubtedly had ceased to be of advantage to the animal, as the normal antlers of its ancestors were, but had become instead just an enormous burden and a handicap. The skeleton of the forepart of the body had to become correspondingly strengthened in order to be able to support the weight of the antlers, and their great span made the former forest mode of existence impossible, compelling the

Fig. 3.124. Reconstruction of *Stegosaurus ungulatus* Marsh, a dinosaur from the Upper Jurassic of North America. About ⅟₆₀ ×. (After O. Abel 1929.)

giant deer to leave its original habitat. All these overspecializations are characteristic of the final members of dying lineages, most of which, at the same time, also display gigantism.

*   *   *

Finally, a few examples from *modern animals:* The Recent *giraffe* is overspecialized in the extreme lengthening of the neck and forelimbs, which is usually assumed to be a special adaptation for browsing on the leaves and twigs of tall trees. The excessive length of the legs is compensated for only very incompletely by the neck, upsetting normal proportions; the giraffe can no longer reach the ground or the surface of a water hole with its mouth when standing in a normal position but must instead spread its forelegs wide apart (fig. 3.126). In addition, the long, slender neck makes impossible the development of extensive weapons on the forehead, as are seen, for example, in the related, relatively short-necked Pliocene *Sivatherium.* For reasons of weight, only short ossicones could develop.

A curious overspecialization is seen in the skull of the *musk ox,* an animal that first appeared in the Diluvian [Pleistocene] and is limited today to the Arctic. Males, in particular, have very sturdy horns, which show a pronounced swelling at the base and lie flat against the temples behind the eyes, sharply reducing vision to the rear. Numerous other ungulates, inhabitants of open country where visibility is good (giraffes, camels, warthogs [fig. 3.160], many antelopes, cer-

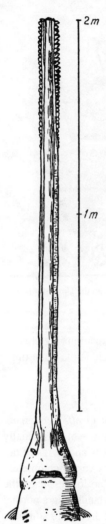

Fig. 3.125. Saw and reconstructed head of *Propristis schweinfurthi* Dames. Upper Eocene of Egypt. (After E. Fraas, from O. Abel 1912.)

vids, and so on) are characterized by more or less prominent eyes, which, protruding beyond the cheek bones, permit unobstructed vision even to the rear. In the musk ox, this situation has become extreme: the orbits are shaped like long tubes, giving a clear view beyond the horns and backward (fig. 3.127).

With this and many similar examples, the question is raised, Which came first? Did the orbits become so long and telescopic to counter the obstructed vision brought about by the development of the thick horns, or were the horns able to arrive at their particular form and thickness without presenting a disad-

Fig. 3.126. Different body positions of the Recent giraffe, *Giraffa camelopardalis* L. (From H. Krieg.)

vantage because the eyes were so protruding? This is the wrong question, in that naturally neither structure was present in its finished form right from the start. Both organs evolved at the same time in close correlation; growth in the horns as in the orbits increased bit by bit and simultaneously. Nevertheless, there are physiological limits here, which, if this evolutionary process were to continue orthogenetically, would sooner or later lead to extinction.

Another frequently cited example is the *babirusa,* a native of the Celebes. In this animal, the upper canines grow upward to pierce the upper lip because they are no longer in contact with the canines of the lower jaw and therefore are not worn down (fig. 3.128). They have become functionless and have begun to spiral. This is an extreme form of a stepped series [*Stufenreihe*] of suids (swine)

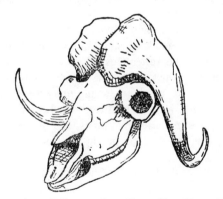

Fig. 3.127. Skull of the musk ox, *Ovibos moschatus* Zimm. About ¹/₁₁ ×. (From H. Krieg.)

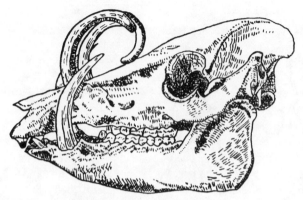

Fig. 3.128. Skull of the Recent babirusa, *Babirussa celebensis* Den. ⅓ ×. (After L. Plate.)

with a positive allometric increase in the size of the tusks (figs. 3.158–60). This, too, is certainly a case of an animal that is not destined for further ascendant evolution.

Phyletic Increase in Size

A special case of an orthogenetic trend is size increase during the course of evolution. This may even be the essential, central process of orthogenesis and one that contributes to and effects at least a portion of the other phenomena.

The corresponding "rule" was proposed years ago by Bronn, Gaudry, and Deperet. It holds that every lineage begins with small forms and that body size increases progressively during the course of evolution. Here, too, as with the orthogenetic development of individual organs, the size increase usually ulti-mately exceeds the limits of what is physiologically and ecologically tolerable; all sense of proportion is lost.

We were able to show evidence of phyletic size increase several times during the evolution of the cephalopods (figs. 3.21, 3.41). Particularly impressive are the examples offered by the vertebrates. Mesozoic mammals, without excep-tion, were dwarfs; not until the later Tertiary and the Diluvian [Pleistocene] were large forms produced. Other examples: as the *titanotheres,* an extinct group of ungulates (fig. 3.129; pl. 31B), evolved from the Lower Eocene to the Lower Oligocene the body size increased five times, and in the Oligocene *Brontops,* which is not illustrated, the shoulder height is seven times that of the Eocene *Lambdotherium!*

The genus *Eohippus,* of the Lower Eocene of North America, which stands at the beginning of the horse lineage, had a shoulder height of twenty-five centi-meters and was the size of a cat. The subsequent forms were, in order, the size of a fox terrier and then of a sheep before gradually attaining the size and pro-

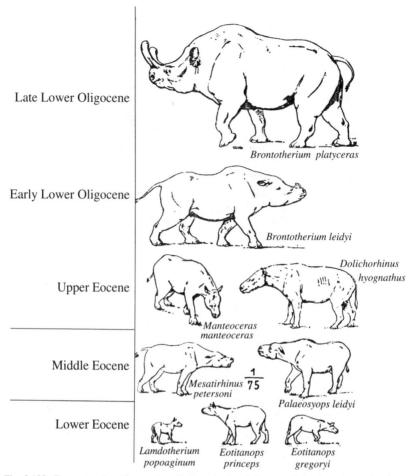

Late Lower Oligocene

Brontotherium platyceras

Early Lower Oligocene

Brontotherium leidyi

Dolichorhinus hyognathus

Upper Eocene

Manteoceras manteoceras

Middle Eocene

$\frac{1}{75}$

Mesatirhinus petersoni

Palaeosyops leidyi

Lower Eocene

Lamdotherium popoaginum

Eotitanops princeps

Eotitanops gregoryi

Fig. 3.129. Reconstruction of a number of stages in the phylogeny of the titanotheres showing a progressive increase in body size. All about ⅟₇₅ ×. (After H. F. Osborn 1929.)

portions of the modern horse. A yardstick for the increase in size is provided by the series of skulls of some horse forms, shown to scale in figure 3.109, and by the reconstructions in figures 3.130–35. A further example is the evolution of *camels,* which also begins with dwarf species about the size of a rabbit, miniatures of the Recent representatives (pl. 27A).

In these examples, the size of the body increases with the approach to modern times, and the Recent forms, as the provisional terminal stages of the lineages in question, are the largest of their respective kinds. And yet, this is by no means always the case; extremely often it is just the opposite, that extinct, ancient ani-

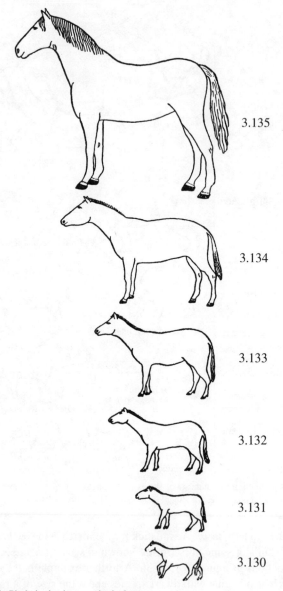

Figs. 3.130–35. Phyletic size increase in the horse.
3.130. *Eohippus*. Lower Eocene.
3.131. *Orohippus*. Middle Eocene.
3.132. *Mesohippus*. Oligocene.
3.133. *Merychippus*. Miocene.
3.134. *Pliohippus*. Pliocene.
3.135. *Equus*. Recent.
All much reduced. (After R. S. Lull, redrawn.)

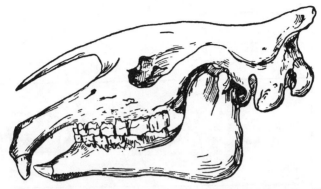

Fig. 3.136. Skull of *Baluchitherium grangeri* Osb. Upper Oligocene of Mongolia. About ¹⁄₁₅ ×. The length of the skull of this specimen, which belongs to the American Museum of Natural History, in New York, is 128.6 centimeters! (After Granger and Gregory, from O. Abel.)

mal forms are characterized by unusual size, and the layman is indeed inclined to imagine these, without exception, as gigantic monsters.

In fact, we know that among extinct tigers, bears, elephants, rhinoceroses, and so on, there are some extinct species that were considerably larger than those living today. A particularly conspicuous example is the mighty *Baluchitherium,* from the Oligocene of Asia, which is assigned to the *rhinoceros* group even though (like most ancient rhinoceroses) it has no horn on its nose (fig. 3.136). The shoulder height of this animal comes to about 5.3 meters, and the length of the torso is as much as 10 meters, making it one of the largest terrestrial mam-

Fig. 3.137. Reconstruction of *Baluchitherium,* to show the enormous size of this animal compared to that of the Recent Indian rhinoceros shown at the left. (After H. F. Osborn, from O. Abel 1939 [*Tiere der Vorzeit*].)

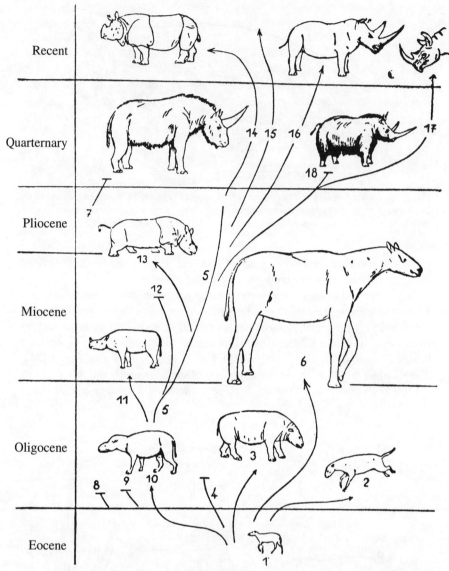

Fig. 3.138. Phylogeny of the rhinocerotids with reconstructions of a few typical representatives: (1) *Hyrachyus;* (2) *Hyracodon;* (3) Amynodon; (6) *Baluchitherium;* (7) *Elasmotherium;* (10) *Caenopus;* (11) *Aceratherium;* (13) *Teleoceras;* (14) *Rhinoceros* s. str.; (16) *Ceratotherium;* (17) *Diceros;* (18) *Coelodonta.* The illustration is intended to show both the phyletic increase in size of the rhinocerotids and the phylogenetic position of the enormous animals of the genus *Baluchitherium,* which is the terminal member of a lineage that died out in the Miocene. (After Osborn and Berry, from W. O. Dietrich.)

mals that ever lived. The enormous size of this animal is clearly seen in the comparison of a reconstruction of *Baluchitherium* with the Recent Indian rhinoceros, both shown to scale in figure 3.137.

These examples, however, by no means represent a contradiction to our rule of phyletic increase in size, for the extinct forms in question are not the immediate predecessors of the smaller Recent species. They are only members of a broader, related group within which they represent the terminal forms of extinct collateral lines (fig. 3.138). To this extent, they thoroughly confirm the general rule that gigantic forms mark the end of evolution.

Unquestionable examples of a once-attained body size being secondarily reduced are almost unknown except in instances where such a reduction is succeeded by a thorough remodeling to a completely new typal structure, which, itself, begins again with small forms. The exceptions occasionally cited are probably only apparent, for in those cases it has not been shown that the forms with the supposed reduction in size really issued from larger ancestral forms of *the same genetic lineage;* only in such a situation would our rule be contradicted.

Accordingly, the evolution of size is, in general, *irreversible.* However, it is immediately clear that gigantic forms are indicators of dying lineages, for ultimately a point would be reached beyond which continued increase in size would be impossible for physiological reasons.

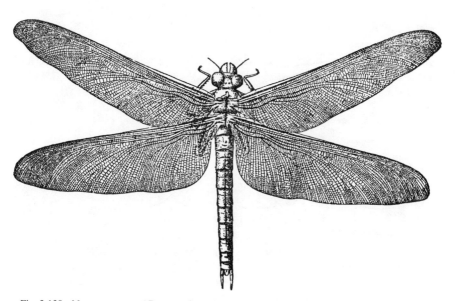

Fig. 3.139. *Meganeura monyi* Brongn., from the Upper Carboniferous of France, a member of the Protodonata, the ancestors of modern dragonflies. The wingspan of this largest of all known giant insects was ¾ meter. (After A. Handlirsch, from E. Stromer von Reichenbach 1909.)

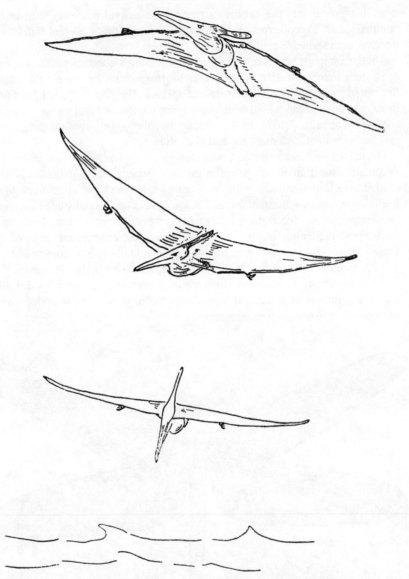

Fig. 3.140. Reconstruction of *Pteranodon ingens* Marsh. Above: In flight. Below: Fishing above the water. (After D. von Kripp.)

The gigantic species of Carboniferous *insects* (fig. 3.139), with a wingspan of 70 centimeters, had clearly not yet reached this upper limit, here established purely mechanically by the relationship between the increasing weight of the chitinous skeleton, which served as the support structure, and the expenditure of energy necessary for flight.

The *internal* skeleton of vertebrates lends greater stability, allowing for correspondingly larger flying forms. The mightiest flying creature of all times is *Pteranodon,* from the Upper Cretaceous of North America (pl. 27B and fig. 3.140). With a wingspan of as much as 8 meters, this reptile by far surpassed the Recent albatross and the great pelican, with their wingspans of 4 and 5 meters respectively. But all of these vertebrates with flapping wings are naturally subject to the biomechanical limits imposed by factors of surface loading, the necessary musculature, strength, and feeding. In *Pteranodon,* these components are still in balance, as has been shown mathematically. Further increases along this path, however, were impossible; *Pteranodon* died out without descendants.

* * *

A certain increase in body size is undoubtedly advantageous for animals. It represents a more robust constitution, increases absolute speed of locomotion, and expands the radius of action, thereby improving the chances for getting food as well as escape from possible enemies; and of course, enemies are also more easily resisted or overcome by greater physical strength.

Furthermore, individuals of large species generally have a longer life span than those of small species. Consequently, they can reproduce more often and can accumulate and exploit more experience in the struggle for existence. Finally, metabolism is more efficient in large forms than in small ones, as Bernhard Rensch has recently shown for various organs (intestine, kidney). In particular, the heat exchange of warm-blooded animals improves with increase in size, for the volume is tripled whereas the surface, where heat is given up, only doubles in size.

Once the development is introduced, however, it never stops but continues unswervingly, and when a particular optimal size has been exceeded, many of the advantages become disadvantages: the animals become ponderous and ungainly, and their mobility decreases. As a consequence of their conspicuous size, they are not able to hide from predators and are no longer able to flee swiftly, with the result that if these slowly moving mountains of flesh have no defensive weapons of their own, they are at the mercy of their enemies. Absolute strength, then, is of less importance than the superior maneuverability of the attacker.

Furthermore, growth is retarded and fertility is reduced. Large mammals have few offspring; in general, only one or two young are born at once, and the length of gestation and of the period of parental care is sharply increased. Ultimately, retarded reproduction causes the reproductive rate to drop below the death rate. Then, too, gigantic forms have enormous requirements for food; should the climate change and the vegetation be affected, herbivores will experience difficulties in getting enough to eat. All these reasons contribute to inexorable decline and seal the fate of gigantic forms.

The enormous *endoceran* shell we mentioned earlier (fig. 3.141*a*), which grew to a length of 5 meters and had to be dragged behind the animal, was undoubtedly a considerable hindrance to movement and had already exceeded the maximum tolerable size. Similarly, *Cretaceous ammonites,* with their shells of 2.5 meters in diameter (fig. 3.141*c*), and the final *eurypterids,* which, having

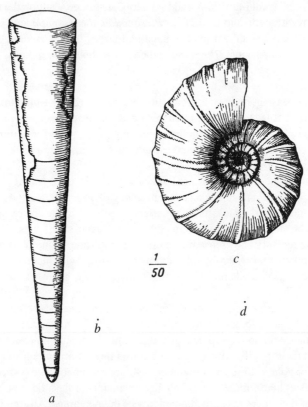

$$\frac{1}{50}$$

c

d

b

a

Fig. 3.141. Phyletic size increase in cephalopods. *a* shows a very schematic Ordovician *Endoceras;* in contrast, the oldest representative of the Nautiloidea (*Plectronoceras*), shown at the same scale (¹⁄₅₀), is represented by the size of the dot above the letter *b* and even exaggerated at that. *c* gives an idea of the size of the giant ammonite *Pachydiscus,* from the Upper Cretaceous of Westphalia, compared to the average size of the oldest Devonian ammonoids, the dot above *d*.

only a chitinous skeleton, reached a length of 1.8 meters before their extinction, represent only an apparent high point in the development of strength; otherwise, the fact that they did not persist would be incomprehensible. Beyond doubt, the inevitable, orthogenetic course of size increase had led them to overshoot size proportions that were biologically advantageous.

*   *   *

A similar situation in the reptilian group of the *dinosaurs* provides particularly impressive evidence. These creatures start right out in the Triassic system with fairly imposing forms and by the end of their evolution, in the Upper Jurassic and Lower Cretaceous, have gone on to become the largest terrestrial animals of all time. In the paleontological exhibit at the Museum of Natural History in Berlin there stands one of the most valuable showpieces of any natural history collection—the almost complete skeleton of one of those giants, from the Upper Jurassic of our former colony of German East Africa [Tanzania]; the specimen was excavated and prepared under the direction of Werner Janensch.

The genus is *Brachiosaurus,* which, according to the largest remains found thus far, reached a length of 25 meters and a height of 13 meters as measured, with the remains in a lifelike position, from the ground to the top of the minuscule head. The specimen on exhibit is somewhat smaller than this; it was evidently not yet fully grown. Its length is 22.65 meters and its height 11.87 meters; its live weight has been estimated at forty metric tons. Our plate 28, which I owe to the kindness of Professor Werner Janensch, shows the skeleton and its proportions in a completed, free-standing mount; an even better impression of the enormous size can probably be gained, however, from plate 29, which shows the skeleton, surrounded by scaffolding, being worked on.

These gigantic dinosaurs unquestionably had the largest appetites of any vertebrates and were probably also the stupidest that nature ever produced; they were completely and disproportionately specialized for size and weight. The skull is so small as to be a caricature, and the brain has the smallest relative volume of any vertebrate; Werner Janensch estimated that in our *Brachiosaurus,* the brain constituted 1/200,000 of the total body weight.

Here, in contrast, are comparative figures taken from some other large vertebrates: living elephants, 1:375 to 1:560; the hippopotamus, about 1:3,000; whales of the genus *Balaenoptera,* 1:10,600 to 1:22,700. The tuna has a very small brain, the ratio being 1:37,400, but this value is many times more favorable than that of *Brachiosaurus.* However, these dinosaurs did have, in a manner of speaking, a second "brain": an enlargement of the spinal cord in the sacrum to provide nerves to the huge hind legs and the tail—in the interests of locomotor functions. This "sacral brain" was several times larger than the brain in the head; motto: if you're not smart, you'd better have good legs!

It hardly needs to be expressly pointed out that the excessive size of the dinosaurs entailed the disadvantages already cited. These were extremely ponderous, clumsy animals and probably did not find it easy to get enough food to fill their bottomless stomachs. Furthermore, their dull wits and lack of defenses made them the underdog against more agile enemies.

Moreover, special measures were necessary to support the mighty weight of the body. Large quadrupeds need a relatively larger skeleton than small ones because the body mass grows to the third power whereas the bone surfaces and diameters required to serve as anchors for the musculature and support for the whole, which determine the static value, increase only to the second power. The expanding musculature of locomotion increased the size of the skeleton, which in turn made a more extensive musculature necessary, and so on: a never-ending spiral! However, since bones, because of their weight, cannot be enlarged infinitely, there is a limit to the weight they can bear and to further increase in size.

Brachiosaurus clearly reached this limit, for many quite conspicuous economies of material and weight in the development of its bones can be recognized. For example, the cervical and dorsal vertebrae, massive in other animals, here consist of coarse-meshed, spongy bony tissue and, in places, of almost paper-thin lamellae. This pneumatization reduced the weight of the bones to a tolerable range, but only at the cost of strength. Thus, disintegration of the bone past the point observed in *Brachiosaurus* is ruled out. It is quite likely that these gigantic forms were no longer able to carry their enormous body weight around on land; they lived like amphibians, submerged in swamps and lagoons, where water lightened the load and made locomotion easier.

A pronounced thinning of the bones to reduce weight also appeared in the gigantic pterosaur *Pteranodon,* but here, the bones lost considerably in resistance. In this connection, we should also mention the extreme porosity of the skull in large elephants, this pneumatization being related to the excessive development of the tusks and the trunk.

*   *   *

With regard to our rule, it could easily be objected that if it were valid, the world of modern organisms would have to consist only of large to gigantic plants and animals, which clearly is not the case. One might perhaps speculate that truly small organisms would no longer exist at all because during the course of their long evolution, they would have grown larger.

However, the only extremely large modern mammals are, basically, the whales and the elephants. There is no doubt that what is true of all gigantic forms is true of them: they are highly specialized groups and stand at the end of their respective evolutionary branches; they do not have a long life ahead of them. Therefore, it would be incorrect to assess the modern organic world only on the basis of the occurrence of *absolute* gigantic forms. The yardstick is only *relative* size—the size of individual species of plants and animals as compared with that in their direct ancestral lineages. From this point of view, the only one appropriate to the rule, it would probably always turn out to be true that Recent species are relatively the largest forms within the lineages leading up to them.

On the other hand, true gigantic forms have, as we know, only a limited life span and belong to collateral lines that will soon become extinct. It is not surprising, then, but rather an obvious consequence of our rule, that forms exhibiting gigantism appeared in much earlier periods and did not continue into modern times; examples are the treelike Carboniferous club mosses and horsetails or the examples discussed above.

Individual lineages evolve large forms *at various times.* Thus, among the ammonoids we occasionally see gigantic forms appearing in individual collateral lines of Paleozoic clymenians, Triassic ceratites, and Jurassic ammonites; such forms represent the typolytic final stages. Nevertheless, the rule of increase in size also holds true here, both for individual lineages and for the entire stock, in that the average size of the Triassic ceratites exceeds that of the Paleozoic goniatites and clymenians, and that in turn, the mean size of ammonite shells of the Jurassic and the Cretaceous surpasses that of Triassic forms.

It can be determined quite generally *that gigantism is tied to an active evolutionary temperament and rate.* This can be seen *within individual lineages:* extremely pronounced increase in size and gigantic forms are found only in the typolytic final phase of short-lived collateral lines, which evolve rapidly, specialize extensively, and then soon become extinct.

In contrast, the conservative lineages, undifferentiated and slower to change, from which those collateral lines derive show only a gradual, insignificant increase in size. The persistent forms from stocks that otherwise evolve more actively, with which we have already become acquainted—*Lingula* (figs. 3.63, 3.64); *Limulus* (pl. 4B; fig. 3.62); *Triops* (fig. 3.69)—and so on, are scarcely larger in the present than they were at first. It is the same with the ancient snail genus *Pleurotomaria* (fig. 3.15; pl. 30A), which has shown only a trivial increase in size since the Triassic; further, the tuatara *Sphenodon* (pl. 31A), the most primitive reptile and sole surviving representative of the old, conservative lineage of the rhynchocephalians, has grown to only a moderate size, not in the least comparable to the extinct, excessively gigantic forms of reptiles.

Thus, the rule for increase in size does not maintain, as is incorrectly interpreted and then applied as a counterargument, that the last representatives of dying groups of animals must necessarily be characterized by unusual body size and extreme specialization. As Ernst Stromer (1944) has correctly pointed out recently, it is often the small, unspecialized forms that survive, relics of ancient animalian stocks that have become sterile.

Contrary to what Stromer believes, however, it should be kept in mind that the overspecialized, oversized lineages of these groups are *always* destined to die out and that it is in these very features that *the reason for their extinction* is to be sought. They die out without descendants; the survivors arise from other lineages that remained unspecialized! This state of affairs does not show up in the schematic tables that Stromer published, which, in that respect, present a skewed view of things.

Furthermore, the phyletic connection determined above is valid with regard to the rate of evolution for the *different major lines of organisms:* only the stocks with rapidly progressing evolution, such as the cephalopods, reptiles, mammals, pteridophytes, phanerogams, and so on, show a pronounced increase in size; these are the lines that produce the actual gigantic forms. The more primitive groups, however, with their slow evolutionary tempo, remain small and exhibit a scarcely perceptible increase in size as they approach the Recent; examples are the foraminiferans, radiolarians, sponges, diatoms, and so on. This is the reason that there are still small plants and animals living today.

There are indeed exceptions. Even the *foraminiferans,* whose average size has remained almost invariable, have produced genuine giants of their kind within certain specialized collateral lines, which underwent highly intensified evolution and brief periods of radiation. As an example, in spite of their *absolute* small size, the old Tertiary nummulites (fig. 2.53), with an average test measuring as much as 120 millimeters in diameter, must be mentioned; or as another example, the cigar-shaped loftusians of the Upper Cretaceous, which reach a length, or better, a width, of 100 millimeters, and whose spiral coils would measure about 1 meter in length if unwound.

The *stony corals* are also one of those groups of animals with a slow evolutionary tempo. Correspondingly, they exhibit no obvious increase in size; the average size of corals living today is probably not much greater than that of the Paleozoic representatives. Furthermore, corals have not produced any forms that are definitely gigantic, either in the present or in the extinct lineages of the Paleozoic pterocorals.

Nonetheless, many Recent stony corals, such as the fungians (fig. 3.142*b*), grow to sizes that are not to be found in the Paleozoic. The more or less round, disk-shaped corallite of this genus, which lies flat on the sea bottom, has a diameter of between 30 and 40 centimeters and perhaps even more. In contrast, Paleozoic forms (*Palaeocyclus, Microcyclus,* fig. 3.142*a*) that are similar in external form and bionomically equivalent show a diameter of only between 10 and 20 millimeters. This is a considerable difference in size. A direct comparison of the two forms, however, is not justified because they are not in a direct line of descent. We can only determine generally that isolated forms of living corals are of a size that we have not yet found in their Paleozoic predecessors.

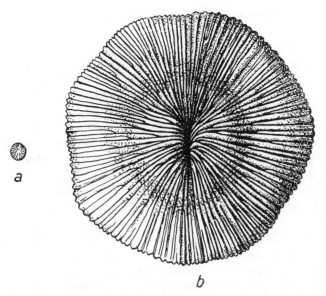

Fig. 3.142. Comparison of the sizes of the disk coral *Palaeocyclus,* from the Gothlandian (*a*), and the morphologically and ecologically equivalent Recent coral *Fungia* (*b*). Both ¼ ×.

The Cause of Increase in Size

From all that has been said, it should emerge well enough *that size increase is a genuine evolutionary phenomenon with unmistakable connections both to the overall character of the individual lineages and to their different evolutionary phases. Thus, size increase and gigantism are not brought about by chance and at random by any environmental factor whatsoever.* If such were the cause, then at particular times gigantism would have to be a widespread, general phenomenon, or at least affect the closest relatives of the gigantic forms. This, however, is not the case.

While dinosaurs were bringing forth real monsters of size during the Upper Jurassic and Lower Cretaceous, the primitive mammals living side by side with them were extremely small. Other examples: In contrast to the titanotheres, which, in the Lower Oligocene, had already reached the end of their evolution with gigantic forms, the precursors of elephants and horses appear during the same age with decidedly small forms; and in the Ordovician, the enormous endoderans lived side by side with the rest of the nautiloid lineages, all of which were normal in size.

It was first thought that *climatic influences* could explain size increase. In modern animals, such climatic relationships are known among *the warm-*

blooded animals. This has been expressed in Bergmann's rule, which holds that the larger forms within individual species, and sometimes groups of species and genera, are inhabitants of cooler regions whereas the small representatives are found in warmer climates. This size differential is usually attributed to climatic selection because the larger forms have a relatively smaller surface compared to body mass and are therefore at an advantage with regard to heat balance. The validity of this rule and its explanation through selection is indeed not unchallenged, but let us assume for the moment that it is justified.

Bergmann's rule could then be applicable not only spatially and geographically for a particular unit of time but might also be applied to a temporal *sequence,* as Bernhard Rensch, in particular, has done. Accordingly, the phyletic size increase of different groups of mammals would be attributable to a gradual cooling during the Tertiary. However, closer examination shows that the various lineages of mammals are by no means uniform in arriving at their maximum size at the end of the Tertiary or during the subsequent Ice Age; this happened, rather, at very different times: for titanotheres, during the Lower Oligocene; for certain collateral lines of rhinoceroses (*Baluchitherium*), also in the Oligocene; for elephants, in the Diluvian; and for horses and camels, in modern times.

One might perhaps object that we are naturally not talking about the *absolute* temperature of the various units of time but the *relative* temperatures in their overall context; it would not then be contradictory that the large forms of the individual lineages appear at completely different times. But even if we assume that, during the Tertiary, within a particular geographic area there was a progressive, undeviating decline in temperatures, the temperature difference between the Eocene and the Lower Oligocene, that is, the period during which titanotheres evolved, was so slight (annual mean about one degree Centigrade!) that the enormous increase in size cannot be attributed to it.

Added to this is the fact that titanotheres, horses, and so on by no means evolved within a narrowly restricted habitat. Rather, their size increase, analogous and linear, proceeded in several parallel lineages living on different continents (North America, Europe, Asia) and in varied biotopes; in other words, each lineage lived under quite different climatic conditions, for during the Tertiary, the climate was not at all the same everywhere on earth.

Therefore, we believe we must reject a causal relationship between the temperature trends during the Tertiary and the evolution of size in mammals; the connection with the Tertiary seems, rather, to consist only in the fact that the evolution of placental mammals took place during this period. The times at which individual growth maxima were reached, however, *were based on internal, phyletic grounds, were dependent upon whether the lineages were evolving slowly or rapidly and whether they were conservative stocks or excessively evolved, short-lived collateral lines,* which entered their typolytic final phase at different times depending on their phylogenetic position.

Within a given unit of time, it might well have been that there was a certain parallel between climate and body size, for example, in races of a single species living under different climatic conditions; the principle of phyletic size increase, however, is supraordinate to that. *It is superimposed upon the individual trivial fluctuations that occur within its framework and is independent of the climatic changes taking place as it progresses.*

Moreover, Bergmann's rule is valid, if at all, only for warm-blooded animals and not for *cold-blooded* ones, making a corresponding explanation for the giants among insects, clams, cephalopods, reptiles, and so on, impossible. In these animals, the opposite is true, for the large forms are mostly found in warmer areas, in tropical lands and seas, although isolated giants also appear in cold climates. Furthermore, in marine animals, the salinity of the ocean also plays a large role; a decrease in salt leads to dwarfism and thin shells in mollusks and to a considerable size reduction in fishes.

During the Late Cretaceous, in the warm mediterranean sea [Tethys], which extended into the region where the Alps are today, the *hippurite* group of clams, highly and curiously specialized, produced shells one meter long and more in a riotous burst of evolution, whereas in the North German deposits of colder waters only isolated, undersized forms were found.

Similarly, the giant specimens of *nummulites* (fig. 2.53) already mentioned appeared only in the warmer Eocene seas; in contrast, the occasional representatives that extended as far as northern Germany are much smaller, with a diameter of two to three millimeters. Here, the main factor is dependence upon climate and other life conditions. But this explains only the *accidental fluctuations in size within individual species and genera of giant forms,* in other words, only the degree of external realization of the gigantism that already lies in the genes; *left unexplained, however, is the origin of gigantic forms themselves and the increasing enhancement of size independent of any climatic change.*

Moreover, indirect climatic effects have been assumed, *by way of the plant world:* luxuriant vegetation is said to have caused the rise of the large herbivores. Unquestionably, however, there is *no causal relationship and not even any conditional one.* Ernst Stromer has pointed out correctly that such large forms as the ostrich and the camel live and thrive today in semideserts and deserts, where vegetation is extremely sparse.

*   *   *

Isolation of the habitat has also been considered as a factor in the evolution of size. According to this explanation, the occurrence of an animal on an isolated island presupposes a lack of enemies and a reduction of the danger from epidemics. Such a habitat would therefore guarantee undisturbed evolution and favor the development of large size. Many examples can be shown: the gigantic birds of New Zealand (Dinornithidae, Moas, pl. 30B) and Madagascar (Aepyornithidae), which became extinct a few centuries ago, the giant tortoises of the Galapagos Islands and Tenerife, the giant monitor lizard from Komodo (pl. 27C), and so on.

These bits of evidence, however, are not conclusive; the situation is by no means one of inevitable, unambiguous relationship and a pattern of phenomena. There are many counter examples and arguments against this view:

1. Huge flightless birds and tortoises are also found in the Tertiary of *mainlands* (Europe, Russia, Egypt, India, China, and so on); for these areas, the advantages assumed for islands would not apply. And monitors even larger than the one already cited have been found in the Pleistocene of Australia.

2. On the other hand, there are also typical *small forms* to be found on islands, such as the Javanese island, or Sunda, tiger, the Bali tiger, the island deer, the dwarf donkeys of Sardinia and Malta, the Diluvian pygmy elephants of Malta, and on and on. R. Mertens has recently added new examples of giants and dwarfs from the same related groups (lizards, skinks) occurring side by side on many islands. If one of the forms has since died out, it is sometimes the giant and sometimes the dwarf. Both increase and decrease in size are thus to be encountered in the same isolated island habitat, and both are favored by rapid, divergent evolution in small populations.

3. The absence of hostile predators, of course, is *not the cause* of size increase or of any other kind of overspecialization and degeneration. In this attempt at explanation, such enemies could only appear as agents of elimination *after* the evolution of gigantic forms had been initiated by other causes. The lack of predators on oceanic islands would only explain why large types have been preserved and could grow larger there and not elsewhere, which is not consistently the case anyway.

4. The *oceanic depths* are also rich in gigantic forms, and for this environment, none of the conditions that are considered critical for the occurrence of gigantic forms on islands applies. There is no limitation of range; the deep-sea fauna is cosmopolitan, and the potential for extensive panmixia exists. Furthermore, there is certainly no lack of enemies, dangers, and pursuit. As F. Doflein, in particular, has shown, the struggle for existence rages at the bottom of the sea with the same intensity as in other realms.

5. Finally, it should be pointed out that for the examples of phyletic size increase presented above, the conditions of island life do not apply at all. *Therefore, those kinds of environmental effects must be ruled out as explanations for size increase;* we must look around for other factors.

* * *

As is well known, growth in individual humans and animals is regulated by the *hypophysis,* an outgrowth of the interbrain. In humans, an enlargement of the main lobe of this endocrine gland by a particular tumor causes, in addition to certain metabolic disturbances, an increase in growth, especially of the distal parts of the body (acromegaly of the hands, feet, lower jaw, nose, and so on). Even childhood gigantism is connected to an enlargement of the hypophysis. This suggests that such a connection should also be sought concerning the size increase of prehistoric giants, and in fact, something of the kind has been posited for the gigantism of dinosaurs.

The *Brachiosaurus* we have already described as being the largest of these dinosaurs has, in fact, a hypophyseal fossa, which, compared with the minuscule brain, is quite large. But with respect to the overall body weight, it is not at all overlarge and therefore cannot be the cause of the gigantism. According to Janensch, its volume is only about $1/1.2-2,000,000$ of the body weight, an index considerably smaller than, say, that seen in the dove, the domestic chicken, and the domestic duck, which show values of $1/23,000$; $1/100,000$; and $1/200,000$, respectively. In addition, not all large dinosaurs have such a large hypophyseal fossa, and in many Recent gigantic animals, the whales of the genus *Balaenoptera*, for instance, the structure is conspicuously small.

But apart from all that, there are many other sources of error in the assumed connection between the size of the body and that of the hypophysis, as Tilla Edinger, in particular, has pointed out. It is not even known to what extent the hypophyseal fossa was filled by the hypophysis and what the relative contribution of the individual lobes was. In humans, gigantism is not connected just to a simple enlargement of the hypophysis but is caused by a very particular kind of tumor in the anterior lobe. Other kinds of tumors and large cysts, on the contrary, can diminish the secretions of the anterior lobe, with the result that an enlargement of the hypophyseal fossa is coupled with dwarfism!

Furthermore, it is doubtful that pathological gigantism is heritable. It does indeed appear that the tendency to hypophyseal disturbances is inherited; but they manifest themselves in different ways in each generation, and only sometimes as gigantism. The fundamentals upon which this whole approach is based are thus as yet quite uncertain and little understood.

It appears to me ultimately that a pattern of phyletic size increase can scarcely be placed on a level with these kinds of pathological phenomena in individuals and with the fluctuations in size contingent upon them within a given species. What we have already said with regard to climatic influences is also true here: phyletic size increase exists at a level superimposed upon innerspecific variability, and it runs its linear, orthogenetic course independent of the lesser phenomenon, outplaying it, one might say. *That this path should be introduced and furthered by the continued selection of pathological individuals is, however, highly unlikely indeed.*

* * *

Finally, there is yet another possible cause of gigantism that should be mentioned: *polyploidy,* an increase in the normal complement of chromosomes in the individual cells. We know that polyploidy (and pseudogigas development, an enlargement of the entire genome, as well) occurs frequently in the plant world, in the giant races of our cultivated plants, for example. Nevertheless, there exists no unconditional, inevitable connection between an increase in the number of chromosomes and body size. There are polyploids that do not differ in size from forms with the normal complement of chromosomes, and there are others that are even characterized by dwarfism.

On the other hand, according to the present state of our knowledge, polyploidy is extremely rare in the animal kingdom and is probably limited entirely to hermaphroditic forms and those with asexual reproduction. Thus, we can hardly assign evolutionary importance to it, at least for animalian evolution. Recently, the supposition has been voiced that certain large Ice Age snails may have arisen through polyploidy. In principle, the possibility that isolated large forms would develop in this way exists; nevertheless, reliable proof has not been forthcoming.

It is highly unlikely, however, that the phyletic size increase of stocks and lineages headed for extinction would be based on this cause. Polyploidy endows plants (and the few animals in which it has been found) with unusual productivity and resistance. In particular, under unfavorable climatic conditions, polyploid races and species grow better than normal ones. Furthermore, it has been shown that the "living fossils" and relics of prehistory to be found in the plant world today are frequently highly polyploid. Through time, such forms have been able to withstand all adverse environmental influences, completely in contrast to the gigantic forms in the process of becoming extinct, in which this capacity is lacking. Because of this contradictory situation, *it is impossible to use polyploidy in an attempt to explain the size increase of gigantic forms, all of which are on the verge of extinction.*

* * *

There remains the further assumption that size increase is caused by a generalized *selection,* independent of all of the factors cited, *of larger mutants,* for which, as already stated, selective advantages may be assumed. This explanation, however, would be valid only for the portion of evolution extending up to the point when optimal body size was attained. Yet the markedly gigantic forms we know of have unquestionably exceeded this stage by far. They grew to such dimensions that the size of the body was not physiologically beneficial but rather carried with it disadvantage and impairment. It is inconceivable that, after this limit had been exceeded and the first detrimental effects of continued size increase manifested themselves, there would be direct selection for the character of body size.

At this point we can take *genetic pleiotropy* into account, that is, the known fact that individual genes contribute to the development of numerous characters and that, correspondingly, any selection, such as that for body size, would entail correlative changes in several features. It could thus be presumed—and this assumption has, in fact, been made—that after optimal growth had been achieved, the continued process of selection for any biologically advantageous character could secondarily entail excessive increase in size, in a certain sense a side effect of the corresponding, selected mutations.

Then, however, it would indeed be extremely remarkable if the selection of the most varied characters in animals and plants—in all of the individual groups of animals, vertebrates and invertebrates, warm- and cold-blooded forms, marine and continental animals of the various special biotopes—was always strictly bound up with a correlative increase in size, and this, in addition, always in the typolytic evolutionary phase, that is, when the organisms in question also show signs of excessive specialization and degeneration in other regards. Such a rigorous pattern of connection cannot seriously be assumed, and *therefore, there must be, in addition to selection, another directive factor.*

We find it in *orthogenesis, the primary trend of evolution,* which at first yields a normal, beneficial size increase and then, once the optimum is achieved, automatically and inevitably exceeds it and finally, having placed the affected animal at a serious disadvantage, leads to its extinction. *This trend is determined by the initial forms of the individual lineages and by the impossibility of reversal once the evolutionary course has been embarked upon.* Any genetic change in the ancestral form sets up requisites, and these necessarily entail a chain of further requisites, which lie along the path already inaugurated. The next section will attempt to analyze the question of factors in more detail.

Factors in Evolution

Now that we have presented the most essential phenomena and rules encountered in evolutionary processes, we shall devote the rest of the book to summing up, to interpreting what has been discussed and placing it in context.

It is true that many attempts have been made to challenge the right of paleontology to make explanations of this kind. Paleonotology is said to be a *descriptive* science and should therefore limit itself to a purely descriptive presentation of the data. This demand is clearly and totally absurd. "Descriptive science" in itself is a contradiction; if the term is understood in the most literal sense, there is no such thing.

The goal of any natural science is the rational understanding of the physical world and the fathoming of its causations; therefore, either it also concerns itself with interpreting its data or it is not a true science and renounces any deeper understanding of its subject matter.

The gathering of observations, directed more or less by chance, and the purely empirical description of such observations are really just *the preliminary stages* of science. True science does not begin until theoretical concepts are introduced, concepts that are not limited to the observations and individual facts at hand but rather tie them together in a generalized way and expand upon them through predictions of situations as yet unobserved. Furthermore, without evaluation, without any kind of theoretical base position, without a guiding concept, the description of an isolated case or even of a consistent, orderly factual situation is impossible. *Unbiased* research is the goal we strive for; but the demand for research that is without *assumptions* and empty of theory is falsely conceived and unrealizable.

In our case, it is true that the possibility for physiological experimentation is closed to paleontology, with the consequence that research into the problem of the factors of evolution cannot proceed inductively. The main task of paleontology is, at first, to establish *patterns of evolution,* thereby providing physiologists and geneticists with actual facts, the causative circumstances of which are to be investigated in living material. However, in doing this, paleontology cannot simply limit itself to the determination of this or that particular line of descent; this would be of no use to further biological research. Rather, individual observations must be tied together through comparison and evaluated theoretically.

The result is a *general, comparative history of lineages,* which must first be evaluated judiciously before the question of cause can be broached. It should be emphasized repeatedly that only paleontology, with its historical documentation, can give information about how things used to be and how the modern organic world came into being. The sciences of genetics and developmental physiology, which build upon Recent organisms, are not in a position to do this because of their nonhistorical character, because the element of time is missing; they can only establish the causes of trivial morphological transformations evoked through experimental intervention in living organisms. These sciences can also make further deductions as to how natural evolution *may* proceed, but can never reveal how evolution actually did proceed through geologic time.

Moreover, perceptive zoologists and botanists have given full recognition to this really quite evident state of affairs. For instance, L. S. Berg wrote, "The deciding vote in the question of the course and causes of evolution belongs to paleontology." And Richard Hertwig stressed, "There should be no conflict between any speculations on evolution derived from the investigation of living organisms and the positive results of paleontology." These words are all the more remarkable in that Hertwig himself considerably overemphasized the incompleteness of paleontological data and, aside from that, did not have a particularly high opinion of our fossil record.

Completely inappropriate is the view sometimes expressed that conclusions derived from fossil material are only hypothetical, whereas genetics has a foundation of verified facts upon which to base its arguments. The evolutionary data of paleontology are the product of comparative historical methods, and to the extent that this way of proceeding is carried out with unimpeachable logic and yields unambiguous results, the data must, as we said earlier in reference to H. Dingler, be considered valid as *historical facts,* as having the same degree of reliability as any other historical evidence. In any event, in matters of the course of evolution, it is only paleontology that can claim *factual material,* the actual vehicle of evolution, as its basis. In contrast, the reliability and conclusiveness of genetic evidence extends only to experimentally analyzed mechanisms of microevolution; to apply these unconditionally to evolution as a whole and to assume that they alone are valid is, however, pure hypothesis.

Many recent authors have spoken of *experimental evolution;* there is *no such thing*. Evolution, a unique, historical course of events that took place in the past, is not repeatable experimentally and cannot be investigated in that way. W. Kühnelt (1942, p. 9), pointed out correctly that even "the development of races is a historic process, one that has never taken place twice in the same way and, consequently, is not reproducible. The only factors that are known and accessible to experimentation are some that, as far as can be anticipated, can lead to the differentiation of races."

Wilhelm Roux (1912) has also seen the matter with great clarity. According to him, the task of *phylogenetische Entwicklungsmechanik* [evolution-oriented developmental mechanics] is to investigate the causes of the historical evolution of organisms. Since, however, this process is not one that "repeats itself," genetics can only study the origin of mutations and so on in the living organisms of the present. It would be better, therefore, to speak of a "causational theory of the transformation of organisms." To what extent and with what limitations or possible amplifications the findings of genetics can be applied to the concrete, material data of fossils is the business of paleontology to investigate.

Therefore, to arrive at an overall theory of organic evolution that will stand up, both types of research, genetics and developmental physiology, on one hand, and paleontology, on the other, must cooperate and supplement one another. In the process, they will gain the opportunity for mutual oversight and enrichment. What paleontology in and of itself has to contribute to the question of factors is essentially of *heuristic,* regulative significance. Nevertheless, it appears for the present that as long as we do not yet have a satisfactory, consistent, overall picture of evolution and its causation, the clues yielded by paleontology are quite important in this area, too. They can be used to point out the directions in which experimental answers are to be sought and also the gaps that still exist in our present state of knowledge with regard to the causative factors of the processes by which lineages evolve. It is in this sense that the following comments are to be interpreted.

Lamarckism

In all attempts to elucidate the driving forces of evolution, environmental influences, that is, the physical and chemical peculiarities of individual habitats, have played a large role. Such an interpretation is obvious and downright impressive for the observer who does not know better. We observe that organisms are adapted, that their organizations are in accord with their environment, and that form and function are in harmony with one another. From this situation, the cause of the processes that led to such results was inferred. And since the organisms could not have transformed their environment, the converse must be true—and this was assumed without further consideration—that it was environmental factors that shaped the organisms.

Environmental stimuli and the organic functions they set off were thus proposed as the driving forces of morphological transformation. Geographic and climatic changes in the habitat forced the organisms to new adaptations that were consistent with these changes; they provided the organisms with the "need" to exercise new functions, and these then led, through the use or disuse of organs, to modifications of form and of the entire organization. Evolution is thus seen purely as a reaction to changing environmental influences, and the organism is understood as the plaything of external factors. We usually call this idea "Lamarckism."

Currently, it is probably only among paleontologists that a few adherents to this view can still be found; in modern biology, it has been almost completely abandoned, for the naive assumption that evolution is affected by functional adaptations and reactions directly and appropriately evoked by environmental influences is in conflict with the results of genetics. It has been found that the kinds of reactions that individuals exhibit when conditions in their habitats change are not heritable. As soon as the particular external influences are removed, the changes of form also reverse themselves and the original appearance is reestablished.

Similarly, the results of the exercising of a function, of the use or disuse of organic structures that have been acquired during the lifetime of an individual, are not transmitted through inheritance. These are only individual changes, which do not enter the genome and are therefore reversible. Conversely, genetic changes triggered by environmental influences, such as are induced experimentally through ionizing rays, temperature, or chemical influence, are not adaptations, for they are not directly appropriate, specific responses to stimuli.

Here is an instance of a situation, described in chapter 1, in which paleontology must submit unconditionally to the biological findings obtained through the study of Recent animals and plants, for there is not the slightest likelihood that in the past, processes of genetic inheritance differed from the laws and patterns in force today. On the other hand, an unbiased examination of the fossil material itself also reveals that absolutely no direct response to environmental influences or appropriate adaptations in the Lamarckian sense must necessarily be inferred. Rather, the phenomena are to be interpreted in a completely different way.

Formerly, in emphasizing the supremacy of the environment, the properties and qualities of organisms were unduly disregarded. Yet it should be obvious that in such chains of reactions and complexes of conditions the objects themselves must be credited with critical significance. When I heat two chemical substances together, it is not the rise in temperature but the composition of the original material that is decisive. The rise in temperature only triggers the reaction; under certain circumstances, it can be replaced by a different physical or chemical action (pressure, catalysts), and the result, determined by the original material, will still be the same. At most, the environment plays only a similar role with regard to organisms; *it can only provoke and set in motion some potential that is already present.* Reactions always assume a material that is capable of a particular reaction.

The Environment and Evolution

We have to limit ourselves here to a few examples and references. First of all, claims are made that *tectonic events* are the causes of evolutionary remodeling, of the appearance of new forms and the extinction of old ones: the dewatering of oceanic areas due to the upward folding of mountain ranges or the uplifting of portions of the Earth's crust, and the reverse, the flooding produced as mainlands subside. The history of the earth is one of continuous change in the proportions of land and water, and the evolution of terrestrial or marine organisms is often seen in this context.

For example, it has been maintained that as a consequence of Gothlandian [Silurian] mountain building [Caledonian phase], marine vegetation was left high and dry and acquired thus the potential for adaptation to life ashore. Conversely, the expansion of oceans caused reptiles and mammals to leave dry land and plunge into the sea, where they became adapted to an aquatic way of life. Further, the reduction in the extent of mainlands caused reptiles to ascend into the air, where they turned into pterosaurs and, eventually, birds. For the greater part of known tectonic events, it is not difficult to find for any evolutionary happening a corresponding geological process, especially if one is not too particular about chronology.

Now, all of these notions are quite naive and hardly need serious refutation. The simple *preconditions* for what went on are here confused with the actual *causes.* Obviously, the presence of mainlands is a basic precondition for the development of a continental flora and a terrestrial fauna; oceans have to exist if there is to be a progressive adaptation of reptiles to a marine mode of life. But there have been mainlands, oceans, and atmosphere, habitats of every kind, at every epoch of geological history for as long as the earth has been inhabited by organisms. Individual habitats may have been more extensive at some times and more restricted at others; but they were *present* at all times and extensive enough both to guarantee the continued existence of faunas and floras with particular life requirements and to serve as centers of origin for new kinds of forms with different modes of life.

Proof that any particular geological course of events was the cause of an observed biological event has not been forthcoming. On the contrary, it can be determined that even *prior to* a tectonic process for which claims are made, folding and the relative expansion and reduction of continental and marine areas had always taken place already without yielding the particular organic results. Therefore, when we observe the appearance of terrestrial plants, of marine or flying reptiles, at a particular time, it is clear that there are other decisive reasons for it, namely, those that lie within the structure of the plants and animals themselves. They simply moved to dry land or into the sea or the air when their time had come, which is to say, when their organization had reached a level such that the necessary specialization could take place.

*   *   *

This is certainly the critical point; for if an organism does not already have the aptitude, or certain preconditions, it will never be in a position to alter its mode of life but will simply succumb to the changed environmental conditions. The American paleontologist Alfred Sherwood Romer once pointed out that the crossopterygians, an ancient group of lobe-finned fishes developed organs and characters—protection for the brain and the sensory organs through bony plates; ossified body skeleton; bony skeleton of the lobed fins (figs. 3.143, 3.144)—that are not essential for animals that are purely aquatic, as for example, the modern cyclostomes (lampreys and hagfishes) and cartilaginous fishes demonstrate. Instead, these features represent an obligatory precondition for land-dwelling quadrupeds, which evolved from these crossopterygians and require a supporting skeleton able to bear the weight of their bodies and for locomotion, since water as a supporting element has been dispensed with. The crossopterygians, as ancestors of the tetrapods, already so thoroughly prefigured them that, because of those peculiarities, which seemed incomprehensible in aquatic, swimming animals, Otto Jaekel got the phylogenetic relationship backward and construed crossopterygians as being derived from tetrapods.

Consequently, when we encounter the first tetrapods in the Upper Devonian, the reason is not that mainlands existed or were enlarged, for land masses had also been available in abundance previously, but rather that, by that time, the precursors of the tetrapods had created the morphological preconditions for life

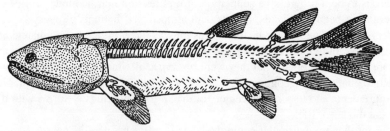

Fig. 3.143. *Eusthenopteron,* a crossopterygian from the Upper Devonian of Canada, with armored skull, bony body skeleton, sturdy, ossified fin supports, and shoulder girdle firmly attached to the roof of the skull. (After W. K. Gregory.)

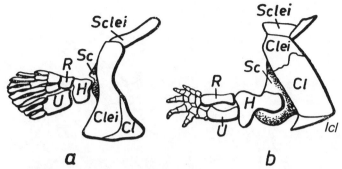

Fig. 3.144. The pectoral fin of the crossopterygian (*a*) as preliminary stage of the forelimb of the stegocephalian (*b*) and of tetrapods in general. An example for *a* is *Sauripterus,* from the Upper Devonian of North America. The diagrammatic model for *b* is *Eogyrinus,* from the European Upper Carboniferous; to make the comparison easier, the limb is shown in a position that would correspond to that in a fish, whereas in reality, it is turned outward and the forearm is bent at an angle against the upper arm. For their crawling mode of life on hand, tetrapods used a structure that had arisen in their ancestors under different life conditions, even though not required for an aquatic mode of life. In a certain sense, therefore, it was a structural advance for use later, for life on land, a mode that only the structure itself made possible. *Sclei* = supracleithrum; *Clei* = cleithrum; *Cl* = clavicle; *Icl* = interclavicle; *Sc* = scapula; *H* = humerus; *R* = radius; *U* = ulna. (After W. K. Gregory and A. S. Romer.)

on land and, remarkably, in an environment different from the one that would later prove to be advantageous and essential for them. Certainly, no mountain-building process of any kind can be considered as the necessitating cause of the transformation.

The same is true for the colonization of land by psilophytes, the most primitive representatives of the ferns. One prerequisite for life on land is the possession of a conductive system to supply the above-ground part of the plant with nutritive fluids. Now, these vascular bundles had already been acquired by the simplest psilophytes, superficially resembling algae, of the type of the *Taeniocrada* shown in figure 3.145, which clearly were still aquatic plants. The rudiments of these conductors of nutrients and their attendant support tissues, so essential later for life and growth, constituted the base that enabled formerly aquatic plants to colonize dry land during the Gothlandian [Silurian] and the Devonian. The incidental enlargement of the continental mass due to tectonic happenings was extremely unimportant to the situation.

Moreover, for vertebrates to persist in life on land (to return to this subject again) lungs are indispensable. These, too, were already prefigured in the swim bladders of fishes. At first, these structures were used as accessory respiratory organs, as the modern dipnoans (lungfishes) demonstrate, and later, when the transition to a purely terrestrial mode of life was made, they were transformed into lungs. In the same way, nasal passages and internal nasal openings (choa-

Figs. 3.145. *Taeniocrada* (formerly called *Haliserites*), from the Lower Devonian of western Germany, as an example of a primitive, leafless, algalike psilophyte, which lived mainly in water but already possessed vascular bundles inside the stem. (After Kräusel and Weyland, from W. Gothan.)

nae), which could serve later in lung breathing, were already developed in the aquatic crossopterygians. *The rudiments of the organs were thus already present before their development for the new function took place.* Another example is provided by Othenio Abel, who pointed out that the ancestors—bottom-dwellers and not yet capable of flight—of certain flying fishes already possessed enlarged fins, which were then used by their descendants for gliding flight.

Moreover, of the originally small "tusks" of elephants, (figs. 3.111 and 3.112) and the initially knobbed "horns" of the titanotheres (pl. 31B, *c* and *d*) and other horned animals—the oldest antlers—none were emplaced *as weapons,* for in their early stages they could certainly not have functioned as such. Therefore, use or disuse was not a factor in their development; they did not become useful until they had reached a certain size. Here, too, the rudimentary morphology was there first, the organ *before* the function, and the function was later exercised *because* the already present organ enabled its owner to do so.

Use does not *create* new organs, for there can be no function without an actual organic base, which means that the organ in question must already be in place before the function can be exercised; the organ does indeed then develop or is transformed concurrently with the new, intensified demands made upon it, and perfected (not, however, according to the naive Lamarckian notion, under the direct influence of use). *Only this perfected organ can be regarded as an adaptation; the generation of a new organ, however, has nothing to do with adaptation.*

Insects and birds fly *because* they have wings; the rudiments of wings, however, are not adaptations for flight. This skewed way of expressing the matter, which one sees occasionally, would mean that the "actual" mode of life of the wingless precursors of insects and birds was flight, and that they strove to attain it without having the necessary equipment. Nor is the appearance of legs an adaptation for walking and running. *The primordia of these organs are present from the outset in the structural design of the animal that will later have them; they are independent of specialization, adaptation, and function.*

In summary, it is universally true that *the organ or, in the broader sense, the form predates and outlives the function: as an unusable primordium or oriment,*[18] *before any function is exercised; and, usually for an extended period, as a vestige, long after functioning has ceased.*

In addition, it should be pointed out that all tectonic episodes—uplift and subsidence, marine transgressions and regressions—do not affect the entire surface of the earth but are always of only limited local or regional extent (fig. 3.146). In the areas where their effect is felt, they of course result in orographic and, consequently, *biogeographic changes,* but nothing more. Naturally, where a sea bottom has been drained, further marine life will be impossible.

Even in such areas, however, marine life, except for sessile plants and animals, seldom dies out. In general, the organisms will have enough time to migrate to unaffected areas far greater in extent than the narrow zones of the earth's crust affected by the tectonics. When an area that was temporarily dewatered is again inundated by the sea a short time later, the organisms that used to live there will migrate back; they will have undergone no changes at all as a result of the tectonic episode and its aftereffects.

Even sudden, local catastrophes have generally had no influence on the course of life, for an abundance of other, intact habitats, where evolution could proceed unaffected, were available. Fundamentally new conditions, such as could and would necessarily provide an impulse for evolution, have not been created by such local events.

The Environment and Extinction

Nevertheless, it is asserted again and again that drastic geological happenings are at least responsible for the extinction of organisms. Usually reference is made to animals living today in a very restricted range, such as the lungfish *Neoceratodus* (fig. 3.147), which lives only in two neighboring rivers in Australia (the eastern coast of Queensland), or the tuatara, *Sphenodon* (pl. 31A), which

18. [Oriment (Abel 1914; Lat. *oriri* = to arise). A structure that is phylogenetically ascendant and will be strengthened orthogenetically. The opposite is "vestige," or "rudiment" in the narrow sense (Lehmann, *Paläontologisches Wörterbuch*).—Trans.]

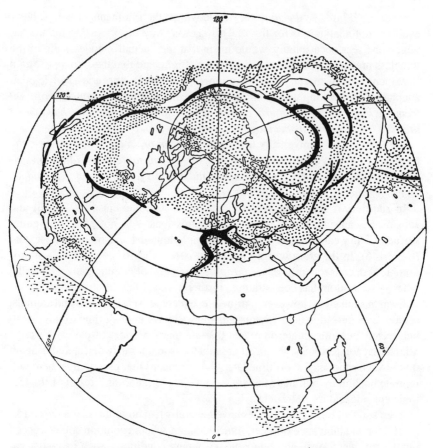

Fig. 3.146. Map of the distribution of land and sea during the Lower Carboniferous (Visean stage), with inclusion of the main Carboniferous folded mountains. Stippled: Ocean. White: Land. Stippled and streaked: Continental deposits of southern lands. Emphasized in black: Mountain ranges. (After F. Kossmat, from Schindewolf.)

Although the Carboniferous mountain-building event is one of the most important of its kind, the changes it occasioned in the extents of land and sea were relatively minor, or, in any event, did not decisively affect the overall picture of the world. Many of the Carboniferous mountain ranges shown here were already in place in earlier times (Gothlandian, Devonian) and were already, to some extent, terrestrial elements. Moreover, the folded mountains are the total result of a whole series of tectonic phases of the Lower Carboniferous, making the increase in the amount of land during the individual temporal units considerably smaller than it appears to be in this paleogeographic map, which summarizes a longer period of time. Individual episodes of mountain-building had, therefore, only a very subordinate effect on the shaping and displacement of life realms overall.

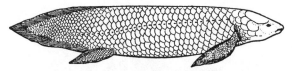

Fig. 3.147. *Neoceratodus forsteri* Krefft. Recent in two Australian rivers (the Burnett and the Mary). Reduced. (From E. S. Goodrich.)

occurs only on a few small islands north of New Zealand. If any kind of geological catastrophe—a draining of the two rivers or, for the islands, a marine inundation or a major volcanic eruption—were to befall these restricted habitats, the last representatives of the ceratodonts, in one case, or of the ancient reptilian group of the rhynchocephalians, in the other, would become extinct.

In these instances, however, it would be thoroughly wrong to say that the extinction of the faunal groups in question was occasioned by such a geological event. To do so would be to confuse the *violent destruction of the last survivors* of those lineages with the *actual process of extinction,* which had already begun much earlier. Both the ceratodonts and the rhynchocephalians were once flourishing groups of forms with, during the Triassic, extensive distribution all over the earth. Then, however, a gradual decline set in—species after species, lineage after lineage, died out, and one distributional range after another was lost—without external environmental factors having been discernibly responsible.

Other examples of a continuous reduction of the ranges of forms that were once cosmopolitan are evidenced by the trigonians (figs. 3.148 and 3.149) and the rhinocerotids (fig. 3.150).[19] The modern, reduced distribution of these forms as well as their representation by a single species, for one, and a few species, for the other, is the result of the process of extinction as it gradually nears its conclusion. A geological catastrophe of the kind mentioned would then be only the external last straw, putting an end to a process that had been inexorably under way for ages for internal reasons. It is the same as when the wind finally topples an old and rotten tree.

This always turns out to be the situation if one examines the existing phenomena more closely and is not satisfied with observing superficial but only apparent correlations. The *ammonite stock* did not become extinct, as is always being asserted, as a consequence of the alpidic orogeny in the Tertiary, for there is no precise temporal coincidence of the two processes. Long before the onset of the major alpidic folding, as far back as the Lower Cretaceous, we encounter extremely varied manifestations of decline (fig. 3.41*d* to *m*), which makes it impossible to regard the tectonic episode mentioned as the true cause of extinc-

19. Another such relict with a sharply reduced geographic range is the conifer genus *Metasequoia,* which was found only quite recently living in central China but which had worldwide distribution during the Tertiary.

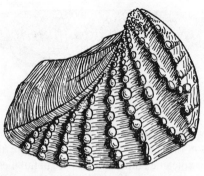

Fig. 3.148. *Trigonia navis* Lam., from the Lower Brown [= Middle] Jurassic of Alsace [eastern France]; a representative of the genus *Trigonia* (group of the Scaphoidae), very common and widespread during the Jurassic but living today only as sparse relicts along the coast of southern Australia (fig. 3.149). 1 ×. (After K. A. von Zittel 1924.)

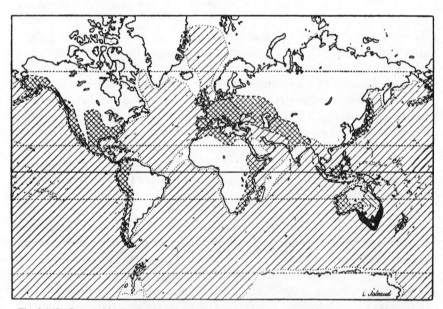

Fig. 3.149. Geographic distribution of the trigonians as an example of a once-cosmopolitan animal group with a modern range that is much reduced. Cross-hatching: Distributional area in the Mesozoic, during which time, beginning in the Lower Lias, the trigonians attained worldwide distribution; but since the Cretaceous, they have retreated increasingly toward Australia and New Zealand. Simple hatching: Oceans connecting the Mesozoic epicontinental life realms. Black triangles: Locality of Tertiary trigonians (*Neotrigonia*) in Australia and New Zealand. Solid black: Modern distributional range along the coast of southern Australia. (After L. Joleaud 1939.)

tion. In addition, it is unlikely that space was a problem, as has already been demonstrated: an important reduction of the area for dispersal has not been observed for Cretaceous ammonites, the last representatives of the stock, and, therefore, it is completely incomprehensible how something that happened in only one region could have had such worldwide results.

Moreover, no corresponding tectonic episode is known that can be held responsible for the widespread mass death of ammonoids at the Triassic-Jurassic boundary. It is a period of crisis for the ammonoid stock analogous to that at the end of the Cretaceous, with the difference that here, a single lineage escaped destruction and went on to evolve further. If tectonics were to be held responsible, one would have to presume a geological constellation similar to that existing later, at the time of the ultimate extinction. Since that is not the case, the

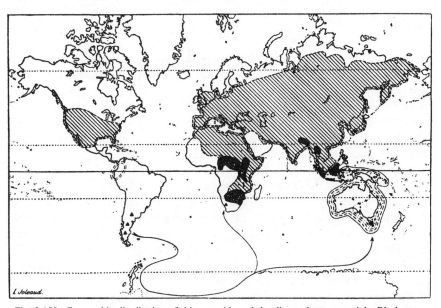

Fig. 3.150. Geographic distribution of rhinocerotids and the diprotodont marsupials. Black areas show the modern disjunct distribution of the *rhinocerotids: Dicerorhinus* and *Rhinoceros* in the Indo-Malaysian region and western Indochina, and *Diceros* and *Ceratotherium* in central and southern Africa. Shown in hatching, for comparison, are the widespread and once-connected areas (Asia, Europe, Africa, North America) of the occurrence of fossil (Tertiary and Quaternary) rhinocerotids, the main and collateral lineages of Recent representatives.

The rest of the symbols refer to the *diprotodont marsupials* (herbivorous marsupials with only two incisors in the lower jaw). The areas outlined in two broken lines (equatorial South America, Australia, and Tasmania) constitute the modern distributional range of individual families, and the Tertiary occurrences from the groups of the Caenolestidae and the Hypsiprymninae are shown as black triangles and squares. The arrows indicate the migration route of the diprotodonts. (After L. Joleaud 1939.)

dying out of the ammonites and the alpidic mountain building are only coincidental, independent, parallel phenomena.

We look in vain for other, discernible influences that the Caledonian or perhaps the Variscan folding might have exercised on the evolutionary course of the trilobites. Trilobites did not become extinct in the aftermath of the Variscan orogeny; rather, long before that, even as far back as the Gothlandian, they exhibit a very gradual decline, one that can be followed step by step. Without any correlation with tectonism, one group of forms after another disappears, and a progressive impoverishment in families, genera, and species can be observed. Thus, the reasons for extinction or continued existence are essentially *internal*—they lie within the lineages themselves. As P. Jensen, C. Zimmer, Karl Beurlen, and other authors believe, the reasons may perhaps be sought in an aging of the germ substance, a gradual loss of function in the sex glands resulting in reduced fertility.

* * *

Othenio Abel has interjected an interesting slant on the question of extinction. The point of departure for his considerations is the very abundant evidence of Ice Age cave bears, which, shortly before becoming extinct, exhibited extremely wide variability and all kinds of manifestations of degeneration. The last representatives showed widespread evidence of severe bone and tooth disease, and of injuries as well, and it has been shown that these conditions affected even young animals, which nevertheless went on to reach an advanced age.

Abel was correct in his judgment that in a species engaged in a strenuous battle for existence, those kinds of vertebral, jaw, and tooth diseases would have worked toward early elimination of defective individuals. Only among animals living under particularly favorable conditions and having no enemies at all could the evident situation occur, that even sick and weak individuals attained sexual maturity and corrupted the species by transmitting their inferior genotype. Abel regarded the degeneration of the species as a consequence of *optimum existence,* and saw it as the major cause of extinction.

This conclusion is quite perplexing, for one would first be inclined to see *unfavorable* living conditions as a source of degeneration and extinction. Abel's interpretation seems to be supported by Nachtsheim's observation based on Recent animals, that when nutrition is good, a higher percentage of pathological individuals appears than when it is poor. On the other hand, the symptoms of the cave bears are strikingly similar to the disease-altered bones that Hansen described from Norman graves in Greenland. There, however, it has been shown that these syndromes are not connected to optimal living conditions but to *adverse* ones.

Consequently, there can hardly be a causal relationship between degeneration and the conditions under which the degenerate organisms lived, and Abel's thinking explains only why the degenerate forms were able to *persist* and to *increase* to an unusually high percentage of the population, and not how they *arose* in the first place.

Here, too, as in our earlier examples, we arrive at the same conclusion, that *the actual causes of degeneration and extinction lie deeper and manifest themselves earlier than any environmental influences whatsoever* that one might at first wish to hold responsible based on appearances. The actual causes reside in the constitution or the genome of the organism in question, in extreme specialization, in a reduction of evolutionary potential, and so on, and take effect at a given time, regardless of whether living conditions are favorable or poor, independent of tectonic episodes and oceanic inundations and recessions.

The Environment and the Tempo of Evolution

Environmental conditions also have no appreciable influence on the tempo of evolution. If changing external factors did constitute the basic, direct driving force of organic evolution, then transformation would have to have proceeded most slowly in those areas of the earth that are the least affected and most constant, namely, at the depths of persistent oceans, whereas in fresh water, where pronounced temperature fluctuations, temporary drying up, and frequent displacement of habitat bring about the most variable conditions with regard to time and space, it would have to have run its course rapidly. This, however, is absolutely not the case; in many respects, even the opposite can be observed.

The fish fauna of modern deep seas includes, in addition to the predominant, highly specialized types of more recent origin, only relatively few ancient forms, whereas in fresh water, there is an abundance of them: *Neoceratodus* (fig. 3.147), *Lepidosiren, Protopterus, Polypterus, Lepidosteus, Amia, Diplomystus,* and other genera to which the first ancient, primitive fishes belong and which, moreover, go back generically unchanged far into the geologic past, some as far back as the Upper Cretaceous; in addition to these, there are representatives of the ancient line of the agnathans. Similarly, freshwater mollusks and turtles have changed little since the Tertiary. For example, according to K. Hummel, the Early Tertiary turtles of the genus *Trionyx* can hardly be distinguished from Recent ones. Many genera of Recent pulmonate snails have been present since the Upper Jurassic; older yet is the malacostracan *Estheria.*

Moreover, we saw earlier, in the freshwater crustacean *Triops cancriformis* (fig. 3.69), a form that goes back to the Upper Triassic, with representatives there that, because they are identical right down to the last detail of form, must be considered conspecific with the species living today. This does not exactly indicate that changing environmental conditions are the cause of the modification of forms and, conversely, that persistent forms are proof of stable, unchanged habitats. *Here, too, there is no doubt that it is the special nature and evolutionary position of the organism in question that is decisive.*

A counterexample is offered by the *reef corals*. Modern coral reefs are tied to

very particular, narrowly delimited life conditions: pure, sediment-free sea water, precisely defined depths, and a closely circumscribed temperature range.
Corals thrive only in areas where these needs are met and will die immediately
if conditions change. Because even in the oldest Paleozoic there is exactly the
same contrast as today between the true, usually colonial, reef corals and certain
solitary corals of other life realms (for example, the oceanic depths, in association with cephalopods), one can probably conclude that even then reef corals
were living under the same conditions as those in which Recent corals are found,
especially since the associated faunas of both ages are similar.

Thus, although it is likely that the environmental conditions for these forms
have remained unchanged throughout time, the forms themselves do not in the
least exhibit evolutionary stagnation. On the contrary, as we saw earlier, they
have undergone a thoroughgoing, albeit slow, transformation of their structural
design and, within this framework, have set out upon numerous different, sometimes short-lived evolutionary paths.

There are, moreover, multiple examples showing that under *the same environmental conditions* evolution has proceeded at *very different rates. Ammonoids,*
throughout their existence in the Late Paleozoic and the Mesozoic, have been
characterized by a heightened capacity for transformation; in contrast, the *clams*
and *snails* associated with them in the same biotope show a considerably slower
modification. This is the reason that certain groups of animals—and ammonoids
are the best example—are used as index fossils in the determination of geologic
time, whereas others from the same beds, that is, from the same former environment, are completely unsuitable for this purpose.

Cetaceans had their period of rampant evolution during the Miocene. Within
this brief period of time, we encounter a surprising abundance of genera and
species revealing broad differences in dentition, skull, and mode of feeding.
Then, from the Lower Pliocene on, the pace of evolution slows considerably. In
contrast, as Othenio Abel has pointed out, the *Sirenia* have evolved very gradually from the Middle Eocene to the Pliocene, at a slow, even tempo, and the
changes that took place during the Miocene are extremely trivial.

Here, too, there are broad differences in the rates of evolution, even though
the external living conditions for whales and sea cows have remained largely the
same since they both began their marine mode of life. *Tempo, extent, and diversity of modification are consequently grounded in internal factors of organic
lineages and are not determined by the environment and its changes.*

The Influence of Climate

Even such drastic climatic changes as those occurring during glacial epochs have
not produced much in the way of far-reaching innovation. The typical Diluvian
[Pleistocene] mammals of Central Europe migrated, some to the far north and

some to alpine regions. However, as W. Soergel has set forth in detail, the adaptations for cold developed by the arctic migrants had already been introduced in the Late Tertiary, although they may have intensified later during the glacial periods.

We owe one special, particularly noteworthy example to the meticulous investigations carried out by H. Wehrli on the alpine marmot, *Marmota marmota,* for which evidence exists as far back as the last interglacial period, which is to say, for about the last twenty-five thousand years. He reports, in summary, as follows:

The modern alpine marmot is the westernmost member of the polytypic species [*Rassenkreis*] of marmots and differs from the related *Marmota bobak,* the steppe marmot of southern Russia, not only in external appearance but also in various osteological characters. Formerly, these differences were generally ascribed to the different habitats (steppe and high mountain). This view was believed to be all the more justified because fossil marmots from the time of the last glacial period do not yet exhibit the osteological differences so clearly. I was able to prove, however, that in spite of that, the marmots of the last glacial are distinctly divergent and that further differentiation did not take place after the retreat to the steppes of southern Russia, in one case, and to the alps, in the other, but during the last glacial, when the alpine marmot extended from the Alpine foreland throughout western Germany and France, in close proximity to the steppe marmot, which occupied Germany east of the Rhine. Therefore, the morphological differentiation occurred in the same habitat, for we may assume with certainty that climate, vegetation, and so on were all identical in the immediate vicinity of both sides of the Rhine. Unfortunately, we cannot say just when alpine and steppe marmots began to diverge, for the oldest fossils from the last interglacial period are very incomplete. They show only that the Alps were already inhabited by marmots at that time. However, since the fossils directly subsequent to those from the last glacial period already show distinct differences, the stage must have been set for the separation during the last interglacial at the latest. Despite the migration from the Alps into the lowlands, the development, once initiated, proceeded and also showed no alteration when the alpine marmot retreated once more to the high mountains. The evolutionary trend, once embarked upon, continued on its inevitable way, in spite of environmental dislocation. For example, during the last glacial period, the Mainz basin surely did not have the same climate, vegetation, and so on, as are found in the high mountains today. It should be noted further that, based on our present knowledge, the osteological peculiarities of alpine marmots (the course of the temporal ridge, determined by a lobe of the tear gland; development of an enamel ridge on the lower premolar, among others) cannot be related to environmental factors.

We must also reject the assertion sometimes made (Matthew) that a dry, cold climate intensifies evolution in terrestrial animals, fostering the appearance of new forms. As Ernst Stromer has already objected, if that were the case, then in times when the climate was warm and equable, during the Jurassic and Cretaceous, for example, slower modification in terrestrial animals would be expected; but in periods with sharp climatic contrasts and temperatures that were often low, as during the Diluvian age, rapid transformation would be anticipated. None of this, however, has been proven: "At least among the dinosaurs of the Middle and Late Mesozoic, new forms kept appearing in abundance, whereas for the Diluvian, we indeed know much about zoogeographic displacements but, apart from the Hominidae, little about the appearance of new species" (Stromer 1944, p. 85).

These last findings constitute the basic reason that I do not ascribe any appreciable significance for evolution in animals to *local* climatic changes. Animals capable of mobility were in a position to evade unfavorable local climatic conditions and to seek out climatic realms that suited them. Such a possibility would not exist if the climatic changes were very generalized *worldwide;* therefore, there is no contradiction at all, as Stromer believed, in the fact that under such circumstances, and *only* then, I concede to climate a certain influence on evolution.

"Adaptations"

All in all, when assessing certain peculiarities of form one must take care not to jump to the conclusion that they are adaptations to particular life conditions and feeding situations. R. Mertens has shown that in the living *Varanus niloticus* there is an ontogenetic transformation of originally pointed teeth to blunt ones, not as a consequence of wear but as an inherited specific character. It was formerly believed, incorrectly, that this transformation took place only in the West African Nile monitor lizard, which eats snails, and that it was an adaptation for eating hard-shelled food.

However, blunt teeth are by no means limited to the *West* African race just cited but are also found in the *South* African race, which seldom, or even never, eats snails. The broadened, stumpy teeth must indeed be appropriate for crushing hard-shelled food, but they still constitute only a general specific character, one that is also found in forms that do not eat hard-shelled prey. If this type of diet had appeared as a secondary feeding mode, it would have involved the use of structures already present, which had arisen under different feeding conditions.

It is generally known that in aquatic tetrapods there is widespread occurrence of *webbing,* skin connecting the digits on both the hind feet and the forefeet, which aids in swimming, such as is seen today in Old World otters, seals, water moles, and the majority of aquatic birds. But there are also animals for which these interdigital membranes have a completely different significance. In the coastal dunes and sand deserts of southwestern Africa there lives a gecko, *Pel-*

matogecko rangei, with pronounced webbing on all four feet. However, this is not a swimming animal and, like all other geckos, has never been one. Here, the webs are useful in desert life, for the gecko uses them as shovels to burrow into the sand.

In the flying frog (*Racophorus*), however, we see the enormous interdigital membranes put to work in the service of gliding. Here, they are probably enlargements of the webbing of aquatic ancestors, and these webs in turn developed from membranous margins that existed before the swimming function was assumed. Thus, we see once more that *mode of life, adaptation, and change in function do not create new organs; rather, they use existing organs or the rudiments of them, which later simply develop or are transformed.*

On the other hand, there are *aquatic birds* that are excellent swimmers and divers but have no webbing (or only extremely negligible indications of it), such as the water rail, the moorhen, the crake, the oyster catcher, and numerous others. Other aquatic birds have well-developed webs but make little use of them, relying mainly on their wings for propulsion in water. Similarly, the webs of some *newts* are not used; locomotion is effected through the wriggling of the tail, with the legs held straight and up against the body. That magnificent swimmer, the water vole, *Arvicola amphibious,* has the same webless front and hind feet as its terrestrial cousin, *Arvicola terrestris.*

These examples teach us that in many animals—because the proper rudiments are lacking—webbed toes are not developed even though the mode of life is aquatic and that, in any event, they are not an absolutely essential attribute; and in other animals, completely developed webbed feet are present but are not used in swimming at all or else function only slightly.

A similar situation exists for *digging feet.* The *mole* possesses shovel-shaped limbs that are outstandingly well developed for a burrowing mode of life. But *rabbits* also have perfected the capability for burrowing without having the same kind of adaptive character on their extremities, which differ only slightly from those of their closest relatives, the nonburrowing hares. In the first pair, the mole and the rabbit, we have the same functional possibilities for different structures, and in the second, the rabbit and the hare, there is, by contrast, the same structure with different functions, which thus have neither required nor caused any particular adaptation at all.

But enough of examples. We run the risk of losing ourselves in individual cases, for their number is legion, and they demonstrate the *insufficiency of a Lamarckian explanation of evolution.*

This last is true, however, only for *dogmatic Lamarckism,* which in various respects has changed considerably from the original theory of Jean-Baptiste de Lamarck. However, even from his original group of concepts we must reject the factors that are supposed to determine the origin of adaptations. A very important insight, one that we endorse, is nonetheless due to Lamarck: he by no means intended to explain *all* of evolution in terms of the needs of organisms or the satisfying of those needs, and of environmental influences. Rather, he regarded the elaboration of organic types and their basic structural peculiarities as an exception. For this process, he proposed a kind of organic law, a development through sudden, discontinuous radical change or upheaval. It was only within these existing organic types that he considered necessity and external conditions to be decisive for specialization, for the final branching of lineages.

Darwinism and Typostrophism

As the second great attempt to explain evolutionary processes, Darwinism stands in contrast to Lamarckism. Its basic explanatory principle is the struggle for existence and its consequence, *natural selection.* Darwin was by no means the first to introduce this concept as such; it is present in the work of a whole series of predecessors (Carl Linnaeus, F. Tiedemann, F. S. Voigt, A. P. de Candolle, W. C. Wells, P. Matthew, E. Darwin, T. R. Malthus, and so on). However, Darwin was the first to make a thorough, consistent study of the idea and to apply it to an abundant, diligently assembled body of evidence.

Using as his point of departure experiments in the breeding of domestic animals and cultivated plants, Darwin became convinced that speciation and, beyond that, all of evolution could be effortlessly explained by the natural variability of organisms and subsequent selection of the most qualified variants, those most fit for life. Nowadays, essential points of Darwin's intellectual edifice have been abandoned, in particular, the Lamarckian element—a considerable part of the concept—of direct environmental influence and assumed heritability of physical features acquired by an individual. Even the omnipotence of selection has had to be qualified in view of the many exceptions.

Here, then, we have only to treat the refined form of Darwinism, as it is professed by the modern science of genetics. Genetics has replaced the vague and ambiguous term "variability" with the concept of "mutability," and we now say that it is *mutations,* minute, sudden, random alterations in the genome, which bring about in offspring a negligible transformation of the former morphology. Selection then operates on these inherited, genetically altered forms.

Originally, in contrast to Darwin's theory, the science of genetics limited itself only to the problem of actual hereditary transmission and the mechanisms whereby races developed—evolution on the smallest scale, or *microevolution,* as it has often been called. (This term is unfortunate from the philological point of view but essentially more justifiable than the term in common use recently, *microphylogeny,* for these small modifications of form that take place before our eyes are not phylogeny in the true sense [the evolutionary history of a lineage].)

At first, the young science of genetics was not interested in the theory of evolutionary descent itself and often even took the negative view. In recent years, however, genetics has expanded its scope and begun to apply the mechanisms of microevolution it has discovered to actual phylogeny, to the prehistoric evolution of higher-ranking lineages, or *macroevolution.*

Modern genetics, if we go along with one of its leading representatives, N. W. Timoféeff-Ressovsky, distinguishes four different *factors in evolution:* (1) *mutability,* which supplies the material for evolution in the form of random changes; (2) *selection,* which constitutes a directive evolutionary factor and effects differentiation over time, adaptation, and higher evolution; (3) *isolation,* which likewise supplies a directive momentum to evolution and, in particular, determines differentiation over an area; (4) *population waves,* which, like isolation, restrict panmixia and the population size, thereby provoking sharp variations in the concentration of individual mutations or combinations of mutations.

However, these factors can scarcely be regarded as being of equal value. In terms of autonomy, isolation and population waves, because they only produce certain exceptional conditions under which selection operates, are unquestionably subordinate to the first two elements. Moreover, as important as the difference between temporal and spatial separation may be in the present or in any other restricted unit of time, it has little basic significance from the historical-phylogenetic point of view, for, from this perspective, spatial displacements also fall within a temporal course of events. For our phylogenetic approach, then, we shall take from genetics the basic pair of factors, *random mutability* and *directive selection.*

These two factors and their mechanisms provide a satisfactory understanding of microevolution, of the experimentally ascertainable modification of forms of lesser rank. The changes observed here are usually confined to species and have nothing to do with innovation, with the creation of new organs, but always only with relatively trivial, gradual changes regarding size, shape, number, color, and so on in *organs that are already present.* For example, the mutations in wing structure investigated experimentally in the fruit fly *Drosophila* (fig. 3.151) do not teach us how the wing structure of *Drosophila,* of dipterids, or even of insects in general arose, but only illustrate the range of development and certain forms that deviate from the norm of this organ so long embedded in the genome.

*   *   *

The question now arises whether those minuscule mutational steps and morphological changes within an existing framework are also sufficient to explain *macroevolution,* the comprehensive remodeling of forms of higher taxonomic categories, which means the elaboration of new organs and structural designs. Many geneticists would answer in the affirmative. They proceed according to the

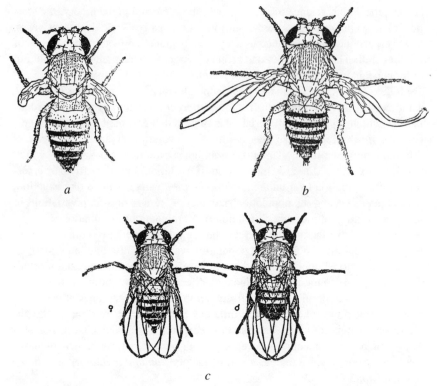

a   *b*

♀   ♂

c

Fig. 3.151. The fruitfly, *Drosophila melanogaster,* the favorite research subject of geneticists. *a, b.* Two mutations with stunted wings. *c.* Two normal specimens (♀ female, ♂ male). (From F. Reinöhl 1940.)

principle of actualism, which is to say that they attempt to explain the processes of the past by the transposition and simple summation of processes that are observable and analyzable in the present. This would mean that microevolution was the preliminary stage of macroevolution; that evolution on a large scale came about through an increase in small morphological differences, through the continuous adding on of tiny, new modificational steps over the long stretches of time in the geologic past. Depending on the degree of restraint observed, it is believed that this explains, either unconditionally or with varying degrees of reservation, all of evolution.

It is clear that this question cannot be answered conclusively through purely intellectual means. Obviously, actualism as a method is clearly justified, but only until one comes up against phenomena that do not easily fit in. Whether or not there will be any such phenomena cannot be determined from Recent material alone.

This is a dubious way of proceeding, to begin by postulating the exclusive validity of microevolutionary mechanisms and to then deduce from them a system for how the evolution of higher phylogenetic units could have taken place, with no empirical evidence at all as to whether or not this picture corresponds to reality. This by no means has to be the case a priori. It absolutely must be taken into account that in addition to phenomena that are observable in the present there were other processes in the geologic past that appeared as relatively rare events at long intervals and, therefore, would have eluded observation during the brief period of time covered by experimental research.

In another context, we have already referred to an instructive parallel that exists with geology, with the science of the inorganic history of the earth. It has been demonstrated that from time to time revolutions and manifestations of force—the upward folding of mountain ranges, for example, or the formation of extensive thrust sheets—must have occurred for which no analogue has yet been observed in the present.

The situation could be the same for the far-reaching remodeling of types we call for, counter to Darwinism. The chances that new sets of characters arising abruptly would be harmoniously integrated into the existing structure are extremely slight, as is shown by the small number of major structural designs—the only ones to exhibit such successful solutions—in the organic realm. Events such as these would not exactly happen every day.

Moreover, the comparison with geology shows clearly that all present-day experiments and all of our knowledge of physical and chemical mechanisms do not suffice to reconstruct actual geologic history, the evolution of the surface of the earth. We cannot deduce from such a basis the historical period during which each universal law was established, or at what magnitude and with what supplementary sources of energy it took effect.

To summarize briefly: we are convinced that the laws governing the modification of forms in the present were also unconditionally in force during the geologic past—there is no reasonable doubt about this—but we hold from the start that it has by no means been proven that all large-scale evolutionary phenomena can be explained in this way.

In a reverse situation, the concepts and principles of classical mechanics are also not applicable to the infinitely small dimensions of atomic, electronic, and nuclear physics; they had to have been extensively supplemented and changed. In the eyes of modern theoretical physics, the application of early experiments, observations, and concepts pertaining to macroscopic dimensions to the enormously different intra-atomic level would seem extremely rash and naive. We are convinced that eventually the same judgment will also be made in the area of evolution—that the indiscriminate, exclusive transposition of microevolutionary mechanisms to macroevolution was a mistake with serious consequences.

An Imaginary Picture of an Organic World Shaped by
Mutation and Selection

The touchstone for the legitimacy of the transposition cannot be just any deductions and rough guesses—nothing less than the historical facts that paleontology provides will do. Let us first visualize, to have something for comparison, a picture of how the organic world would appear if it were shaped only by the aimlessness of random mutability and by selection.

1. Since mutations are minute transformational steps, and since selection also works continuously and very gradually, a *continuum of forms* would have to arise *in which there would be no pronounced gaps.* Even a breakdown into species would be lacking; the concept of species would no longer be valid in an evolution that flowed so smoothly. Even less would there be a separation into sharply defined, higher categories. All differences in form that somehow or other appeared would be linked together in long, continuous, uninterrupted series of transitional forms.

2. Further, if mutability intervened here and there purely accidentally, affecting the most varied of characters indiscriminately, the result would have to be a *disorganized confusion of forms diverging in all directions.* A clear graded progression in the diversification of form as reflected in structural designs that encompass one another stepwise and in sustained, orderly series of forms could not arise in this way. We would have an anarchic chaos of forms, not a hierarchically constructed cosmos.

3. Nor would a sharply delimited temporal sequence of individual structural types be achieved. Evolution would consist solely of the *differentiation and accumulation of adaptive characters,* those characters that offer selective advantage with regard to a particular mode of life. It is indeed true that various levels of organization could be arrived at in this way, but instead of being sharply discrete like floors in a building *they would become polyphyletic, permeated by the entire network of forms, and their boundaries would be transcended.* There would exist even on a large scale, beyond the category of species, a network of relatedness, which many authors have held to be altogether logical.

4. Finally, under the conditions posited we could not expect to see a pronounced periodicity in evolution; we would have rather *a steady, very gradual, slowly advancing modification of forms in extremely small steps.*

Comparison with Reality

Clearly, this picture agrees only slightly with the one we see reflected in the fossil record, portions of which we have become acquainted with in previous chapters.

*   *   *

1. Our first point, that there would have to be a consistent, uninterrupted continuum of merging forms, which does not offer the possibility for taxonomic

organization, is already contradicted by the modern organic world. Nor did Darwin overlook this, of course. To explain the undeniable divisions between the individual taxonomic categories, he hypothesized that the existing gaps were once filled with consistent series of transitional forms, which, however, became extinct. The boundaries were simply the results of chance, brought about by the collapse of former bridges.

For the modern plant and animal world, such an explanation might, in a pinch, appear conceivable; but it is no longer tenable once we take the fossil material into account. We would have to find there all the transitional forms and links that are missing in modern classes, orders, and families; but this is not the case. Even in groups that are entirely extinct, we always see, even when material evidence is extremely abundant, the same picture *of a sharp separation and discontinuity between the individual typal categories.* Earlier, we explored in detail the reasons why this situation must be a primary, natural phenomenon and not an accident of preservation.

*   *   *

2. Our second point, that random mutability working together with selection could produce only a disorderly tangle of forms and not well-defined, orderly series, must be somewhat qualified. The objection can be raised, and rightly so, that the minute mutational steps with which the geneticist works cannot even be observed by the paleontologist. In fossil material, they fall beneath the threshold of the visible or within a range of variability, and do not appear for the paleontologist until after they have been subject to selection for some time, have taken hold, and are manifested as a progressive modification. Mutations may be completely haphazard, setting off in the most varied of directions and resulting in an unstructured network of mutants of every kind, but unnoticed by the observer, a particular orientation for separate evolutionary lineages could nonetheless arise.

What is important is the circumstance that the evolutionary lineages that do in fact exist evolve *linearly over long periods of time,* even though it can be proven that the conditions for life and the environmental circumstances changed repeatedly during that time. Once their course is set, the individual lineages pursue it steadily, regardless of such changes. There is no wild zigzagging; evolution does not weave back and forth continuously in every direction, solely contingent upon mutation and selection of this or that character, and upon adaptation to ever-changing environmental conditions. *There must be, then, a progressive evolution of characters that is not directly subject to selection and that differs in kind from the processes of adaptation regulated by selection.*

Above all, however, random mutations, which arbitrarily strike at the most varied characters and organs, would never have yielded the *hierarchy of structural designs,* that is, the sequential grades of taxonomic categories, each encompassing stepwise the previous one. To put it another way, within one form, individual characters and sets of characters are *decidedly unequal in value and*

show a hierarchy according to their variability and the amplitude of their scope.
As Karl Ernst von Baer has already clearly perceived, a sharp distinction must
be made between (1) the attributes of types, (2) attributes that characterize the
level of organization within a type (the subtypes), and (3) adaptive characters.

Many readers will see this way of putting it, this referring back to von Baer
and the using of terms such as types, typostrophes, and so on, as an intellectual
relic of idealistic morphology, which is now outdated. On the contrary, I am of
the opinion that comparative morphology (as I prefer to call it, leaving out the
dated metaphysical ingredients from the epoch of idealism), as the vital foundation of biology and the only means of expression accessible to phylogeny, is and
will continue to be indispensable—there is no way in the slightest that it can
become outdated. More will be said about this in chapter 4; we have already
discussed the concept of types earlier in this chapter. You can look at it any way
you like, but the basic facts are inescapable: considered purely phylogenetically,
the individual characters of organisms show a graded variability; there are characters that remain constant or quasi constant over a more or less lengthy period
of time and, correspondingly, are indicators of a more or less extensive group of
forms, and other characters that are extremely labile, and their stamp at a given
time characterizes only the most minor groups of forms. These latter characters
are those of races and species, and the others are the attributes of the supraordinate categories, of the structural designs that are more comprehensive step by
step—in short, of the types.

Correspondingly, each lineage shows a large number of more or less far-
reaching transformations, through which these sets of characters are elaborated
at different times, one after another, thereby introducing evolutionary trends that
permeate the entire lineage for shorter or longer periods of time. The evolution
of a lineage thus consists of various series and stages of modifying processes,
which run in parallel, side by side, and are graded according to their appearance
in time, their scope, and their order of magnitude.[20]

Let us go back once more to our favorite example, the *ammonoids;* there, in
the reversal of the *septal bulging* and the *septal necks,* we have become familiar
with two extensive evolutionary trends, which set in immediately after the line
appeared during the Lower Devonian, permeate every individual lineage, and do
not terminate until the Triassic and the Cretaceous respectively (fig. 3.152). In
addition, there are the progressive development of the *suture* and the increasing
enhancement of the *prosuture* and the primary suture through the addition of
lobal elements, which go steadfastly on their way independent of the other evolutionary trends cited previously.

---

20. A similar rhythm and assemblage of processes of different duration also exists in individual
development. We refer, using mammals as an example, to the twice-repeated cycle of development
of the dentition, the annual replacement of antlers, the semiannual molt, the monthly reproductive
cycle, the daily metabolism, and so on.

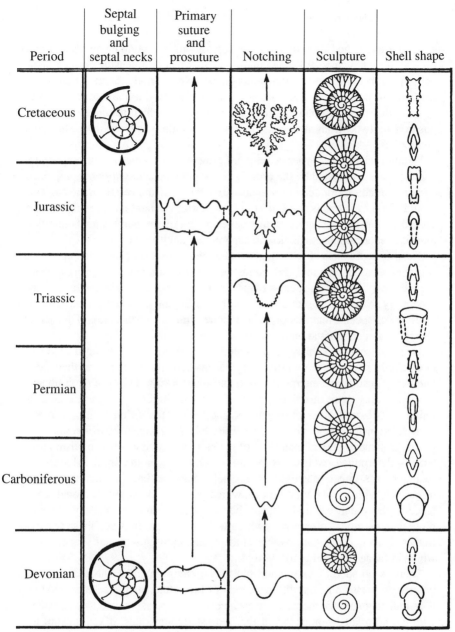

Fig. 3.152. Parallel processes of restructuring in the ammonoid stock running side by side at different rates of speed and with varying scope. The septal bulging and reversal of the septal necks indicated in the first column are shown in more detail in figure 3.36. With regard to the primary suture and the prosuture in the second column, see the more precise drawing in figure 3.38 for comparison. Here, only the prosuture (below) and the first true suture (above) of the initial and terminal stages of evolution are shown. In the last column, "Shell shape," a number of shell cross sections have been selected at random from forms that were subject to continuous modification and typify the shortest evolutionary cycles within the framework of sculptural evolution. The space available here is quite insufficient to reproduce completely the lineages of the least breadth, most of which usually exhibit gradual, but sometimes also proterogenetically abrupt, remodeling of the shell type (see figs. 3.77– 82, 3.87).

Later, during the Carboniferous, the *notching* of the suture is introduced, and regardless of any special living conditions, this feature intensifies orthogenetically from the simplest beginnings to the point that finally the entire suture is toothed and notched. At the Triassic-Jurassic boundary, a transformation takes place in the mode of notching (unipolar to bipolar), which results in the juxtaposition of two different types, the Triassic ceratites and the late Mesozoic ammonites.

Within each of these types, a number of lineages and superfamilies segregate, and these are dominated by the orderly development of the *sculpture,* described earlier. Already introduced by the ancestral forms of the ceratites and the ammonites into their respective lines (and, similarly, at an earlier level also by the goniatites and the clymenians), this development of sculpture runs its parallel, independent, orthogenetic course in each individual lineage.

The different stages of sculptural development form the genera (or, depending on the evaluation, also the subfamilies), and it is within that framework that we see the most diversity in special shapes of the *shell:* wide or narrow umbilicus, differently shaped cross sections of the whorl, different markings on the external side, trivial special features of the sculpture, and so on; the rank of genus or species is assigned based on the scope of these features.

There are several paths here for evolution to follow, which differ in kind and in extent, in rank and in duration; they encapsulate one another, and their morphological results are interpenetrating and interlocking. Of these evolutionary processes, only those that shape the shell over the short term, representing evolution at its lowest order of magnitude and least extent in time, have a direct, close relationship to the mode of life. The swimming capability of ammonoids, given the same organizational level of the soft parts, that is, the propulsion system, is determined exclusively by the outer shape of the shell, which consequently has direct selective value. Shells that are flat, high-apertured, and discuslike, with a narrow umbilicus, which present only slight frictional resistance to the water, are undoubtedly better for rapid movement from place to place than plump, swollen forms with a wide umbilicus. Between these two extremes there is every conceivable kind of gradation and special shape, all of which influence the ability to swim.

There is no such close relationship to the environment connected with the orderly progression of the differentiation of the suture. Its increasing crinkling does indeed guarantee increased resistance to water pressure, and to that extent, with the correlated refinement of the hydrostatic apparatus, it signifies an increase in biological efficiency. Consequently, the younger ammonoids with differentiated sutures were, on the whole, better divers than their more primitive ancestors. But the less specialized types were also viable, and we have no indication that they occupied habitats that were essentially different.

More than anything else, however, it is the extreme slowness of the transformation of the suture that must be considered. The addition of a single new lobe or a few teeth, the elaboration of which required lengthy periods of time, would have brought about no significant increase in the ability to dive and swim. Much more critical for that is the shape of the shell, the various models of which are repeated at every single phylogenetic evolutionary stage of the suture.

The same conclusions can be drawn for the orthogenetic reversal of septal bulging. The convex curvature of the septa, which occurs in parallel in the various lineages of ammonoids, brings about an improvement in the diving apparatus, as we pointed out earlier. But this advance, too, achieved only after a long period of evolution, seems relatively trivial in comparison with the important differences in ability to swim brought about by changes in the shape of the shell, changes that overlap the progressive bulging of the septa. Most important, however, is another factor, that convex septal bulging becomes an advantage *only when it is finished,* whereas the flattened, intermediate stages between concavity and convexity are distinctly disadvantageous.

The same thing holds true for the orthogenetic evolution of sculpture in the different cycles of ammonoid evolution. There, too, the increasing folding of the shell increases its strength and, to that extent, signifies an improvement. Nonetheless, there is no connection to a special mode of life, except perhaps the negative one, that very sturdy, protruding sculptural elements would interfere with rapid swimming. There is absolutely no coordinate change in mode of life or in environmental conditions corresponding to the orderly evolution of sculpture. It is not as if, for the different sculptural cycles, evolution began with shallow-sea forms and then moved gradually into deeper oceanic realms; rather, we have in combination with each type of sculpture the most varied of shell shapes, which are the primary determiners of swimming ability.

* * *

We encounter basically the same situation everywhere. In the *agnathans* and the *fishes* there is a broad evolutionary trend toward progressive degeneration of the armor. The earliest representatives had thick, heavy dermal armor consisting of large plates and shields, which was thought to have served as protection against enemies, whereas its breakup and degeneration in phylogenetically younger forms was interpreted as an adaptation for greater mobility.

It seems, however, that this orderly orthogenetic reduction in bony tissue has yet another, deeper physiological significance; for it affects not only the external armor, for which the above-mentioned interpretation would be understandable, but the internal skeleton as well, the endocranium, for example. Moreover, as E. Stensiö and E. Jarvik have brought to our attention, we see this happening in *all* groups, not just in those that have in fact acquired increased mobility but also in those, like the dipnoans, the crossopterygians, the coelacanthids, and the ganoids, that maintained their modes of life and mobility largely unchanged until they became extinct or, for those that are still in existence, on into the present.

In amphibians, too, similar reductions in ossification have been demonstrated.

They are not adaptations occasioned by selection, even though in forms that assumed a really active way of life selection may have accelerated the course of armor disintegration. Watson postulated that internal factors common to all members of the lineage caused the degeneration of cartilage bone and other orthogenetic evolutionary trends in the labyrinthodonts; only the special refinements of any given orthogenetic evolutionary stage were shaped with the help of selection.

Earlier, in another context, we became acquainted with a number of orthogenetic evolutionary trends in the group of fishes known as the *actinopterygians* (restructuring of the tail and the endocranium, degeneration of the scales, loss of the clavicle, increasing ossification of the centra of the vertebrae, and so on), which, in numerous parallel lines, permeate the orders Chondrostei, Holostei, and Teleostei, which replace one another in temporal sequence. These changes are also not adaptations for a particular mode of life, nor are they related to a consistent shift in the habitat; rather, they are phenomena that are superordinate to spatial and temporal adaptive processes, taking place in both marine and limnetic forms regardless of their particular mode of life.

The actual adaptations are more like quickly changing images, attached, so to speak, only to the periphery of the different morphological platforms created by long-term transformational processes. We find at the level of the Chondrostei as well as at that of the Holostei and the Teleostei, and thus subordinate to the individual stages of the comprehensive evolutionary trend, the most varied of adaptational types with regard to external body shape, length and shape of fins, development of dentition, and so on: long bodied, torpedo- or arrow-shaped, fast-swimming fish; high-backed, short-bodied animals, adapted to calm water or to life on a coral reef; predatory forms with pointed teeth, shellfish eaters with tooth plates, and so on.

This does not mean, naturally, that initially there was, for instance, not yet any body shape, no special adaptation of the dentition, of the extremities, and so on, or that the individual groups of characters appeared at strictly different times. Obviously, all these processes ran in parallel, side by side and all together. We maintain only that there are different categories of transformation of characters and that these carry on more or less independently of one another, follow separate directions, proceed at very different rates of speed, and exhibit very different durations.

* * *

In *amphibians,* too, we see evolutionary trends of different rank and duration overlapping one another and closely interwoven in individual representatives to form an integrated structure. The first to be emplaced, as J. Piveteau and F. von Huene have brought out, are the structural characters that typify the various major lineages, the features with the broadest scope, each of which dominates evolution for long periods; the more finely tuned adaptations appear later, however, and the most particular are the last to develop.

The basic structural elements that characterize the various lines of amphibians are related to the *vertebral column* and the *skull.* Both are typical entosomatic organs in the sense of Sewertzoff, that is, those that are not directly related to environmental conditions. In the line leading from the embolomerous labyrinthodonts to the Recent *anurans,* to cite one example, the structure of the vertebrae evolves from the embolomerous through the phyllospondylous to the notocentral form of the anurans (figs. 3.71, 3.153, 3.155). This last type of ver-

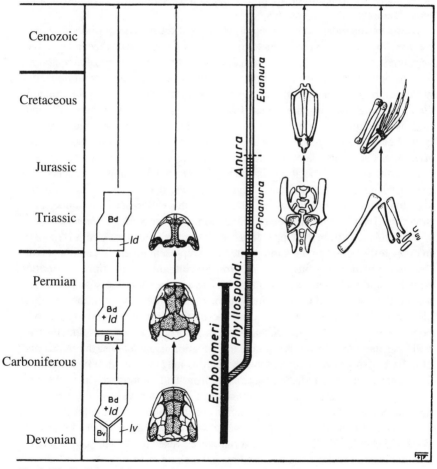

Fig. 3.153. Evolution of the vertebral column (composition of the centra) and the skull from the embolomerians to the anurans, and transformation of the pelvis and the hind extremities in the anurans, as parallel evolutionary trends. (Using illustrations from D. M. S. Watson and J. Piveteau.)

tebra and, in addition, the decrease in the number of vertebrae typical for anurans, was attained during the Lower Triassic. The structural variation thus produced in the vertebral column was unquestionably not an adaptive character, for the column's physiological value to the axial skeleton lay only in its strength, with the kind of elements making up the centra (ventralia or dorsalia) being irrelevant.

Similarly, the skull, with its much flattened, parabolic shape, the reduction of the cranium far advanced, the large eye sockets, the elongate frontoparietal, and so on, had also attained the developmental level of the anurans in the Lower Triassic. Even the structure of the endocranium, the foramina for the cranial nerves, and the structure of the middle ear and the columella are identical with those features in Recent anurans.

However, the oldest Triassic representatives of the order, the genus *Protobatrachus,* still lack the typical adaptation for jumping and the change in the pelvic vertebrae that are peculiar to modern anurans. Thus, the *organs of locomotion,* as ectosomatic structures with direct relation to environmental conditions, did not develop until *after* the anuran vertebral column and skull had been remodeled. In *Protobatrachus,* the bones of the forearms and lower legs are still separate, as yet unfused; carpus and tarsus are ossified. Further, the three vertebrae of the pelvic region are still separate, not exhibiting the fusion to form the urostyle that is so typical later. In addition, there are at least three free caudal vertebrae still present. The ilium is indeed longer but has not yet reached the extreme length of that in Recent representatives; it is connected to the axial skeleton by cartilage or ligaments. Because of this difference from modern anurans, the Triassic type has been assigned to its own suborder, *Proanura.*

Thus, it is only from the Upper Jurassic on that the structure of frogs is modern in every respect, and so the appropriate fossil representatives are combined with Recent forms in one group, the *Euanura.* Since that time, the hind extremities have become long jumping legs: the thigh, and especially the shank, is much longer; the tibia and fibula have fused to become a single element; the proximal elements of the tarsus, the tibiale and the fibulare, are structured as two long bones joined at the ends; the distal connecting portion of the tarsus remains partially cartilaginous. In the forelimbs, the situation is analogous: the radius and ulna have also grown together, and the carpus may remain largely unossified. The ilium is much lengthened toward the front and connected securely to the sacral vertebra. The structure and makeup of the vertebrae have been retained unchanged; nevertheless, further reductions related to the adaptation for jumping have taken place in the length of the vertebral column: the number of presacral vertebrae is reduced; the postsacral vertebrae are fused to form the characteristic dagger-shaped urostyle; caudal vertebrae are lacking.

Within the Euanura, these characters, once acquired, remain constant, and starting there, other evolutionary trends appear affecting the anterior and posterior faces of the vertebrae (opisthocoelous, procoelous, and so on), the degree of degeneration of the ribs, the elaboration of the transverse processes on the sacral vertebra, the composition of the shoulder girdle, and the development of the teeth. Some of these characters are definitely adaptive, the dentition, in particular, and are subject to selection. This is even more true for the last external elaborations, for the generic and specific characters, such as the development of webs, claws, or suction disks on the toes, brood care, coloration, and many others.

*   *   *

The examples just presented reveal that to some degree, *selection concerns and affects only the most superficial layer of characters.* However, the phylogenetic position and affiliation of a form based on structural design is not conditioned by the environment or by selection, but is determined independent of those factors by the stage of each large-scale evolutionary trend dominating the phylogeny and finding its expression in individuals. *It is not increasing divergence of racial and specific characters* that produces the supraordinate categories and structural designs; rather, these are added on to the structural characters of the higher typal units, in a certain sense, only as capstones; they introduce the last variations and arabesques—indeed the most important features for specialized fitness for life—into the existing framework of the structural design.

This was already recognized almost fifty years ago by Otto Jaekel. According to him, superficial adaptational forms represented only "a local deviation from the main evolutionary trend." In the old comparison using the tree, the trunk and the branches represent the real phylogenetic evolutionary trends—the sequence of major types—with only the leaves corresponding to species. "Just as leaves are a temporary phenomenon on the slowly growing tree, so the species are rapidly changing portrayals of the evolutionary state of each branch as it takes shape through contact with the environment and temporarily assumes a fixed expression."

Further, Edwin Hennig has recently said it beautifully: "Adaptation, as a toll extracted by nature, makes it possible for the organic world to express itself according to its own internal laws, as, when winding one's way through a dense forest, one makes way for trees and obstacles and still manages to hold steadfastly to the course."

Of critical importance for actual evolutionary advance is the production of new, general structures, which will determine the base for extensive groups of forms, and the introduction of the far-reaching orthogenetic evolutionary trends that will govern the lineage. *These processes are independent of the mode of life of their bearers and do not directly involve the factor that is decisive in the formation of races—selection. Only the adaptive characters within the framework of the individual structural design are selected, and not the design itself.*

Selection, then, through gradual elaboration, places its stamp only on the final, peripheral subdivisions of the evolutionary cycles and types.

The structural designs themselves, however, are emplaced or transformed suddenly in a temporal sequence that corresponds to and determines their rank. The first to be developed is the typal complex of the comprehensive unit, such as the class. This set of characters governs all of subsequent evolution and forms the foundation, the deepest layer, upon which, later and at ever more advanced levels, the features of the subordinate types, orders, families, and so on, develop. Thus, the organizational features of the higher typal categories arise, contrary to the Darwinian interpretation, not through an increasingly intensified and progressive divergence of specific characters and generic peculiarities, but the other way around, through a *descending disassembly,* a breaking-up of the higher typal units into the lower ones.

Consequently, evolution does not proceed from the particular to the general, but in reverse, from the general to the particular, until finally a new, far-reaching transformational step casts off the specialized forms and creates another broad base for a new point of departure. *Evolution does not proceed by constructing the major types synthetically, cumulatively, out of individual building blocks collected over a long period of unfolding; rather it follows the path of direct, complete transformation of types from class to class, and within the class types, from order to order, and then, descending, from family to family, and so on.*

This is the core of our *theory of typostrophism,* which differs fundamentally from the Darwinian view and interpretation. Only this theory yields an understanding of how the types encapsulate one another, a phenomenon that, as grades in morphological diversification, had attracted the attention of systematicists early on but that cannot be thought of as having arisen by way of just any arbitrary change in characters. Moreover, the conspicuously low number of major structural types in the plant and animal kingdoms speaks for the fact that their production must be attributed to very rare, radical events in the overall course of evolution. If the major types had arisen simply through the accumulation of numerous small remodelings of characters, which went on endlessly and randomly in all directions, they would not be so sharply limited in number and kind.

A very interesting representation of evolution in the animal kingdom based on typostrophism (which is to say, on the principle of the descending breakup of types, on one hand, and the early ontogenetic development of the comprehensive types, on the other) was published a few years ago (1938) as a magnificent wall chart by A. Heintz, with the help of L. Störmer. In figure 3.154 we reproduce a much reduced version of this chart, for its "evolutionary bush" offers the only scientifically unobjectionable illustration we know of the situation as it really is. The line drawing, however, gives only a slight impression of the large, brightly colored wall chart.

The ideas developed here are in touch with earlier explanations by A. E. Parr, according to which *phylogenesis* [*Phylogenese*], or true evolution, and *adaptogenesis* [*Adaptiogenese*], the processes of adaptation, are two completely different things. The distinction made by A. N. Sewertzoff (1931, p. 136ff.) between *aromorphosis* [*Aromorphose*], the organizational modifications of very general significance with regard to increased efficiency, and *idioadaptation* [*Idioadaptation*], adaptation and specialization on the general morphological base developed through aromorphosis, is of the same tenor. Also, the contrast, emphasized again and again by H. G. Bronn, between *concentration* [*Konzentration*] and *centralization* [*Zentralisation*] or *unitation* [*Unitation*], on the one hand, and *differentiation* [*Differenzierung*], on the other, means the same thing and ultimately refers to the two processes in the development of structural designs and their adaptive, differentiating elaboration.

For further historical references, we add that even many older authors, such as O. Hamann, O. Heer, and A. Kölliker, have supported ideas of an abrupt transformation of the organizational structural design. In so doing, Kölliker, as well as H. Spencer, later, had strenuous objections to the idea that the higher organic types arose in the Darwinian sense, through selection.

* * *

3. With regard to our third proposition, presented earlier, we have already pointed out that phylogenetic advance does not consist of increased differentiation and superficial accumulation of adaptive characters, but rather of profound transformations in the characters of basic organization. The set of characters constituting the type behaves more or less neutrally toward special modes of life, and in general, changes in those sets of characters do not confer directly tangible advantages with regard to fitness for life and, therefore, are not directly subject to selection. The different developmental paths taken by the sutures in ammonites (fig. 3.75), the divergent structural designs of the septal apparatus in corals (figs. 3.46 and 3.74), the different types of vertebral structure in the tetrapods (fig. 3.155), have no direct relationship to mode of life. Typal characters are on a different level from adaptive characters and are part of different trends; they are reshaped abruptly and apart from the refinements of specialization.

The origin of a new structural type lies, consequently, not in the continuation of any adaptive trend whatsoever but comes about *in an undifferentiated lineage* of the ancestral type, which constitutes the sole connection, whereas all the highly specialized branches of the old type die out and do not carry over into the specially adapted lineages of the new one. The result is a sharp separation and demarcation of the types as they succeed one another in time; *the highly specialized groups of the different types do not overlap, and consequently the types could not have arisen by way of adaptation and selection.*

To the plant kingdom

Fig. 3.154. Evolutionary bush of the animal kingdom. (After a colorful wall chart drawn by A. Heintz and L. Störmer, from A. Heintz.) Here, contrary to custom, the evolution of the animal kingdom is not shown as a tree, with a main trunk and a series of branches that issue from the trunk at various heights, but as a bush or shrub with numerous equivalent, sometimes parallel branches, many of which take off immediately above the roots. The starting point of the evolutionary bush is the fertilized egg, from which the protozoans—which remain unicellular throughout their existence—derive (shown on the right), and which forms the point of origin for the evolution of the plant world, shown on the left. Evolution is posited as issuing from the fertilized egg in stages analogous to ontogenetic development: stages that correspond to the morula, the blastula, and the gastrula of ontogeny.

At the gastrula stage, multiple branching begins. First, the sponges (Porifera) and the medusae and polyps (cnidarians from the phylum Coelenterata) diverge. Moreover, this spot is also the site of the bifurcation that results in the two great lines of the deuterostomians and the protostomians. The deuterostome embryo is the starting point for the echinoderms and the chordates—essentially, the different groups of vertebrates. On the other side, by way of the protostomian germ with persistent blastopore and the trochophore (larva), the annelid worms, arthropods, and mollusks arise. The Tentaculata (brachiopods and bryozoans), whose position with regard to these two lines is still unclear, are traced directly back to the gastrula stage, a neutral position.

In the upper part of the illustration there is a wavy line, representing the surface of the waters. The animals shown beneath that line are aquatic, and those above are terrestrial forms. Secondarily aquatic animals, such as ichthyosaurs and whales, are shown in the illustration as diving back into the water. This interesting depiction illustrates clearly the significant preponderance in number of marine types over terrestrial types. We cannot go into more detail here about the 125 pictures, each of which shows a typical representative of each branch. Most of them speak for themselves.

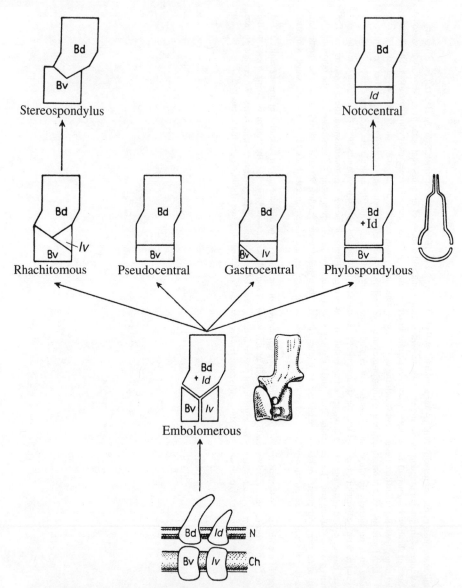

Fig. 3.155. Evolution of different types of vertebrate vertebrae, borrowed from O. Abel, F. von Huene, and other authors; schematically shown. The original emplacement of a vertebra (lower picture) consists of four pairs of arches: the basiventral (*Bv*) and the interventral (*Iv*), which enclose the notochord, or chorda dorsalis (*Ch*) (chordal arches), and the two elements lying dorsal to these, the basidorsal (*Bd*) and the interdorsal (*Id*), which surround the hollow nerve cord (*N*) of the spinal cord (neural arches.) Through nonsynchronous development and ossification of the individual ele-

Continued

This is as true on the small scale as on the large. We have already emphasized that there are no lineages connecting reptiles adapted to swimming or flying with correspondingly specialized mammals, which there would have to have been if selection and adaptation to the mode of life were a decisive factor, superordinate to the evolution of the structural design. Whales are not the descendants of ichthyosaurs (marine reptiles, fig. 3.110), and flying reptiles are not the ancestors of bats. These specialized types from the reptilian realm left no descendants when they became extinct.

Rather, what developed first, from reptilian stock that had remained unspecialized, was the general mammalian type, not as an intensification of any special

---

ments, the centra come to have very different makeups, giving rise to a great number of different types of vertebrae, which are very diagnostic for the individual lineages of amphibians and reptiles.

The oldest terrestrial vertebrates had *embolomerous* vertebrae. In each metamere there are *two* centra, consisting of the equally developed and ossified basiventral and interventral. Of the two original dorsal arches, the posterior one (interdorsal) has degenerated and fused with the basidorsal to become the major component of the single upper bony vertebral arch. The interdorsal has disappeared completely from all the other vertebrae except for the phyllospondylous and notocentral types, of which it is still a component. Next to the diagram of the embolomerous vertebra, a presacral vertebra form of the stegocephalian *Cricotus* (suborder Embolomeri), from the Upper Carboniferous of Texas, is shown, to illustrate how the structure actually looked.

Because in certain descendants of the embolomerids the anterior ventral arch developed at an accelerated rate, and the ossification of the interventral, in contrast, was delayed and finally did not take place at all, the *rhachitomous, stereospondylous, pseudocentral,* and *phyllospondylous* types arose, each developed in a different way. In the latter type, the basidorsal extends as much as half way down the thickness of the chorda dorsalis, where it comes into contact with the thin-walled, socket-shaped, only superficially ossified ventral portion of the vertebra (see the figure next to it, which shows the phyllospondylous vertebra in cross section, without the apophyses). The *gastrocentral* centrum is built just the opposite, consisting exclusively or almost so of the interventral. Finally, in the *notocentral* vertebral structure, which issued from the phyllospondylous type, ossification of the ventral pairs of arches is completely suppressed; sheathing for both the hollow nerve cord and the chorda is thenceforth provided by the two dorsal arches.

The changes in the structure of the vertebrae take place orthogenetically in stegocephalians and reptiles; the processes can often be followed step by step throughout the extended period of time encompassed by the Carboniferous and the Permian, and to some extent, the Triassic. Obviously, during this long period of evolution all kinds of changes took place in tetrapod environments, but they had no influence on the controlled evolutionary process. Since the transformations in the structure of the vertebrae begin in early ontogenetic stages, they are absolutely and fundamentally not in direct contact with the environment.

Furthermore, with regard to function, it would seem to make no difference whether the centrum consists of the basiventral or the interventral; there is no question here of adaptations to different modes of life. The only critical factor is the strength of the vertebrae and of all of the supporting structures of the vertebral column, and this can be guaranteed as well in one way as in another; thus, selection does not have a role in the elaboration of the two different types of vertebrae. Clearly, the evolutionary trends were introduced by random mutations—initially, a slight lagging behind of one or the other element during ontogeny—and then advanced in a controlled manner consistent with the foundation thus created, regardless of environmental conditions and selection.

reptilian adaptation but as something thoroughly new, something independent. Only after the new base had appeared did specialization begin again, first in the aplacentals and then in the placentals, sometimes in exactly the same directions.

The first representatives of a new type, however, are undifferentiated and still without any special adaptations, although, if they are to live and persist at all, they must naturally bring with them certain aptitudes, or *preadaptations,* for life in one environment or another. They then seek out habitats appropriate for their structure, where further adaptive adjustments will take place.

Thus, the mode of life and the superficial appearance, the results of selection in combination with the environment, are *secondary* to the transformation of the typal structure that took place first, and what is more, that typal structure is not altered by these adaptations. Reptiles that became adapted to life on land, in the ocean, or in the air remained always and essentially reptiles, just like the primitive forms of this type; horses, whales, and flying foxes are and will always be mammals. Within the mollusks, the structural designs of clams, snails, and cephalopods did not come into being through selection and adaptation to various habitats; rather, all are present in the same environment and, since they were originally all benthic, bottom-dwelling marine animals, also arose under the same external conditions.

Correspondingly, at the lowest scale, the development of races takes place only within the framework of and on the base supplied by the general specific characters, which are common to all the races. Some one strain of one species is transformed into a new species, which, in radiating from its center of origin, once more breaks up into different races. Some of these races may exhibit adaptive characters similar to those of their predecessors without being directly connected to them developmentally. The specialized form does not issue from the corresponding specialized form but rather from a more general, intervening form.

With humans, what came first was the development of the species *Homo sapiens,* that is, of the genetic structure common to all human beings. The splitting into races set in only after that; however, the individual races did not arise side by side directly from special races of species at the *pre-sapiens* level and then only secondarily and polyphyletically develop the new specific characters common to all of them. Interpretations of this ilk by Weidenreich, according to whom there is a direct relationship between Mongoloids and *Sinanthropus,* a prehuman stage, have been well and solidly refuted.

The detailed investigations by Richard Goldschmidt on the gypsy moth, *Lymantria,* have demonstrated that races do not exceed the boundaries of species and do not form the preliminary stages of species; we must accept that there is a direct transformation from species into species. According to Goldschmidt, the origin of new species and higher categories comes about through systemic mutations [*Systemmutationen*].[21] The index species with worldwide distribution owe their stratigraphic usefulness as zone fossils only to the circumstance that speciation took place independently of ever-changing, local environmental conditions and selective pressures. Only within the individual species did local races, geographic forms, and modifications influenced by facies take place.

Thus, specific characters and all supraordinate structural designs for organization are primary and antecedent; only within their framework does further special development take place. *Adaptation controlled by selection therefore forms the conclusion of individual evolutionary cycles but never their beginning.*

This behavior also stands in profound opposition to Darwin's idea that progressive evolution took place by way of gradual differentiation through selection of the best-adapted individuals, with the less specialized being weeded out. The rule of unspecialized descent teaches, rather, that it is not the most specialized, not the most differentiated within the framework of the individual structural designs, but precisely *the simple form that builds the starting point for true evolutionary advance.*

Moreover, as has already been mentioned, only the fact that adaptive specialization takes place only within the scope of individual, mutually exclusive types and does not exceed the boundaries of the type explains why, today, the more primitive types still exist side by side with the advanced ones, for example, why protozoans still live alongside mammals, why algae exist with flowering plants. Otherwise, if there were no such typal restrictions on selection, the more primitive lineages would long since have been replaced by the more advanced ones or been absorbed by them.

* * *

4. Just the fact that evolution does not proceed at a steady rate but unevenly, rhythmically, could, with the help of certain auxiliary assumptions, be understood on the basis of random micromutability and selection. What this view does not explain, however, is *the definite periodicity permeating evolutionary cycles, consistently breaking them down into three phases with very different evolutive behavior.* This periodicity presupposes that evolution proceeds according to certain laws that go beyond random mutational and selective processes.

And the *varying evolutionary tempos of individual lineages* are by no means determined exclusively by the paired factors of mutability and selection. We should expect, based on the Darwinian view, a proportional relationship between

---

21. [*Systemmutation* (= *Makromutation;* Goldschmidt 1940), a transformation of the structure of the chromosomes, which entails radical changes in the entire reactive system. According to Goldschmidt, *Systemmutationen* with strong effects lead to the origin of new types of higher taxonomic rank (Lehmann, *Paläontologisches Wörterbuch*).—Trans.]

the speed of the succession of generations and that of phylogenetic change. Rapid successions of generations would establish the preconditions for an increase in the rate of mutation and for heightened effects of selection. As Ernst Stromer has already pointed out, however, we observe the opposite: in mammals, for example, we see phylogenetic change that is on the whole *rapid,* although the generations are particularly long and correspondingly fewer in number within a temporal unit. Nor has any pattern of difference whatsoever within the mammals with generations of differing lengths been determined, as George Gaylord Simpson and F. E. Zeuner, in particular, have recently shown. Elephants, for example, with their extremely long generations, are among the mammals with the most rapid phylogenetic evolution. They are transformed much more rapidly and thoroughly than, say, insectivores or rodents, whose generation times are only a fraction of that of elephants.

Instead, the critical factor in the speed of transformation is the more or less aggregate appearance of *far-reaching, abrupt, large-scale transformations, which characterize the typogenetic phases of lineages and cannot be explained solely by micromutations, the only kind thus far given credence.* For the former, we would have to assume abrupt genetic transformations of a higher order of magnitude taking effect early in ontogeny and bringing about complex transformations.

* * *

All in all, we come to the assessment that Darwin's view of evolution, or rather, dogmatic Darwinism, has put the cart before the horse. The essence of evolution does not lie in the evolution of races or species, not in differentiation and adaptation; what is *decisive for evolutionary progress and the sustaining of life is the production of structural designs of a higher order, of new types, which regularly get rid of overspecialized adaptations, protecting the lineage from the common death of old age.* Where such a rehabilitation of the lineage has not taken place, the affected groups either became entirely extinct, or survived to the Recent in only a few, relatively unspecialized lines, whereas all the rest, having lost their flexibility through overspecialization, died out.

Precisely those characters to which Darwin's utilitarian doctrine primarily applies are therefore either insignificant for evolution or even detrimental to it. In an earlier context, we spoke of the digging feet of the mole and compared them with the functionally equivalent but not specially adapted limbs of the rabbit. It can be concluded from this that for the mole, the acquisition of digging feet has not meant any significantly greater functional efficiency than that which might possibly have been achieved by its ancestors without adaptational feet.

Progressive adaptation and the selection of adapted forms only places limits on originally multiple possibilities for use, establishing a one-sided commitment to the exercising of a single function, which may be advantageous if the mode of life is extremely specialized and competition is fierce but is detrimental to further evolutionary possibilities. In this respect, every refined specialization verges on overspecialization, which is not conducive to evolutionary progress but inimical to it and, at the slightest change in living conditions, leads to extinction.

Macromutations

In our interpretation, the sudden production of comprehensive typal organizations demands macromutations with complex consequences affecting the organizational structure itself and going beyond micromutational adaptational processes. Macromutations are the determining factors of evolution. Genetic research has recently presented models of macromutation, as we described earlier. In magnitude, it is true, they extend only to the level of generic characters, but for the time being, more cannot be expected for the reasons already given.

The main thing is that such macromutations, which bring about complex consequences and are independent of adaptational characters, have been proved in principle, and that, moreover, there is no basic doubt about also linking them to the more extensive mutational hops that seem to call for them. For there is, as we have emphasized again and again, a graded series of typal transformations, which differ in rank but exhibit the same behavior (fig. 3.156).

Moreover, it may be that macromutations taking place, say, at the beginning of a class or order, inaugurating them, do not even need to be of a greater magnitude than the ones already demonstrated experimentally. The difference in behavior lies essentially only in the scope and the evolutionary longevity of the organization transformed by the macromutation, and in the extent of each of the groups that develop from this base. But these are manifestations that do not appear until later, and they are not dependent perforce on the primary scope of the mutations that supplied the impulse. Moreover, it is of prime importance that these mutations, unlike the usual micromutations, intervene in the basic structure of the organism, opening up definitively new evolutionary paths, which is why W. Ludwig has recently termed them, appropriately, *key mutations* [*Schlüsselmutationen*].

Paleontology will have fulfilled its mission when the evolutionary processes it has deduced are successfully attributed to mechanisms that are also observed in Recent organisms and can only be studied further physiologically by means of them. The question of the nature and mode of operation of macromutations lies beyond the capacities of paleontology and must be left to experimental genetics to answer. This branch of research will have to discover what is responsible for far-reaching, discontinuous evolutionary transformations—whether

they are due simply to a genetic mutation with pleiotropic, complex consequences, to a "repatterning" of a chromosome or of the entire genome in the
sense of Goldschmidt,[22] or to some other mechanism.

* * *

Similarly, it can scarcely be decided with certainty based only on fossil material
whether and to what extent such *macromutations are triggered by environmental changes* or were so triggered in the geologic past. That such effects
(climatic influences? ionizing rays? chemical influences?) existed could be concluded from the fact that during certain geological periods organisms exhibit
an especially active remodeling and development of forms, whereas during
others, they follow their evolutionary paths more slowly and peacefully. Particularly at the turning points between the major epochs of geological history—
between the Paleozoic and the Mesozoic and between the Mesozoic and the
Cenozoic—is the organic world characterized by profound transformation.
These important upheavals have long been recognized and used in the classification of geological time.

22. Richard Goldschmidt laid out his intellectual edifice in 1940 in an extensive, thoroughly
provocative work entitled *The Material Basis of Evolution,* with which I was not yet familiar when
I prepared this manuscript. His earlier communications on this subject have had considerable influence on my thinking or have strengthened it, but in essence, the concepts described here grew out of
my own analysis of paleontological material. All the more surprising and pleasing, then, is the broad
agreement in our views. "Schindewolf's theory is practically identical with that of Goldschmidt," as
D. D. Davis (1949) observed recently based on my 1936 publication. I regard this convergence of
views arising from extremely different premises as a welcome sign that we are on the right track.
    Indeed, Goldschmidt goes further than I and is in a position to support his phylogenetic conclu-
sions genetically. He holds that microevolution through the accumulation of micromutations is a
process that, in adaptation to the environment, leads only to a diversification within the framework
of species and does not exceed the boundaries of species. "Subspecies, therefore, are actually neither
incipient species nor models for the origin of species. They are more or less diversified blind alleys
within the species." According to him, macroevolution would require a different evolutionary
mechanism, one that would create the decisive transformational step from species into species, from
one higher category into another. It would not take place through a series of atomistic alterations but
by way of a far-reaching transformation of intrachromosomal structures. This *repatterning,* or *Systemmutation,* is attributed to cytologically provable breaks in the chromosomes, which evoke inver-
sions, duplications, and translocations. A single modification of an embryonic character produced in
this way would then regulate a whole series of related ontogenetic processes, leading to a completely
new developmental type. Accordingly, gross anatomical differences between two taxonomic groups
would not have to evolve through the simultaneous selection of numerous small mutants as determiners for each individual organ but could arise through a single evolutionary step.
    This explanatory attempt by Goldschmidt has aroused much opposition among other geneticists.
Paleontology has no right to intervene on this dispute. From my personal point of view, I can add
only that Goldschmidt's inferences completely meet the challenge that fossil material appears to me
to pose, and that he, a leading geneticist, has presented a complete interpretation that does justice to
the tangible, historical phylogenetic data.

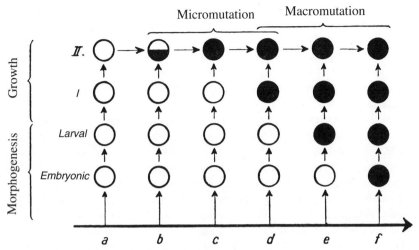

Fig. 3.156. This diagram representing a series of ontogenies arising from a continuous germ line is intended to illustrate that changes from the previous ontogeny can occur at very different stages, during the second (II) as well as during the first (I) growth period and even during the early stage of morphogenesis, at larval or embryonic stages. Correspondingly, the transformations vary in extent and, because the further consequences they trigger differ in scope, in complexity. Consequently, there is no fundamental difference between micromutations, beginning with the quite negligible differences between races in *b*, proceeding through the more pronounced transformations in *c* and *d*, to distinct macromutations, which appear in early stages and are distinguished only by the profoundness of their effects. Circles indicate unaltered stages; those filled in with black represent altered structures.

It is typical for such epochs that the representatives of almost all of the extant lineages enter a phase of more active evolution (see, for example, our information on the Permian system in chap. 1, p. 35ff.) and that the inhabitants of all biotopes—land, sea, and air—consistently show radical transformations. Moreover, it is noteworthy that these transformations do not display any common, uniform tendency at all. For example, there appear no concurrent adaptations to climate that display one particular trend, no uniform gigantism, say, no similar evolutionary trends in the development of skeletons or shells, and the like. Consistent with the ideas developed by research on mutation, it may well be that only nonspecific impulses can be held accountable, impulses that acted on the genetic structure and led to evolution in the most varied of directions.

One difficulty with this, however, is the fact that the turning points in the individual lineages—their typogenetic phases—are not strictly contemporaneous; they are always spread out over a long period of time. One would have to assume, then, that perhaps individual evolving lineages reacted at different rates to the environmental processes affecting them, that there were, one might say, incubation periods of differing lengths.

Perhaps, however, the relationships are much more intricate, proceeding through a complicated network of influences, consequences, and interactions among the separate lineages. There is no doubt that there were such reciprocal influences. The proliferation or extinction of a lineage upsets the existing ecological balance and affects the life conditions of other organisms. It is plausible that in this way, events in one lineage brought about other events in other lineages.

However, such relationships and contingencies are probably also often overestimated. For example, the view has been repeatedly expressed that only the plant world exhibits what one might call primary evolution and that the animal world unfolded passively, secondarily, contingent upon it. It certainly cannot be denied that the prerequisite for animals was plants, for plants are the basic source of food for animals, providing substance and strength. But it is not apparent that there is a particular temporal relationship between the major turning points in plant and animal evolution, nor is the nature of any causal relation clear.

In the oceanic biotope, such influences on the plant world are almost completely ruled out, for marine flora displays hardly any radical changes of any kind throughout geological history. For terrestrial organisms, however, it may be supposed that far-reaching changes in the plant kingdom changed the conditions under which selection operated in the animal kingdom; *but selection and adaptation can only begin to act on organizational systems that are already in existence.* These structures, however, which are to be radically changed, did not develop in relation to plant evolution. *The significance of the plant kingdom for that of the animals is unquestionably not causative but only conditional.*

Selection

Now let us turn to the second factor in evolution—selection.

In the development of structural designs and basic organizations of higher rank, selection is of no appreciable importance. To summarize what has already been said in this connection, we observe the following:

1. The features of basic organizational structure arise independently of the mode of life and differ in nature from the environment-related characters of race and species, which are directly subject to selection and have selective value.

2. The duration of explosive typogenetic phases is much too brief to give slowly working selection time to have an influence.

3. The long chain of linking elements that would have to be assumed between individual types, upon which selection would have operated, are not, in actuality, present. The step transforming one family into another or one order into another lies, rather, between two individuals or two species.

4. The profound transformations of type take place suddenly and discontinuously in early ontogenetic, and often even embryonic, stages, where the effects of selection can scarcely be felt or affect only embryonic and larval characters.

After any new type has arisen through a major mutational transformation, a controlled, *orthogenetic remodeling and elaboration* begins to work on the newly developed base. *And in this process, only a modest role, or in any event, not an all-powerful one, can be conceded to selection.* One reason is that in many cases these elaborations have nothing to do with adaptation but instead proceed in a manner indifferent to the mode of life. Another reason is that the objections already raised with respect to Lamarckism—to the assumption that morphological transformation is directly conditioned by the environment—are also valid here, for selection is driven by external factors and their permutations and is thus only a vector of the earthly and cosmic environmental forces that affect organisms.

The linearity and strict orientation of orthogenesis would then presuppose either an immutable environment or one that changed with matching linearity; this cannot possibly be assumed for long periods of time. Indeed, it is typical for orthogenesis that in many, many cases it proceeds along its controlled path in parallel lineages under the most varied, broadly fluctuating environmental conditions, causing the same sequence of character transformations to unfold.

The same is true of *iterative morphogenesis,* the repetitive orthogenetic copying of evolutionary trends that had already appeared once at an earlier stage of evolution. Ruled out in these instances is the assumption of an identical succession of environmental changes—of the conditions under which selection operates—of exactly the same constellation of circumstances appearing twice or several times at widely different times.

This circumstance, that orthogenetic evolution proceeds regardless of continually changing environmental conditions, was taken by Othenio Abel as the basis for an attempted interpretation that is quite indicative of the nature of the phenomena. He believed that he could attribute orthogenesis as well as the irreversibility of evolution in a purely mechanistic sense to the physical principles of inertia and the path of least resistance. These laws of physics imply that a given body (1) remains at rest or in uniform motion in the same straight line; and that (2) when acted upon by different forces, it proceeds in the direction of the least resistance. Undoubtedly, there is a remarkable parallel here, and the application of appropriate physical laws provides an outstanding illustration of evolutionary phenomena.

It should not be overlooked, however, that the law of inertia applies only to the *mechanics of solid bodies,* and that, consequently, to extrapolate inertia to the situation in the nucleus of a cell is to exceed its realm of possibility. To that extent, the law Abel formulated, the *law of biological inertia,* is only a different way of describing the situation, one of persistence in an evolutionary trend already begun as the result of the conservative, preservational principle of inheritance and the opposing one of mutability. The comparison with the physical principle of inertia, however, is pure metaphor; to attribute orthogenesis causally and mechanistically to this law is out of the question.

*   *   *

We return to selection and observe that even in instances in which orthogenetic evolution follows an adaptational trend, it is very difficult to assume that only selection is responsible. Such a view appears justified and immediately obvious only as long as differentiation proceeds in an *ascending* course and brings improvement and increased efficiency, thus giving the specialized form an edge over its competitors.

But orthogenesis by no means comes to a halt once a biologically more favorable peak is reached. It exceeds this optimum and leads to the typolytic phase of evolution, in which unquestionably disadvantageous forms with excessive gigantism and overspecialization of individual organs develop. *This evolutionary decline is difficult to understand in terms of selection.*

It has been said that natural selection cannot keep specialization from ultimately being forced in a disadvantageous direction, for it only operates in the present and cannot foresee the dangers of the future (Bernhard Rensch). This latter is, of course, correct but does not relieve us of the difficulty of seeking an explanation for the fact that in any particular "present stage," *after* the optimal peak of differentiation has been exceeded, after the future has become the present, selection for the overspecialized, deleterious mutants must, in fact, have taken place, whereas the less highly specialized, that is, the biologically more favored, forms were eliminated.

Here we have the opposite of what Darwin postulated as the consequence of natural selection, that detrimental structures and forms would be suppressed and organs becoming harmful would slowly be reduced. After the antlers of the giant elk had begun to be a hindrance, making life in the forest difficult, one would expect that mutants with lesser antler development would have been selected. Further, why, in the elephants whose tusks were beginning to grow excessively large, was there no selection of forms that were "fitter," more appropriately equipped with smaller and less curved teeth; or in forms afflicted with gigantism that were already past the point of optimal size, why was there no selection of the more advantaged smaller individuals, which would have prevented further increase in size?

In such cases, one aid to understanding is offered by the *pleiotropy of the genes,* the fact mentioned earlier that a gene determines not just a single character but is at the same time involved in a multiple effect on the development of several different characters. When, therefore, a mutation of *one* of the characters influenced by a gene turns out to be useful or to have preservational value for the system, there is subsequent selection for *that* feature, which then, under certain circumstances, might have as a consequence an unfavorable evolutionary trend in another of the pleiotropically contingent characters.

Such an explanation might be appropriate for many phenomena of overspecialization of individual characters, especially for those structures whose special morphological development is without appreciable influence on the physiological performance of the affected organs. In any event, the advantages on one side must outweigh the disadvantages on the other, which, in view of the often striking detrimental effects of overspecialization, is not always realized.

A generally valid explanation for the entire complex of orthogenesis is thus not achieved. The argument fails to explain excessive gigantism, and many authors hold that it is precisely orthogenetic increase in size that constitutes the primary cause of progressive differentiation and ultimate overspecialization of individual organs. We have already pointed out in the discussion of this matter that excessive increase in size takes place in the *most widely varying* plant and animal groups in *all* life realms during the typolytic phase of decline, and that it must be extremely unlikely that selection of the most diverse characters and adaptations, which time and again displayed different trends, would always have brought about a correlative increase in size during exactly that phase.

But even for an orthogenetic increase in size without such excessive exaggerations, or rather, for the initial stages of that kind of controlled evolutionary development, it is difficult to posit selection as the only cause. Let us analyze in somewhat more detail the already discussed orthogenetic size increase in the horse lineage. From *Eohippus* in the Lower Eocene to *Plesippus* and *Hipparion* in the Upper Miocene and Pliocene, the shoulder height increased by about one meter. This size increase extends over a period of fifty million years, or about twenty million generations if we use the situation in the modern horse as the basis for calculation. (This figure is probably too low, for in the small, early forms, the generation length could well have been less.) This would mean, then, that the size increase for each generation would, at the most, amount to about 1/20,000 millimeter, or 0.05 micrometers!

Should selective value be ascribed to this ultramicroscopic difference in size? In itself, unquestionably not; one would not impute any biological effect to that kind of difference. It may be assumed, however, that within individual generations there were greater fluctuations in size that much exceeded that imperceptible amount. From this range of variables, the largest individuals would then have been selected over their somewhat smaller competitors. But is it conceivable that selection would operate on values as small as the averages cited here?

* * *

On the other hand, there are cases in which a mysterious *"anticipatory" selection* seems to have actually existed. A typical example of orthogenetic evolution correlated with a simultaneous increase in adaptation is the *evolution of the horse,* already discussed here several times, upon which Othenio Abel primarily based his law of biological inertia.

The modern one-toed horse with its high-crowned teeth is beautifully adapted to life on the plains, to the hard ground and abrasive grasses of its habitat. For rapid locomotion, it is advantageous that the contact surface between foot and ground is as small as possible. In all fleet-footed animals, therefore, the foot has raised up; the original flat-footed gait has given way to walking on the toes (fig. 3.157), just as running humans involuntarily move along only on their toes. Another advance pertinent to the reduction of the frictional surface came about through reduction in the number of toes (figs. 3.93, 3.108).

To this extent, the one-toed horse must be regarded as the ideal running animal of the plains. Its early Tertiary ancestors had four digits on the front feet and three on the hind feet, and low-crowned cheek teeth. Since, in the later Tertiary, an expansion of plains at the expense of forests has been observed, this change in environmental conditions and the consequent change in the mode of life has been represented as the cause of a linear, progressive selection leading up to the modern horse.

However, in the formulation of this view, not enough consideration has been given to the fact that the evolutionary trend of reduction in the number of toes had already been introduced *long before the plains were occupied* in the early Tertiary by the precursors of the horse; these inhabited dense scrub, meaning that they lived in an environment where the reduction of the primitive five-toed protoungulate foot was not an advantage at all. In the descendants, then, the rest of the lateral toes degenerated and the teeth grew longer step by step, the advance being orthogenetic in the sense that development once initiated was continued— and all this regardless of mode of life, which (as Othenio Abel, in particular, has described in detail) fluctuated repeatedly, with habitats switching around among forests, savannas, shrubby plains, tundra, and so on.

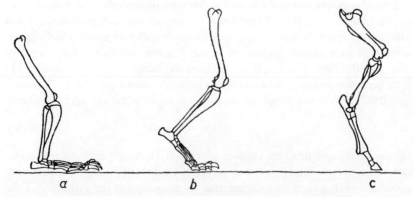

Fig. 3.157. Comparison of the hind limbs of a plantigrade (*a,* bear), a polydactyl (*b,* tiger), and a one-toed digitigrade (*c,* horse), to illustrate the different positions of the foot and the progressively smaller contact surface between the foot and the ground.

If selection alone were decisive in this specialization trend, we would have to ascribe to it *a completely incomprehensible purposefulness,* which aimed from the very outset at the one-toed, high-crowned horse, for the evolutionary trend was unquestionably disadvantageous and nonbeneficial for the primitive inhabitants of the scrub forest. It did not offer any advantages until much later, after the transition to life on the plains.

It should further be considered that the orthogenetic evolutionary trend of the horse does not at all lie in the direction of speciation and the emergence of races controlled by selection. The individual species that diverged within the evolutionary stages defined by the genera are not based on particular degrees of degeneration of the lateral toes but on various other skeletal peculiarities. However, these differences that we discern in the skeleton and upon which we base the "species" indicate units that in scope probably exceed by far what we call species in living forms. In Recent equids, the characters subject to selection and indicative of species and subspecies are the coloration and patterning of the hide, the shape of the ears and the tail, the size of the hoof, and other similar external features, which are to a large extent independent of the form taken by the skeleton and unavailable to paleontological observation.

Thus, the phylogenetic evolutionary trend of toe reduction and tooth lengthening is an independent, autonomous transformation, superordinate to speciation and raciation. It is not directly affected by the selection of those racial characters mentioned, and further, it cannot seriously be assumed that it was brought about as a pleiotropic side effect of mutation and selection for those variable peripheral features, which represent completely different trends. We see here the same behavior that concerned us earlier, namely, that progressive evolution consists of a whole cluster of paths of unfolding of different morphological and temporal magnitudes, which proceed more or less independently of one another, and that their combinations at any given time characterize the different stages of the lineage.

Based on this, it appears that another factor, or at least a supplementary one, is responsible for the linear persistence of evolutionary trends once initiated. A Lamarckian explanation, of course, is also completely inadequate. The earliest forms in horse evolution did not live on the plains, and therefore, this milieu did not trigger any corresponding needs and reactions. Moreover, the subsequent repeated change of habitat and feeding habits could not generate linearity. Rather, on the premise of a direct or a selective effectuation, such changes would have to have led to a zigzag course of evolution.

This shows that, in spite of all changing external influences and independent of the selective conditions thus established, the evolutionary trends introduced in ancestral horses manifested themselves as if through inner compulsion. In any event, the main features of the path evolution would follow had already been decided before the expansion of the plains took place. The younger horses, however, lived and continue to live on the plains *because* their specialization, which proceeded for other reasons, fitted them especially well for this mode of life and, consequently, forced them into the only habitat in harmony with their given form and its functional possibilities.

The causes for this specialized trend are to be sought within the horse lineage itself, in the genetic makeup of the ancestral form of the horse type. Its genetic structure was different from that, for example, of the tapirids and rhinocerotids, which, closely related and descended from the same root stock, lived with the precursors of the horse in the bush and were subject to the same selectional pressures, but which, because of their different predispositions, did not follow the same kind of evolutionary path. They differentiated in another way and held to that course just as unswervingly, in spite of the attraction offered later by the plains, to which they, too, must have been exposed.

What is more, neither the Lamarckian nor the selectionist view explains why horses became perissodactyl (one-toed) and not artiodactyl (two-toed), a foot structure which in some cases makes it possible to run just as fast across the plains; the antelope is an example: with its two toes and delicate foot structure, it can even surpass the one-toed horse in speed. Earlier, in another context, we saw that the cause for this development lay in the mesaxonia, acquired through mutation, of the root stock. This leads to the conclusion that *the main features of the evolutionary trend were laid out right from the start with the abrupt, discontinuous production of the type, and with evolutionary potential being restricted right from the start to certain paths.* We have here Goethe's "imprinted form, which, living, develops according to the law under which it first appeared." [23]

Selection by itself absolutely cannot create something new directly, but can only shape and develop what is already in existence. To this extent, selectionism cannot explain why *anything arises* and displays a given peculiarity, but only why, as a consequence of elimination, something else that is inherently possible, *does not arise.* Selection is only a negative principle, an eliminator, and as such is trivial: it asserts and explains only the nonexistence of forms that could not possibly exist under the conditions in question.

The rudiments of an organ, which are not yet really an organ and are not functional as such, are not yet of any use to the bearer, let alone decisive in matters of life and death. Therefore, a direct involvement of selection is to be assumed only *at more advanced stages* in the phylogenetic development of the organ. The example we discussed earlier also shows this: selection could only intervene after the horse, *using the specialization it had already acquired,* occupied the plains. Only there did meaningful preconditions exist upon which

---

23. Goethe: ". . . geprägte Form, die lebend sich entwickelt nach dem Gesetz, nach dem sie angetreten."

selection could operate, and this happened only during the *concluding portion of horse evolution.*

Put the other way around, other facts make it appear difficult to imagine a selective value for characters during the *later* phases of controlled evolution, as we attempted to show earlier with regard to phylogenetic excessive size. L. Döderlein once referred to the following case: the elephant lineage, from the lophodont (teeth with transverse ridges) mastodons to the Ice Age mammoth, shows a constant increase in the number of lamellae that make up each tooth, from originally three or four to twenty-four to twenty-seven. As long as the numbers are low, the addition of a new lamella is important and undoubtedly is selected for and proves advantageous to the bearer. However, after a certain number was reached, and the tooth consisted of about eighteen transverse ridges, the increase by one additional ridge—the enlargement of the tooth, or the increase in the number of lamellae, by 1/18—was barely noticeable or at least did not carry with it such a life-enhancing advantage that the forms without these additional lamellae were at a disadvantage in the struggle for existence and were eliminated; finally, through consistent increase, species with twenty-four to twenty-seven folds arose. The fitness for life of individuals probably depended far more on other features than on the number of lamellae in each tooth.

Allometric Growth

Before we come to any more conclusions, we must try to penetrate more deeply into the nature of orthogenetic evolution. We have already mentioned that in the view of many authors the orthogenetic differentiation and ultimate overdifferentiation of individual organs can be interpreted as being a simple, automatic consequence of phylogenetic size increase. This question remains to be examined and the insights we obtain thereby must be evaluated.

Almost all orthogenetic transformations can ultimately be traced back to differences in the *relative growth rate of organs,* to accelerations or retardations in development. In the development of the individual, which we take as our point of departure, the simplest case, although one probably only seldom realized completely, consists of all of the organs increasing in size evenly, or *isometrically.* The growth of the organism as a whole is in proportion; all growth stages retain their geometric similarity.

In other cases, however, the relative speeds of growth of individual parts and dimensions vary; because of such *allometric* (or *heterogonous*) growth, the dissimilar growth [*Unähnlichkeitswachstum*] of W. Roux, the adult organism is not simply a proportionately larger version of the juvenile form. Certain organs may perhaps be overdeveloped in the juvenile and then, during later growth, not show the same relative rate of increase as others. Hence, the growth of the body as a whole has a composite nature.

Well-known examples that we see every day are the large heads of small children and the long legs of newborn colts. Through a later retardation in size increase of the organ in question, through *negative* allometric growth, as it is called, the divergent proportions of the adult are established. An example of *positive* allometric growth is seen in humans in the development of the extremities, which grow at a faster rate than the rest of the body.

The individual growth gradients are established genetically and coordinated through the workings of the genes. Mutations affecting those genes or groups of genes could bring forth far-reaching changes of form, which are passed on from the ontogeny to subsequent phylogeny. The effects of such changes in the relative growth rates will then be particularly enhanced if the lineage in question further experiences an increase in overall body size. In major mutants displaying accelerated or retarded body growth, organs subject to positive allometric growth experience a particularly extensive development (figs. 3.158 to 3.160.)

The surprising extent of the transformations that changes in relative growth rates can produce in the geometric configuration of organs and in entire organisms can be illustrated by simple changes in a system of Cartesian coordinates into which a given organism is placed (fig. 3.161). This graphic technique shows that the differences in shape within a group of related forms or of forms that have issued from one another are based only on changes in proportion. And this is the actual situation for the vast majority of orthogenetic transformations. The reduc-

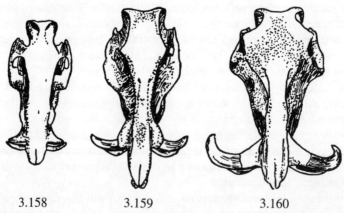

| 3.158 | 3.159 | 3.160 |

Figs. 3.158–60. Positive allometric growth stages of the tusks in the skulls of various Recent swine. Reduced.

3.158. *Potamochoerus.*

3.159. *Hylochoerus.*

3.160. *Phacochoerus* (warthog).

(After Brehm and Böker, from B. Rensch.)

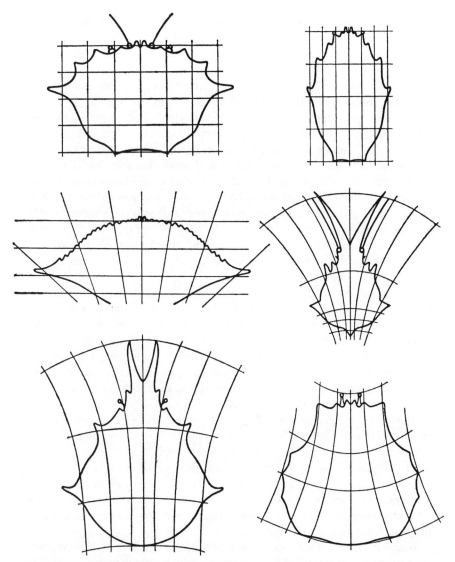

Fig. 3.161. Carapace of the marine crab *Geryon* Kröy (top left) and the changes in shape in related genera, represented by different configurations of the system of coordinates corresponding to the allometric growth of individual parts and regions. (After d'A. W. Thomson.)

tion of toes in the horse lineage is ultimately nothing more than the result of negative allometric growth of the lateral toes. Further, the extremely important evolutionary and taxonomic structural differences in the types of vertebrae in the different lineages of amphibians and in reptiles are, as we saw earlier (fig. 3.155), attributable to shifts in the timing of the emplacement and ossification of the individual elements—hence, also to allometric growth processes.

Moreover, through precise calculations, A. H. Hersh could show that the phylogenetic evolution of the horns in early Tertiary titanotheres (pl. 31B) could be understood as hereditary changes that caused the entire body to grow larger and the growth coefficients of the separate parts of the skull to shift. The oldest representatives of these ungulates (pl. 31B, *a, b*), still lack any horns at all; they do not appear until the skull has reached a certain size, and then only as tiny, buttonlike swellings (pl. 31B, *c*). Then, as the entire skull grows larger, they quickly increase in size allometrically (pl. 31B, *d–f*). As long as the size of the horns was not such that they were useful for anything, they had no selective value in themselves, and their size increase was based, rather, on a selection of mutants with greater overall body size.

It is obvious, however, that the rudiments of the horns must have been present right from the beginning; otherwise, even the largest skull would not have been able to produce them. As soon as the horns had attained a sufficient size to serve as weapons for attack or defense, the animals with the largest horns—*as a consequence of already having them!*—were at more of a selective advantage than their small-horned competitors. From that stage on, the horns themselves acquired direct selective value, and Hersh thought he could observe that from then on, the horns evolved more rapidly than before.

Moreover, one can understand that, as a consequence of a progressive increase in size, overspecialization of certain organs with positive allometric growth can ultimately occur, and that the direct selection of horns, antlers, tusks, and other similar structures is succeeded later by a general selection of the larger mutants. Two things, however, remain unexplained: (1) why the overall increase in size persisted even after the optimal body size was exceeded; and (2) why positive allometric growth of a given organ was retained even though, after a certain state had been reached, it was detrimental.

Why was there no counterselection of the smaller mutants or of the forms with lesser growth gradients? This persistence along an evolutionary path already established is precisely what defines orthogenesis and what selection by itself cannot sufficiently explain. *Here, the workings of selection must, to a certain extent, be overruled or overrun by a linear evolutionary process that is contingent upon different factors.*

*   *   *

We observe, moreover, that allometric growth and changes in the proportions of organs are by no means always tied to an increase in overall body size, that there is no inevitable linkage here. Disruptions in the timing (*heterochrony*) of the emplacement of individual parts belongs here, too, and these may take place *in early stages of ontogeny, when the body is still minute,* and, through phylogenetic enhancement, may bring about extensive changes in the degree of development of individual organs.

We know from our earlier reflections that the most primitive, simplest suture in ammonoids consists of three lobal elements (I, L, E) and that this basic number is later augmented by umbilical lobes issuing secondarily from a splitting of the inner saddle. Whereas in the oldest ammonoids, these umbilical lobes are not produced until later growth stages, in younger representatives, accelerated ontogeny causes their emplacement at ever earlier developmental stages (fig. 3.38).

Thus, in Devonian ammonoids, a first umbilical lobe (U_1) appears on the average between the twentieth and twenty-fifth suture; but in Carboniferous forms, its emplacement has been shifted on the average to the second or third suture. In Triassic ammonoids, continued orthogenetic enhancement sees the emplacement of the lobe in question, U_1, already in the primary suture, in the first true suture of the ontogeny. Jurassic forms show a different progress: there, yet a second umbilical lobe (U_2) is laid down in the primary suture, ventral from U_1, to join the first; in the precursors, this second lobe did not appear until about the twelfth or sixteenth suture. In later stages, a lobe (U_3) is installed between them, and sometimes other lobes (U_4, U_5), and so on, are added in a pattern of alternating, sequential emplacements, ventral or dorsal from the most recent umbilical lobe (fig. 3.37).

Another peculiarity we have not yet discussed is treated here briefly: In younger ammonoids, in one of these umbilical lobes there usually appears a *sutural lobe,* which, in fact, appears in a lobe lying on the umbilical seam (= sutura), on the kink in the line separating the concave inner and convex outer zones of the whorl and dissected by it. It may be in lobe U_1, U_3, or even U_4, the particular lobe naturally being firmly fixed in each group; in the majority of Jurassic and Cretaceous ammonites, it is lobe U_3. This so-called sutural lobe (S) is formed when the umbilical lobe lying on the seam breaks down into a number of subsidiary elements: it is split by the arising of a saddle; at the crest of this saddle, then, another lobe drops down, which in turn is divided by a saddle, and so on (fig. 3.162*d, e*). In this manner, the lobe in question is extensively subdivided and usually comes to cover quite a considerable area.

In one line of Jurassic ammonites, in which lobe U_3 has been elaborated to become a sutural lobe, a striking acceleration in the emplacement of this lobe now sets in. Whereas normally in Jurassic ammonites the primary suture contains only lobes U_1 and U_2, with lobe U_3 coming on the scene only at a much

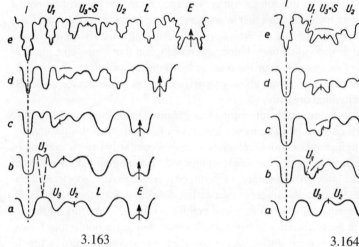
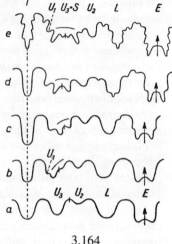

3.163   3.164

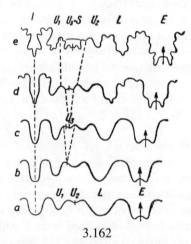

3.162

Fig. 3.162. Development of the suture in *Perisphinctes convolutus* (Qu.), from the Lower White Jurassic of southern Germany, with the normal sequence of emplacement of the umbilical lobe and with sutural lobe development (S) in the last lobe to be formed, U_3.

Fig. 3.163. Lobe development of *Cardioceras cordatum* (Sow.), from the Lower White Jurassic of England. Heterochronous, delayed emplacement of umbilical lobe U_1 contrasted with lobes U_2 and U_3 already present in the primary suture: U_3 becomes a sutural lobe. In the mature suture (*e*), U_1 reaches normal size and appears as an independent lobe.

Fig. 3.164. Lobe development in *Pseudoperisphinctes rotundatus* (Joh. Rmr.), from the Upper Brown Jurassic (Upper Middle Jurassic) of northern Germany, with extreme heterochrony and suppression of the umbilical lobe U_1, which even in the mature suture line (*e*) no longer attains the extent and form of an independent lobal element.

(After Schindewolf.)

later stage, here, lobe U_3 is taken up earlier than usual, right in the first suture, and is already distinguished there by significant depth (fig. 3.163a).

Without knowledge of the connections and further development, one would be inclined to think, because of its size, that it was the first umbilical lobe (U_1). However, the broad saddle separating it from the inner lobe is conspicuous. At a later stage (fig. 3.163b), a lobe appears in this saddle that, based on its position and later behavior, can only be homologous to lobe U_1 of the normal forms. Thus, lobe U_3 is here extremely accelerated compared to U_1 and even U_2, as can be seen from the start by its considerable depth and width.

We may connect this early emplacement of U_3 with the extensive differentiation it experiences later, through the development of sutural lobes, compared with the other umbilical lobes. A well-known rule says that the more complex the consequences the earlier the emplacement. (The actual causal relation is, of course, the opposite: the greater complexity is a result of the early emplacement, or rather, the two phenomena are fed from the same source.) In many extreme forms of the ammonites in question, the heterochronological emplacement of U_1 and U_3 is even more pronounced, and the proportions of the two lobes shift even more at the expense of the former. In further retardation, lobe U_1 does not appear until after the sutural lobe has already developed in U_3 (fig. 3.164b). It takes shape as a shallow dip on the ventral slope of the inner saddle and even in age does not attain the size of a normal lobe but looks rather like a somewhat more extensive notch in a crenelation (fig. 3.164c).

The phylogenetic developmental path of the suture is thus substantially determined, on the one hand, by the orthogenetic acceleration of individual development, that is, by the increased advancement of the emplacement of lobes to earlier and earlier ontogenetic stages, and, on the other hand, by heterochronies, shifts in the normal timing of the sequence in which the lobal elements are emplaced.

That selection plays a decisive role in the abbreviation and condensing of the course of ontogeny is hardly to be assumed, for as we have already pointed out in ammonites, fitness for life and ability to swim depend on external characters of shell configuration and not on the possession of one lobal element more or less. The resistance to pressure of the minute shells was, in any event, more than sufficient, meaning that the earlier or later addition of a crimp in the suture could provide no additional advantage. The increasing heterochrony, however, cannot be understood at all in terms of selection. The group of forms characterized by this phenomenon is in no way advantaged over lineages with the normal sequence of emplacement of the umbilical lobes, for their life span is shorter and they become extinct earlier than the others, without showing recognizable signs of any other distinctly disadvantageous specialization.

Orthogenetic evolution in the lineage of the stony corals is characterized by an increasing appearance of radial developmental norms instead of the original bilateral standard of the pterocorals [Rugosa]. In the latter, and still in the older cyclocorals [Scleractinia] as well, the protosepta are emplaced in pairs *one after the other* and thus display a purely bilateral stamp. In the younger cyclocorals, in contrast, the emplacement of protoseptal pairs II and III is accelerated, making them appear *simultaneously* with pair I as an integrated, radiate circlet (fig. 3.46). The remnants of the former bilaterality are here forced back to the preseptal stage and are still manifested only in the bilateral sequence of origin of the protomesenteries, which precede the septa. A similar concentration of originally separate emplacements appears in the first cycle, and possibly also in later ones, of the metasepta.

Quite generally, in the primitive forms, the developmental times of the individual categories of septa represent phases in which the bilateral aspect appears dominant. As soon as each developmental stage is complete, the septal apparatus displays radial features. As the originally widely strung out emplacements of septa are advanced, these bilateral phases become more and more abbreviated; the separate developmental events are finally compressed into a single event—they become simultaneous—with radiality thereby gaining the upper hand. Thus, it is ultimately the progressive abbreviation or acceleration of ontogenies throughout the entire course of the phylogeny that is responsible for the increasing radiality of the corals.

What the two examples just described (many similar ones could be added) have in common is that *orthogenetic evolution comes about through the acceleration of growth, of the emplacement of individual parts of organs, and that these processes are not tied to an increase in body size.* The ammonites or corals under discussion do not grow to be any larger than their predecessors; we saw, rather, that the transformations do not take place first at an advanced age, at larger stages, but *early in ontogeny, when the organism is still minute.*

Hence, allometric growth can be realized phylogenetically in two different ways: (1) either through a *lengthening of ontogeny,* which is carried on phyletically from individual to individual and produces an increase in size; and (2) through a progressive *acceleration of the course of ontogeny,* through the compressing of emplacements and increased acquisition of new organizations, without extending the period of individual development and without increasing the size of the individual organisms. In the first case, there is an increase in the *duration* of growth, and in the second, an increase in the *speed* of growth.

Allometric growth—heterochronies, accelerations, and retardations in ontogenetic development—is therefore basically independent of the particular size of individuals; *at the least, the allometric and ultimately excessive development of organs is not uniformly and causally coupled with a phylogenetic increase in size.* The joint appearance of the two is either accidental or is attributable to a common cause.

Again, paleontology can only speculate openly about these causes, referring to certain ideas put forth by genetics, according to which the speeds of development of individual organs are determined by the quantity of the corresponding individual genes or groups of cooperative genes regulating the length of time required for the chains of reactions they have triggered to run their course (Goldschmidt). Quantitative changes in these genes cause dislocations in the synchronization of reaction speeds; an increase causes accelerated development of the controlled organs. Allometric growth of *individual* organs would be based, then, on quantitative changes in a small number of genes, only those carrying the trait in question; uneven growth, paired with an increase in body size, however, would be attributed to progressively disparate quantitative changes in the totality or in a large portion of the genes.

Orthoevolution, Autogenesis

The concept just described would presume, however—because, according to what has already been said, the assumption of selection for the larger mutants does not offer sufficient explanation—that mutations are directed, or at least not completely nondirected. And this, too, is not actually the case. We recall the *homologous variability* in related species and in species of related genera, and the restriction of mutability through special *mutational potencies,* which, without the involvement of selection, bring forth similar mutations and parallel races (Vavilow's series) in different species only because of the similarities in the constellation of the genes (*Paripotenz* of V. Haecker).

The observations of S. Wright and Th. Dobzhansky are along the same line: only a tiny portion of the possible combinations of genes are and were realized in organisms. Moreover, according to the authors, the combinations of genes that do exist are by no means distributed purely randomly and indiscriminately over the entire range of possibilities; rather, they cluster in particular groups, separated from one another by pronounced discontinuities. A similar idea is the basis of the gene complexes or radicals of L. Plate.

Even back in his day, August Weismann was already using the typical image, that the number of tracks radiating from any one life form along which possible "variations" travel may be large but is not infinite. Between them lies a trackless area, where advance is impossible, where unfeasible mutations dead-end, their fate sealed.

We refer further to the already described observations and interpretations of Stubbe and von Wettstein, according to whom a large mutational step is succeeded by a series of smaller micromutations, which influence the same set of characters and reduce the original breadth of developmental possibilities. Such an adjustment would have even more effect on the recently declared *factor-specific mutations* (these are specific, but not adaptive, responses triggered by particular influences), if in fact their existence is confirmed.

If it is correct that quantitative changes in the genes determine differences in body size and relative growth gradients, then there are only two paths open to mutations: an increase or a decrease in the number of genes. Perhaps—and we attempted to substantiate this earlier—mechanisms of developmental physiology continue to intervene here, causing a particular trend to be designated and then persistently adhered to.

Such a controlling factor *must* exist, since (1) orthogenetic allometric growth is not inevitably tied to a phylogenetic increase in size, hence, cannot be explained in this roundabout way; and (2) the [initial] increase and subsequent excessive increase in size cannot be adequately understood in terms of selection alone.

Parallel lineages and iterative morphology are particularly convincing in demonstrating that what is decisive for their orthogenetic evolution is not the environment and the workings of selection, which are dependent upon the environment, but rather a limitation of potential established right from the start by the genotype. To solve particular biomechanical problems, there are always only a few paths available to the structural basis of a given organ, and the taking of the first differentiating step means a commitment to one of these possibilities; further progress can take place only along this path.[24] *It is not environmental conditions but factors residing within the organism itself that are decisive here;* they determine the kind and, as we have already observed, also the extent, the diversity, and the tempo of evolution.

Radiation, the spread of organisms, is also a factor that by itself sets no new evolutionary trends in motion. The colonization of any new area can only take place *if organisms that are suited to it are already in existence.* They must take with them the organs and features that enable them to live in the new habitat, qualities they have already acquired independent of the new environment. If the appropriate prerequisites are lacking, an organism in a new environment or one attempting a new mode of life will die.[25]

24. The mammalian canine tooth, for example, undergoes either enlargement or reduction. One of the few ways in which enlargement and enhanced functional demands can be achieved consists in a lengthening of the tooth to an ultimately saberlike shape. Thus, we see the sabertooth-type appearing independently in certain marsupials (fig. 3.100), creodonts, and fissiped carnivores (figs. 3.101, 3.120*c, d*), a parallelism that is probably based on an original limitation and progressive paring down of potential.

25. Wilhelm Busch has expressed this delightfully in terse pictures and verses, which are fun to recall in this context. He is commenting wryly on a frog that has climbed a tree and then, hoping to emulate a bird, attempts to fly and breaks its neck in the process:

Wenn einer, der mit Mühe kaum
gekrochen ist auf einen Baum,
schon meint, daß er ein Vogel wär',
so irrt sich der!

[When someone who has barely managed to climb a tree thinks that makes him a bird, he's mistaken!]

Selection undoubtedly plays a role in further improvements of the features in question and enhanced adaptation to the altered way of life. In the example of the horse lineage described earlier, this did not happen until the final portion of its evolution, *after* the horse, having experienced many changes in mode of life, had colonized the plains, an environment for which previous evolution had, as it were, predestined it. *Here, therefore, selection only enhanced a transformation that was inaugurated under other environmental conditions.*

Similarly, we saw in the evolution of the titanotheres that direct selection for horns does not set in until an advanced stage of evolution, *after* the horns have already attained a certain degree of development. In any event, selection can only influence those organic structures for which at least the potential already exists within the organism.

Moreover, in addition to special adaptive selection there is also a general *orthoselection* operating independent of any particular habitat; examples are the enlargement and improvement of the sensory organs, the central nervous system, and so on. These orthogenetic transformations, however, are probably based more on an *orthoevolution* than on orthoselection.

It also seems that *isolation* is not an appreciable factor in evolution. It is unquestionably important for raciation, for it breaks up populations, thereby eliminating panmixia and favoring various kinds of differentiation. *But on the larger scale of macroevolution, it does not seem to apply.* Otherwise, isolated islands would be characterized by a particularly rich and diverse fauna. But New Zealand, for example, and the islands nearby were isolated throughout the entire Tertiary and still did not produce an endemic fauna, either of mammals or of modern reptiles. On the contrary, their Tertiary and Recent faunas are conspicuously meager.

Furthermore, an abundance of races and species does not by any means have to be equated with a large measure of transformation. There is, as Ernst Mayr (1944) has recently pointed out correctly, sharply divergent evolution without the appearance of numerous new species, and conversely, there is considerable speciation without significant evolutionary divergence. Evolution behaves in such a way that the greatest diversity of divergent morphological types occurs at the beginning of the history of a lineage. Later, there is a decrease in the number of types through the extinction of numerous collateral lines. On the other hand, lines that escape extinction experience a progressive consolidation of structural characters, but in exchange produce mostly an increased number of individuals, minor adaptive forms, and species. In such cases, then, the sterilization of evolutionary potential is coupled with abundant speciation. Finally, we have already mentioned that isolation on islands absolutely cannot be regarded as the cause of gigantism, with which it is often connected.

Most important is the leap in the transformation of types that is accomplished apart from any direct relation to the environment, and it is this kind of transformation that, in our view, is by far the most significant process in evolution; it brings about the far-reaching remodeling of the genome and creates the basis for evolutionary potential, which is then pursued by orthogenetic differentiation. The two processes are different in essence, as must be emphasized repeatedly, and as even Goethe divined: in connection with his theory of plant metamorphosis, he spoke in his *Vorarbeiten zu einer Physiologie der Pflanzen* (1797) of a "twofold law" [*doppelten Gesetz*], one "whereby the plants are produced" [*wodurch die Pflanzen konstituiert werden*] and a second "whereby the plants are modified" [*wodurch die Pflanzen modifiziert werden*]. We also find the beginnings of this kind of distinction in Lamarck, as we have already mentioned.

The production of basic structural designs takes place independent of the environment and of the mode of life: the neomorphic organs are at first neutral with regard to their function and appropriateness; at first, their significance is completely incidental, and they do not attain value—and this usually in an environment different from the original one—until after they have acquired a certain degree of development and, with that, the capacity for being useful. In the early stages of the existence of characters and organs, in our typogenetic phase, environmental factors are influential at most as unspecific stimuli that trigger already existing potentialities; *the quality and quantity of neomorphs, however, are determined exclusively by the organism itself.* To put it bluntly: you can't make an elephant out of a mosquito.

We are far from seeking to deny the influence of selection. It is, as we have already observed on another occasion, a commonplace, a self-evident truth, and unquestionably of great significance for raciation. Evolution, however, does not proceed in the same way as raciation and does not appear to us to be a product of selection, but more the result of *orthoevolution.*

This does not at all mean that evolution is prefigured right from the start; that kind of inevitable limitation applies only to the later portions of the typostatic periods and for the typolytic final phases. The typogeneses, however, are continually throwing open new gates, creating new, unfettered possibilities for form. Hence, there is a continuous alternation between determination and release from the bonds of determination, that is, epigenesis.

Likewise, there are also close *reciprocal relations between the organism and its environment,* as just metabolism and the continuous exchange of substances shows. It is impossible to conceive of life without environment. One might, for example, even hold firmly to the opinion that in sessile animals, whose interface with the environment is the same on all sides, some sort of close relationship between their often radial structure and their mode of life, or external factors, exists. That radiality appears only in very particular animal lineages and always in a thoroughly specific way, and that all of the lineages of echinoderms (which originally grew on a stalk) were five-rayed, the polyps of the hydrozoans and scyphozoans four-rayed, the hexacorals six-rayed, and the octocorals eight-rayed—*these peculiarities are bestowed by the basic structural designs and organizations of the organism in question.*

*   *   *

The workings of evolution thus do not present themselves to us as ectogenesis, as a purely random process driven and dominated by environmental conditions, but primarily as *autogenesis*.[26] This group of concepts holds that the organism is not simply the plaything of external factors but is more extensively *in command* (unconsciously, of course) *of its own evolutionary destiny* because the responsiveness to environmental influences and the functional forms that appear as a consequence are implicit in the organism itself.

This is not a mystical interpretation, and it should not be confused with finalism, purposivism, or perfectionism [*Vervollkommnungstrieb*]. The ideas we have arrived at based on fossil material are indeed quite different from the Darwinian explanations that are common today, but they can nonetheless be traced back to *experimentally substantiated facts of genetics and developmental physiology,* as we believe we have shown. There is absolutely no element of special, supernatural legitimization in our interpretation, nothing that lacks a true physiological basis.

Our explanations are premised only on the constituent of life, of the organismic. In the same way, they are also based on the Darwinian concepts of efficiency, the struggle for existence, selection, and adaptation, which, lord knows, are anything but mechanistic principles, contrary to repeated assertions. But the phenomena of life—the capacities for development and for responsiveness—must provisionally serve as "first causes," as a remainder that, for the time being, cannot be further reduced and, because of its nature, is not completely comprehensible. All of the special terms that have been introduced as an "explanation" (the life force, *élan vital,* the will to existence, the monad, entelechy, the dominant, totality, morphoidea, the creative urge, and so on) are only descriptive, not new understandings, not clarification of the facts of the matter.

This introduction of an unexplained elementary factor is by no means peculiar to biology, and there should be no objection to it as being unscientific. Completely analogous is the use in the discipline of advanced theoretical physics of the terms force fields, elemental quanta, energy, and so on, graphic concepts and descriptions of phenomena whose actual essence and real cause are not known. What we do know is only that there *are* such things as field effects and gravity and how they manifest themselves, but not *what* they actually are, *why* they exist, and *whence* they come. A discussion of these questions must remain the province of metaphysics.

Descartes reproached Galileo, saying that his law of falling bodies was unfounded; he should have first determined *what* gravity was. Today we know that such a demand is pointless and unrealizable; it does not open the way for a successful attempt to treat the subject. As Newton had already acknowledged,

26. Or *automorphosis,* as Pfeffer put it, by which he meant those evolutionary processes that are contingent upon internal factors and take shape out of inherited potential.

the only problems that can be studied productively and with scientific exactitude are those of the effects of gravity, the conditions under which it operates, its distribution in space, the strength of its force, and so on.

We present the concept of gravity only as a derived image, without speculation on its nature, and this is justified, for the introduction of such a term serves to simplify and unify our concepts. We insert it and analogous physical terms into the laws of causality, which clarify the appearance of certain changes for us; we do not use the terms to say anything about the essence of the forces in effect. *Likewise, in biological fields, we must take the basic phenomena of life into account and use them in our deductions,* even though for the time being we cannot determine their nature more precisely or explain their mechanics.

4   Basic Questions in Biological Systematics

The expounding of any field of science is accomplished through a specific sequence of work procedures, the formulating of problems, and findings. All research basically follows the same path. At the beginning, there is a *stage of assembling the material.* This, in a certain sense, is a prescientific stage, during which casual, isolated observations are compiled and described. In general, there is no particular goal connected with this; whatever chance provides is described at random and without assumptions having been made.

After a certain body of observed phenomena has been accumulated in this way, there comes a *phase of ordering,* and this is where actual scientific work begins. One tries to find an ordering principle for the heretofore unconnected individual facts and attempts to bring them into a coherent system. This is the period of classification, of developing concepts and systems; not until these intellectual processes have been carried out is it possible to have a clear view of the material, to manipulate it, and to master it.

Upon this basis, then, arises the *interpretive stage,* in which the research is no longer satisfied with simple observation and ordering of facts but asks questions about reasons and causes, endeavors to discover the laws underlying the observed body of facts. Now, isolated empirical data and purely conceptual systems are no longer the final goal, an end in themselves, but only a means and a point of departure for deeper questions of cause.

It is not as easy to fit the *period of application,* or better, the *applied aspect* of the science—the attempt to draw practical conclusions from the theoretical results—into this sequence of steps. It does not coincide closely with any particular developmental stage of the research. Practical applications for technology and daily life, for culture and world view, are naturally all the more productive and assured the further along the science is in its quest for the causality of its material.

But it is entirely possible that even the first, purely empirical observations, as yet unorganized and unevaluated theoretically, may yield certain useful, practical applications. We saw an example of this in the application of fossils to stratigraphy, which began at a time when a useful system did not yet exist, and there was no insight at all into the evolutionary context of life. This practical application was based only on the simple existence of fossils and their differences as they appeared in each of the superimposed layers, and it did not matter whether they were the product of creation or of evolution.

Moreover, the stages of knowledge just described are obviously not distinctly separate in time; they do not simply appear in succession but in some cases proceed side by side, with only a shift in emphasis in mode of observation and research goals. In biology and, especially, in paleontology, with its inexhaustible wealth of material, the period of description is far from over and will never be over. New, as yet unknown organisms are always being discovered and must be described; new viewpoints are always arising, and methods are being refined, forcing renewed investigations and greater detail in descriptions.

Indeed, such modern descriptions at the level of our current state of knowledge differ substantially from those written during the early days of science. The choice of facts or characteristics considered important enough to be described depends on certain theoretical concepts and varies with changing times in research. Further, since we have an already developed organic system at our disposal, a modern description cannot be satisfied with a simple account of the raw empirical data but is constrained to place these within the existing contextual order and to work out the relationships to neighboring bodies of fact.

Even today, however, really naive, lone descriptions sometimes appear, which, without any objective, without placement within the larger context, provide only rough, isolated contributions. They represent a reversion to a scientific era long out of date. Naturally, this by no means intends to disparage pure, scrupulously objective observations and descriptions. These are *indispensable* and, if they are carried out carefully, have enduring value and hold up longer than the intriguing, fancy hypotheses that are often based upon them. "What is important is the solidity of the foundation, not the gay colors of the tapestries on the walls" (Friedrich Hebbel).[1] Nonetheless, it is necessary that all descriptive work be based on a sound understanding of the entire body of knowledge to which it belongs and that it not lose sight of the relation to the whole.

The foundations and methods of systematics are likewise continually being refined, meaning that the discipline progresses and develops continuously. Contributing to this process, moreover, is the close reciprocal relationship between each historical era of science and its methods of observation. As we have said, the availability of the system influences the kind of description; but the system itself makes use of the advances brought about by the investigation into causes, in our case, then, advances in the theory of evolutionary descent and in genetics.

In this book, we did not observe the order of the historical steps toward knowledge just described: At the beginning, the practical application of paleontology

---

1. [Friedrich Hebbel, German dramatist, 1813–63. "Auf die Solidität der Grundfesten kommt es an, nicht auf die Buntheit der Tapeten, womit die Wände behängt werden."—Trans.]

was described; then, we took up the causality of form—evolution and its factors; and now, we turn to systematics, or better, *taxonomy,* which is the special, more precise term for the classification of organisms. This sequence was chosen to show at the outset how the classification of geologic time was arrived at and to explain the basis upon which our phylogenetic deductions rest. Knowledge of the regular patterns of phylogeny is in turn the prerequisite for a complete understanding of taxonomy as it is currently conceived.

Paleontology has important contributions to make to both the theory and the practice of taxonomy.

THE NATURE AND TASK OF SYSTEMATICS

The general character of systematics is purely that of *scientific organization;* its goal is a conceptual understanding and logical classification of objects, processes, and products either found in nature or produced by the human hand or mind. Thus, systematics and classification are by no means limited to the biological sciences; rather, they are a characteristic component of all other natural sciences and also of those humanities that treat a multiplicity of complex phenomena. From these emerge the general principles of logic that are the basis of systematics, and these, naturally, are also relevant to the system of organisms.

Since the human mind is not capable of taking in and mastering the multifarious physical world in all its many-layered diversity and complex relationships, it requires *simplification through concepts* that strip away all diversity to reveal enduring likeness and commonality of features. One begins with observation, with the detailed comparison of the objects to be classified. The *comparative method* shows what the objects under investigation have in common with a larger group—the characteristic, or *essential* features—and the special characters that are *unessential* because they are variable and inconstant; these could also be described, by way of contrast, as being *relatively invariable,* or *essential,* and *relatively variable,* or *accidental.*

By ignoring the latter and retaining only the identical, constant features, one arrives at notions of objective essence, and the increasing elimination of groups of distinguishing characteristics leads to more advanced, higher, more general notions. In this way, a clearly organized, hierarchical conceptual structure arises in which each individual phenomenon of the reference context in question takes its logical, appropriate place. The system offers a simple and unambiguous way of understanding the position of any given object.

The particular task of biological systematics or taxonomy consists in *classifying the diversity of organic forms and reducing it to a clearly arranged system of concepts.* Systematics arranges the forms through which the organic world manifests itself according to their similarities and gradational differences; *it traces diversity back conceptually to certain basic forms.* To do this, we apply

the process of abstraction, by which characters are disregarded or eliminated,
to the individual we actually see and, using this methodical generalization, arrive
at taxonomic categories ranging from those including the lowest and most similar forms (races and species) to the higher units, in which the degree of resemblance diminishes progressively and there are fewer common traits (genera,
families, orders, classes, and ultimately, the kingdoms of animals or plants and
living organisms in general).

For example, if we carefully compare all of the dogs we know of, we observe
that they have a great number of features in common, but also a series of differences, in hair, coloration, size, special shape of the head, the limbs, the tail, and
so on. We disregard these various characters, which typify only isolated individuals, varieties, or races, respectively, as being unimportant, and considering only
the features that are common to all dogs, arrive at the *concept of the species*
domestic dog (*Canis familiaris*).

To the domestic dog, we then add the animals that most resemble it—the wolf,
the jackal, and so on, which have a considerable number of features in common
with the dog. Combining them conceptually yields the supraordinate *genus
Canis*. Analogously, we have developed the genera *Vulpes, Otocyon,* and so on,
which are connected with our genus *Canis* through a series of morphological
relationships. By once more eliminating the differences, we reveal the morphological commonalities and are left with the *family Canidae;* by adding similar
families, we have the *superfamily of the Arctoidea,* and then, going further, the
suborder Fissipedia, the *order Carnivora,* and so on, hence, the more inclusive
units of structural design; within this framework there is a pronounced gradation
of morphological resemblance.

* * *

There are two reasons that such a classification is at all possible:

1. One critical factor is that the organisms we are working with do not present themselves as a continuum of forms or in series of evenly spaced morphologies; rather, separating the forms there are *discontinuities,* gaps, and leaps *of different magnitudes,* which
mark the boundaries of larger and smaller groups.

2. The other is the critically important fact that organisms are not random, chaotic,
incoherent agglomerations of unrelated characters; rather, there is a *strict, stair-step kind
of structuring based on degrees of generality and similarity,* which permits the categories, each with its own particular combination of characters, to be separated out and either
coordinated with or subordinated to one another.

This is what Darwin called a "wonder," and Lotze called a "fortunate fact,"
but it really falls quite generally within the scope of "nature's comprehensibility" [*Begreiflichkeit der Natur*] (Helmholtz), which is the assumption upon
which all of natural science, the understanding of the external world through our

intelligence, is based. According to B. Bauch, this comprehensibility is rooted philosophically in the logical, functional nature of the concepts. It is our belief—and we shall return to this presently—that in typostrophism and the descending breaking down of types we have perceived the actual processes constituting the objective causes of the interlocking of types and categories.

Systematics is based exclusively on a clear understanding of reality, of the actual circumstances existing independent of any theory or interpretation. Our taxonomy, in particular, provides—and this is its first goal—a purely conceptual, formal resolution of the diversity of organic *forms;* its categories are *morphologically* based concepts and have no relation to space or time. This system does not ask about actual relationships existing among the objects it arranges. It does indeed try to express the natural order and kinship inherent in the organism itself, but says nothing about actual origins. *To this extent, the mission of taxonomy is not one of explanation, of finding causes;* its intent is only to compare, to make observations, and to classify.

The value of systematics is primarily a practical one: the conceptual system imparts an overview of the organic world that is not attainable in any other way; it provides a catalog of animal and plant forms that is indispensable for all biological work; in it, the most important features of each represented species can be read. To possess this system means that we are able to compare any individual we have at hand with all known species, to insert it easily and surely into the conceptual scheme of species, genera, and so on—in short, to "decide what it is." If we are dealing with a new form, the system allows it to be placed unambiguously wherever it fits best based on morphological relationships, without constraints and with a minimum of diagnosis and description.

THE NATURE OF TAXONOMIC CATEGORIES

It has often been maintained that the *species* (taxon) is an exception among the taxonomic categories; its nature is *real,* as opposed to the supraordinate, higher units, which are purely abstract concepts. The basis for this reasoning is that the concepts of species or the diagnoses as species can be extrapolated and applied directly to given individuals.

This view, repeatedly upheld, is clearly based on a widespread error in logic. *Only individuals are actual and real, tangible and corporeal, perceptible through the senses. But a single individual never bears and embodies all that goes into the complete picture of a species.*

We can perhaps see this most clearly by referring to the *holotype of a species,* which plays a role in the nomenclature of animals and plants—in the naming of the concepts developed by taxonomy. We select one concrete individual as the holotype and attach the name of the species in question to it. It often happens that a later author, because of more or better material, arrives at a different view of a species proposed by an earlier author. Newly acquired viewpoints or differences overlooked by the original creator of the species—differences within the group of forms as the first author understood it—may make it necessary to split the former species.

In this case, it is important that there be a type specimen of the species to which the permanent species name is attached for all time: the name continues to be valid for those forms of a species that has been split that are the *same* as the type specimen, or holotype, whereas the form that has been separated receives a new specific name. Thus, the significance of the holotype lies only in its function as documentary evidence for nomenclature: it is the bearer of a name; it is clear, however, that, as an isolated individual, it can never itself objectify the universal type of the species—the conceptual ideal of the species.

The holotype is usually an adult specimen and either male or female. The species concept, however, is more inclusive; it takes both males and females into consideration, disregarding the various characters of the often very pronounced sexual dimorphism. Moreover, in its diagnosis, it takes no notice of all individual, habitat-related, racial features of the individual specimen that go beyond the general features of the species, but rather encompasses all this variability. Furthermore, the idealized species includes not only the fully grown, mature stage of plants and animals but also their ontogeny and the different stages of, perhaps, a metamorphosis, an alternation of generational cycles, seasonal dimorphism, and so on.

The supraindividual concept of species only emerges, therefore, through the *conceptual consolidation of a multiplicity of individuals,* and because, as a generalization of empirical reality, it contains only selected characters, it cannot be directly extrapolated to discrete, tangible individuals. To characterize an individual unequivocally, a whole array of individual features must be included that would be disregarded in establishing the concept of species. However, other features that are essential for the complete picture of the species but that are lacking in the individual for reasons already given must be left out. It is the same with the type *species,* which, from the point of view of nomenclature, forms the backbone of the *genus* but likewise does not embody the general concept of the genus as such.

Thus, the concepts of species are developed according to the same abstracting methods of generalization and stripping away of characters as are the higher taxonomic categories. Consequently, the species concepts have the same logical structure as the others and, like them, are *not actualities and do not characterize tangible objects;* rather, they construct inferential, generalized concepts directly from the thing itself. The species can never be equated with a particular individual; with its essentially longer existence and broader scope, it rises above the reality of ephemeral, always changing, discrete individuals, which are bearers of the concept of species only when taken collectively.

But even in cases in which the species is established only on the basis of a single known individual—which happens occasionally for extremely rare forms—its character is still not real. It is only that one individual that can be identified; under such circumstances, it is completely impossible to develop a well-grounded conceptualized species, and by the same token, there is also no question of a "real conceptual species."

* * *

Now and then, intellectual fuzziness causes curious flowers to bloom in this field. In groups that are poorly represented or little studied, it sometimes happens—analogous to the cases just mentioned—that a genus or even a family or suborder is based on only a single species. Since the species is understood as being real, it is assumed that the concepts of genus, family, and so on, are also real!

To counter this assertion, often made in all seriousness, the following objection can be made: There is indeed thorough justification for the establishment of a separate genus, family, and so on, on the basis of a single species, provided that it differs sufficiently from other species; the specific, generic, and familial descriptions in question, however, should absolutely not read the same, but must, here as elsewhere, maintain an increasingly general formulation. In such a case, obviously, definitive descriptions cannot be developed at all. They would have to await the discovery of other species of the same genus and other genera of the same family with which the original specimen could be compared; only then could a tenable concept be abstracted. Until that happened, the most that could be proposed would be suppositions as to which characters should be construed as specific, generic, or familial and should be omitted in the development of the next higher categories. *A concept of family based on a single species is therefore never identical with the species concept and consequently is even less real, that is, even further from empirical reality, than the species concept itself.*

Furthermore, one should bear in mind that each genus, family, order, and so on, was once, at the time of its origination (phylogenetically speaking), provisionally represented by a single species. In general, a new category makes its appearance with only a single species or, to put it more precisely, with only a few individuals. But does that mean that every taxonomic category would be in statu nascendi, a reality during its early historical stages but subsequently losing in reality as other "real" species appear—be it through progressive phylogeny or only expanded knowledge—to become an unreal, abstract concept? This is clearly nonsense; the notions in question are disposed of on their own terms.

All systematic categories, the species as well as the higher units, are, rather, of a general conceptual nature; they have no tangible, material actuality but are solely metaphysical realities. It is only a matter of degree inasmuch as the species concept is the most objective and, correspondingly, the richest in characters, and from there, the abundance and specificity of common characters diminishes step by step on up through the supraordinate categories. The number of individual characters to be added to the concept of genus, if one wants to begin there in the characterization of the individual, is correspondingly greater than it would be for the concept of species. For the rest, however, both have the same logical structure; they are *purely conceptual,* and contrast with the discrete reality of individuals.

*   *   *

To counter this explanation, it could be argued that there is something special about the case of the species, that there is a real bond between the members of a species inasmuch as they recognize themselves as belonging together, they interbreed and reproduce among themselves. This is not, however, unconditionally true; there are distant races of a species that no longer mate, just as the reverse is true, that there are well-founded different species and even genera that, at least under laboratory conditions, produce fertile hybrids. In the case of reproductive isolation, it used to be that only the breeding community of the race or subspecies in question was considered real and the supraordinate species was deemed an abstraction, which further demonstrates the purely relative and untenable nature of this whole way of looking at things.

Moreover, it should be pointed out that the purely conceptual nature of the species does not at all exclude those kinds of life patterns. Obviously, the concept of species is not a purely intellectual artifact unrelated to any object—this would be useless in scientific research, which is always exact and object-oriented; rather, it circumscribes the tangibles that fall within its scope, it is the quintessence of the reality upon which it is based. And this reality upon which the concept of species is founded consists of a group of individuals with extensive morphological similarities and which form an interbreeding community.

Similarly, the higher categories are also based on objective bodies of facts. What they contain is marked out and defined by nature through discontinuities; they are by no means fictions developed arbitrarily by the human mind, not empty conceptual schemes, as commonly used to be thought. In the previous chapter, we explained in detail that the diagnostic features of types from the levels of genus, family, and so on, arise through saltational evolutionary transformations (typostrophes). Hence, the sets of characters circumscribed by the categories in question have their real basis in the particular evolutionary processes by which they developed. By the same token, they are historical products, as are the characters of species, which in turn shows that the *logic behind the categories of different magnitudes does not differ in essence but only in degree.*

On the other hand, if the Darwinian notion of a generally flowing transition from species to species were correct, the concept of species would actually be eliminated; under such circumstances, neither species nor higher categories would be differentiated or discernible. Numerous authors, beginning with Darwin and going on, most recently, to Ernst Mayr (1944, p. 153), are of the opinion, therefore, that it is only the gaps in the fossil record, hence in our knowledge, that draw the lines between the sharply discrete fossil species. Analogous to this, unknown, unavailable links are assumed to connect Recent species. According to this view, the boundaries of species would rest largely with purely superficial coincidences. In spite of that, such sharply defined species are said to be founded objectively and are regarded as natural realities, whereas in complete, closed descent communities only a subjective classification of species is possible!

This view goes on to say that the genus, however, lacks the objective reality of the species, for most genera are not separated by large, well-defined gaps (Mayr 1944, p. 281); this, however, can scarcely be acknowledged as being generally true. It is particularly the supraspecific categories that are usually separated by wide gaps. Furthermore, since according to the same author, there are very different understandings of the extent and definition of genera, families, and orders, they are held up as subjective categories.

The only thing that is *subjective,* however, is the evaluation and naming of the categories, a relatively minor problem. It is much more important that in nature there exist groupings of various rank and extent separated by distinct, natural discontinuities, and that these are *objective realities* and are the basis of our categories no matter what they are called. This observation also shows that in principle there are no essential differences in the logic of the individual categories.

For reference, we insert here a summary of the most common supraspecific categories, with the optional taxonomic units enclosed in parentheses to distinguish them from the obligatory ones: kingdom — phylum — (subphylum) — (superclass) — class — (subclass) — (infraclass) — (superorder) — order — (suborder) — (infraorder) — (superfamily) — family — (subfamily) — (tribe) — (subtribe) — genus — (subgenus). Other optional categories are the cohort, usually inserted beneath the infraclass; the section and subsection, beneath the subgenus; and so on. In some cases, however, these terms are used in a different sense and position.

Above the level of genus, all the categorical names are plurals. The name of the subtribe is formed by adding the ending -ina to the root word of the genus serving as the type. The tribe uses the same root and adds -ini, the subfamily ends in -inae, the family in -idae, and the superfamily (although here there is no generally recognized rule), most expediently for the sake of brevity, in -ida. The order and its different levels are either designated by names derived from a type genus or bear their own names based on characteristic features.

The type for each unit is named for one of the next lower obligatory categories: the genus (which here is also the bearer of a name) names all categories up to the superfamily; the family, up to the superorder; the order, up to the superclass, etc. It seems to us appropriate and necessary to typify the higher ranking categories, in contrast to George Gaylord Simpson (1945, p. 32), who believes that the infraorders and higher groups have no types. For practical questions of classification by type, of nomenclature, and so on, which lie outside the scope of our subject, we refer the reader to the excellent, lucid description by Rudolf Richter (1948).

*   *   *

With regard to the *species concept* in particular, we add that *common descent* is often stipulated as the determinant of the forms that fall within the category. Such a requirement cannot be met either in paleontology or in modern biology; for how could even a zoologist determine the specific kinship of an isolated find of an individual that is not in the habit of carrying its pedigree around with it! This criterion has nothing to do with pure systematics. The taxonomic concept of species can be based only on characters that are directly observable in individuals; the individual can only be evaluated based on what it is and what it presents.

However, what can be investigated objectively in living subjects, provided there is sufficient material available for research, is the *genetic basis* of variants; identical genetic makeup in representatives of a species is a valid criterion for common descent. On that basis, many researchers define the species as a group of individuals with identical groups of chromosomes and genes, which enables the individuals of a species to interbreed freely with unlimited fertility, whereas in the occasional hybridization between idiotypically different species the result is low fertility. However, such a definition of the species category is, at least for now, essentially a theoretical postulate, for the internal genetic causes of specific characters are known in only an infinitesimal number of Recent species.

In any event, in practice, regardless of any academic requirement, even zoological and botanical species were originally founded and defined almost entirely on the basis of morphology, and only later, in a few instances, was testing for the possibility of interbreeding carried out. In addition to the morphological features, there are subordinate chemical and physiological criteria (type of constituents and metabolism, composition of proteins, ecological behavior, particular manifestations of life) that come under consideration, some of which, however, are of only limited application to fossil material.

In paleontology, therefore, for the most part only *morphological* interpretation and characterization of species is possible, which, in spite of all claims, is just about the case for the founding of species in the biology of the Recent [neontology], too. Accordingly, fossil species can be defined as *a group of individuals that are identical in all features deemed constant and genetically predetermined,* and presumably, as in Recent species, represent a freely interbreeding community. Recent species examined for their genetic behavior provide clues and models for the scope of the morphologically defined species and for the allowed range of variability.

This is by no means, however, a matter of constants, for the degree and magnitude of the differences in the species of individual genera and families vary widely. For fossil forms, and also often for Recent forms, there is consequently no objective standard by which to choose the scope of the species; so in this regard they are largely like the higher categories, with their fluctuating boundaries and reevaluations. No proof can be furnished that the fossil species consistently or at least on the average corresponds in rank to Recent species.

On the other hand, from the point of view of genetics, fossil species do *not* necessarily and fundamentally have to be different from Recent species, as Simpson (1945, p. 19) asserted. According to his interpretation, there is a far-reaching difference in that in contemporary species, which are defined as interbreeding populations, an exchange of genes throughout the entire group is possible, whereas this does not hold true for fossil species as segments of historical evolutionary series, for the latter representatives of the species cannot mate with the earlier elements, which are already dead. But this is not a matter of *actual, realized* physiological mating but only of *potential* mating. Even in Recent species, populations of different years, like fossil species, are prevented from interbreeding by the barrier of time, and the same limitations are imposed on representatives of a species that live far apart. Nevertheless, based on their genetic structure and their physiological behavior they are potentially able to interbreed, and this is the decisive factor.

In any event, it is a fact that particular fixed evolutionary patterns are at the root of *all* of the more significant inherited differences in form. Consequently, categories set up on the basis of phylogeny have an objective ground. Of less importance is whether fossil categories correspond consistently to the Recent entities of the same name, or whether, in some cases, they are not equal in rank.

In modern biological taxonomy, the "new systematics" of Julian S. Huxley, the species has been moved out from its central position. In its place, *races* and *subspecies* have been installed as primary entities. In this system, the species definition does not emerge from a comparison of single individuals but of entire populations and series. The morphological definition of species is basically replaced by a biological one, in which ecological, geographic, ethological, and genetic factors are at the fore. But ultimately, in the recognition, discrimination, and definition of forms, species, and subspecies this method must also use diagnostic characters that are morphological in nature but more or less insignificant from a biological point of view.

This methodology is of very limited application with regard to fossil material. Character differences of the level that serves to identify Recent races are, as has already been emphasized several times, scarcely susceptible to preservation and lie below the threshold of paleontological proof. The lowest categories that paleontologists can establish on the basis of skeletal structure and other easily preserved hard parts are therefore probably consistently of higher rank than the categories established by Recent biology.

NATURAL AND ARTIFICIAL SYSTEMATICS

As a result of what we have presented so far, we observe, in summary, that taxonomy, like all other systematics, is not theoretical and is based purely empirically on given facts, and that it attempts to classify organisms primarily on a morphological basis, that is, to establish an idealized, conceptual system. There are different ways of doing this, of breaking things down—different principles of classification, to use the logician's term.

For example, books can be arranged according to their size, their thickness, or the kind and color of their binding. But these principles of ordering would not make it easy to find a particular work in a large library. On the other hand, the books can also be arranged alphabetically by author. This ordering principle would indeed be unambiguous and easy to figure out; but it would not say anything about the singularities of the books, their content, their intrinsic importance. From a practical point of view, this crude, superficial principle of classification would indeed serve the purpose of making the books easy to find; but in a library arranged like this, works from the most varying fields of science would all be mixed up together. The procedure would be inadequate and inappropriate for the material.

A natural classification can only be gained if one arranges things based on *relevance,* takes the *natural relationships* among objects into consideration. For example, if one assembled literature having to do with railroads, and then works on automobile transportation, shipping, and air transport were added to that, all together they would form one large category of transportation. (It is not important that the usual procedure is the other way around, starting with the larger heading and subdividing it progressively. Such a descending classification is also often carried out in taxonomy, in zoology, for example, where even in prescientific times, one began by characterizing the higher categories—while simultaneously excluding the lower ones; with plants, however, the system was primarily built up from the lower categories to the higher ones.)

The alphabetical arrangement of our books corresponds taxonomically to *artificial systematics,* or raw *diagnostics,* which works schematically with a biased system of ordering based more or less arbitrarily on superficial, conspicuous characters without doing justice to the natural, inner morphological relationships of the organism. The most well-known example of this is Linnaeus's sexual system of classification of the flowering plants (*Systema Naturae,* 1735), which is based on the number of sex organs (stamens, pistils) and the mode of their fusion, which is indeed excellent for determining the forms but often disrupts the natural morphological context.

Artificial systems are purely *diagnostic* or *analytic* inasmuch as they are solely concerned with carefully working out *differences* and pay no attention to existing *similarities of form.* There is no longer room for such systems in modern biology; we still find them occasionally in the form of identification keys, which are based on an arbitrary choice of easily recognized characters and intentionally serve purely practical ends.

Otherwise, artificial systematics has been replaced everywhere by *natural systematics,* which, through detailed consideration of inner structure, of the organism as a whole and all possible morphological relations, works out the commonalities and resemblances of the objects studied, having in that respect a *synthetic character.* In constructing its system, it follows the natural order that is inherent in the organisms themselves and leads to the segregation of groups of related forms that display degrees of similarity based on common basic structures. Their categories take into account all recognized features of the forms they subsume, even, as has been expressed in seeming paradox but very aptly (J. von Pia 1914), characters "in certain cases not yet even observed."

Morphology is also the basis of this natural system, but in correspondingly more depth, for here the individual basic organizations and the transformations they are subject to can be discovered. The prerequisite is a detailed comparative-morphological and evolutionary analysis of the materials to be classified, a careful weighing of all of the separate characters, and the recognition of commonalities in spite of the differences in individual realizations; however, actual differences occasionally masked by extensive superficial similarities must also be recognized. Hence, the important thing is to separate the true, typical similarities of form, or *homologues,* from similarities that are only apparent, from convergences or *analogues,* which often appear as surprisingly similar structures and functions.

* * *

For example, a taxonomic concept that would subsume *winged animals* would be a false construction because the wings in the individual classes of animals are not at all homologous, morphologically equivalent organs in similar organizational contexts. They have, however, the same function and, further, show many of the same features in their external structure. The wings of bats and insects[2] are similar in that both are broad, lateral appendages, both consist of thin membranes and have internal supports spanned by the flight membrane. In each, however, the internal structure is fundamentally different.

2. In presenting this example, we follow H. H. Karny (1925).

The supports of the bat wing are formed from bones, and these correspond completely in form and arrangement as well as in histological structure to the bones in the forefeet of other mammals. The bat, as its other features show, belongs to the mammalian type; its wings are homologous to the forelimbs of the mammalian organic structure. They do indeed deviate considerably in external form and function from the walking legs of a four-footed mammal, the digging legs of a mole, or the prehensile hand of the human, but it is easy to trace them back to the typical mammalian forefoot; the changes in the basic structure consist only of a lengthening of the elements of the toes and the addition of the flight membrane; the rest is the same.

The superficial similarity to the insect wing is much greater. Nevertheless, there exists here only an analogy and similarity of function. The anatomy and tissue of the wing as well as its overall structure, the way it is put together, is completely different in insects. The support structures, the veins, have no bones, and so on. Furthermore, the origin of the insect wing cannot be traced to a leg because wingless insects have the same number of legs—six—as the winged ones. Only the equivalent parts of the same organizational type are homologous; they show the same positional relationships, develop according to the same plan, and have evolved from the same rudiments.

Furthermore, as we now know, the concept developed by Cuvier wherein the Radiata, including the protozoans, coelenterates, and echinoderms—three fundamentally different structural designs—were lumped together, is untenable. Other unnatural units are the marine mammals, or *thalassotherians* (whales, fur seals, manatees); the *Pachydermata,* or thick-skinned animals, which included the elephants, hippopotamuses, rhinoceroses, tapirs, and swine; and the *edentates,* which joined the sloths of South America with the aardvarks of South Africa. The classification of birds into *carinates* and *ratites* is outdated, for the latter is a heterogeneous group of forms with very different characteristics, the only feature in common being the secondary loss of the capacity for flight.

All these categories, now abandoned by modern natural systematics, are based on characters that were chosen incorrectly and given undue importance. In general, classification used to be based mostly on the immediately apparent, superficial likenesses among external adaptive characters such as are found in animals with the same mode of life but very different anatomical structure. Thus, whales were originally grouped with fishes, and bats with birds. Their corresponding adaptations always affect only a few characters, however, and therefore do not express natural kinship, which is demonstrated not through isolated characters chosen at random but through the similarity of as many characteristics, primarily of internal structure, as possible.

Thus, in setting up a natural system, the task is to discover those general organizational features that are typical not only for groups of specially adapted forms but also for the comprehensive basic structural designs that encompass them. It is those features that are relatively very consistent and have the broadest scope that are thus regarded as typical. The determination of these characters does not lie solely with enumeration but rather with carefully considered evaluation.

Looked at purely superficially, the marsupial wolf, according to an example given by George Gaylord Simpson (1945, p. 8), has a greater absolute number of characters in common with the wolf than with the kangaroo, which, in turn, resembles the jerboa in many external features. But if we do not just count the characters here, but evaluate them according to their distribution and scope, it turns out that the anatomical and structural likenesses between the marsupial wolf and the kangaroo, on one hand, and the wolf and the jerboa, on the other, are more basic and more typical than the superficial adaptations, which point in a different direction. It is the former that characterize the larger, broader group of forms and are the basis for true kinship, whereas the others are purely analogous structures built up on different foundations.

Before a system is set up, therefore, it is indispensable that a careful evaluation of each available character and of the different possible ways of classifying them be carried out. In this process, the characters that will decide what is to be combined or contrasted, the *index characters,* as we might call them, must be determined separately for each group, so that, in contrast to artificial systematics, each category receives its own specific, natural principle of classification. This is necessary because the taxonomic value of individual characters varies considerably from unit to unit: characters that are not important for one category may be extremely significant for another one.

The index characters used in contrasting the larger units of the system do not, therefore, have to express consistently and unconditionally an absolute higher degree of morphological difference. For example, the differences in form and morphological disparities among the genera of certain families may, under certain circumstances, be smaller than those among species in other families. Naturally, within any particular unit it is important that the genus be characterized by greater, more numerous, and more far-reaching differences than the species, that families reflect more important and more general characters than genera, and so on. The process should not be schematic; only experience and sensitive observation guarantee an objective system.

In contrast to the sloppy artificial system, which works with only one or a few characters, the preliminary result of establishing a natural classification will be, first, a hierarchy of classes of characters and, then, a hierarchy of taxonomic categories. In general, in establishing the levels of characters above the species, one proceeds in descending classification, from the top to the bottom, that is, determining first, by comparing the collection of forms at hand, the most comprehensive, most general features and then, step by step, the more restrictive ones. Thus, the insertion of a species into this conceptual system means that the combination of characters in the species includes all of those characters that are typical for the sequence of categories supraordinate to it.

The artificial system is subjective and dogmatic in character. In it, principles of classification are established and rigidly adhered to that are inadequate for the nature of the object to be classified. In contrast, the natural system is empirical. It begins with a detailed understanding of the objects and, based on this experience, attempts to discover for each group those characters that appear to be the most suitable for portraying the objective order inherent in the objects themselves.

Against our view of the *system as being objective,* it could be argued that the choice and evaluation of the index characters and, correspondingly, of the diagnoses of the categories are to a large extent a matter of subjective discretion. This is certainly true, especially so for fossil material, which always provides only a portion of the totality of characteristics. Consequently, we have no guarantee that we are apprehending the characters that are the most important, the most basic for any given types and categories. The choice we make and the order of rank we postulate may be related only to derived features that are of relatively minor importance for the structural design in question, representing perhaps incidental phenomena attendant upon the actual first-ranking characters.

As a practical matter, however, this is of no consequence for the development of the natural system. The boundaries of the categories and the order occurring naturally in the diversity of forms are set down for us through the discontinuities among the multiplicity of forms, and every choice of characters that gives proper expression to those wide gaps is objectively justified and portrays the existing order appropriately. It is, in fact, the mark of the natural classification as, one might say, a reconstructive conceptual system mirroring nature that it does complete justice to the totality of structural design features and that, consequently, subdivisions on the basis of different characters lead to the same result. We saw proof of this earlier in another connection, with regard to insects and snails, for which classification on the basis of different classes of morphological and anatomical features led to systems that were nearly identical.

*   *   *

With regard to the *practical development of the system and the diagnoses* we should add just a few brief remarks. The categories should be characterized precisely and logically so that the forms can be diagnosed and assigned without ambiguity. The definitions of the categories, therefore, must reflect the diagnostically most important index characters—those used in demonstrating contrast—and only those. On the other hand, the diagnoses are not descriptions and should not be concerned with anything but the major criteria establishing the distinctions between the coordinated categories.

Nevertheless, this obvious logical stipulation is often misunderstood and the silly charge is made that taxonomy should not be based on only one or just a few characters but must consider them in their totality. This demand, in itself, is naturally quite legitimate, but it is out of place in this context. It goes without saying that a careful examination of all features is essential; it takes place, however, as we have seen, when the material is first being analyzed, *before* the diagnoses and the system are established. On the contrary, it would actually be a mistake of method to weigh down the definitions of the categories with a recounting of all of the observed features; this would contradict the reasoning behind the system, behind the overview and the saving of labor it strives to achieve. The few characters indicated in the diagnoses are valid then pars pro toto, as an abbreviated expression of the totality of distinguishing characteristics.

Another frequently committed error in logic occurs when, in the diagnosis of one category, all of the characters are repeated that are given in the definition of the higher, supraordinate categories and are implicit for all the groups subsumed in the higher category. Thus, a diagnosis should contain only those features that indicate the differences between taxonomic units of equal rank and that constitute the basis for contrasting them.

*   *   *

The natural system, as we have already said repeatedly, yields a faithful image of nature; not only does it provide an overview and the possibility for identification, but (ideally) it reflects natural, true morphological kinship. Comparative morphology (including the metric and statistical variation methods), however, upon which the natural system is built, is a nonhistorical discipline, one not concerned with time. It consists only of a simple comparison of the different morphological states. It can give no information about the temporal orientation of transformation and, therefore, about the sequence in which the morphological steps should be arranged. In this point, an important one even though insignificant for the development of the categories themselves, comparative morphology lets us down, and it is *here that paleontology makes its essential contribution, for its historicity allows the introduction of the temporal component.*

To use one example, among the mammals there are forms with one, two, three, four, and five toes. These characters are important for constructing the system and are used in the ungulates, among others, for classification. We can also (taking into consideration, of course, all the rest of the characters accessible to us) build corresponding lineages and systematic categories; however, Recent material gives no information about the sequence in which these categories should be arranged and understood.

Theoretically, the following five possibilities are conceivable:

1. gradual development from one-toed to five-toed forms;
2. progressive reduction of toes from five- to one-toed forms;

3. initial forms with an intermediate number, two or three toes, and linked to these, subsequent lineages, some with an increase, others with a decrease in the number of toes;

4. remodeling of one-toed to five-toed forms as in (1), but in some lineages degeneration to forms with fewer toes;

5. and finally, evolution as in (2) from five toes to one, but in a few groups reestablishment of a greater number of toes.

The last two possibilities would be extremely unlikely because of irreversibility but should not be excluded entirely.

Occasionally we can indeed find in ontogeny clues to the existing morphological relationships. In an ungulate with few toes, for instance, early in ontogeny vestiges of other toes can be observed which never go on to develop completely. This would lead to the conclusion that *this* species derived from forms with more toes. But for the totality of mammals, this individual case would not prove much. It would be entirely possible that the subject was an isolated regressive lineage, which in general, was nonetheless following the transformational path of mammals advancing from forms with one toe.

In another context, we have already become acquainted with proterogenetic development, in which the juvenile stages of an individual do not reflect the morphological states of their precursors but bring forth novelties, which only in descendants attain their full development in the mature stages as well. Thus, the early stages of ontogenies are ambiguous; they provide no sure clues as to the direction of morphological development.

Unimpeachable conclusions are obtained only from fossil material through the direct observation of actual, temporal successions of individual morphological states, and fossils prove unequivocally that the oldest, most primitive mammals had five toes and that all forms with fewer toes are later, reductive transformations of them. Only in this way do we have reliable grounds for determining the direction of the transformation in time and for setting up taxonomic categories in an order that reflects nature, all of which enhances the naturalness of the system, but does not, however, seem necessary for its practical utility.

*   *   *

On the other hand, we assign no importance to the temporal factor provided by paleontology in the *relative ranking of the categories.* The suggestion has been made (by G. Säve-Söderbergh in a brilliant attempt to revise the system for vertebrates) that the time at which a group originated, or better, split off phylogenetically (in connection with the grade at which branching occurred) be made the criterion for its rank. A low point of splitting off for a group of forms, which would mean that it appeared early, would cause its corresponding category to receive a high taxonomic rank.

Within the Choanata, that is, the vertebrate group characterized by internal nostrils (choana), the splitting into the two fish lineages of the crossopterygians (lobed-finned) and dipnoans (lungfishes) took place earlier than the separation of the mammals (fig. 4.1). To the former, then, the rank of class would be assigned, whereas the mammals, as a later branch of a lower grade, would have to be satisfied with the rank of order. Crossopterygians, dipnoans, and urodeles (long-tailed amphibians) are treated the same as the Eutetrapoda (all tetrapods except the urodelids). The orders of the stegocephalians are on a level with the birds and mammals, formerly considered as classes but now only tiny buds and appendages in the phylogenetic diagram created by Säve-Söderbergh based on his system.

An objection to this (which has already been expressed by W. Gross) is that the temporal origin of lineages and the duration of their existence does not in any way determine the extent of morphological differences, the sum of diversity of form achieved, and the range of the assemblage of forms in question, criteria that are so critical to taxonomy. There are ancient animal lineages that have scarcely changed at all, or only extremely slightly, over long periods of time—among these are the form-poor crossopterygians and the dipnoans—whereas others, the placental mammals, for example, produced profound changes and a great abundance of forms in a short period of time. Hence, there is *no proportionality between the temporal factor and the grade of morphological divergence,* just as, on the other hand, contrary to the well-known "age and

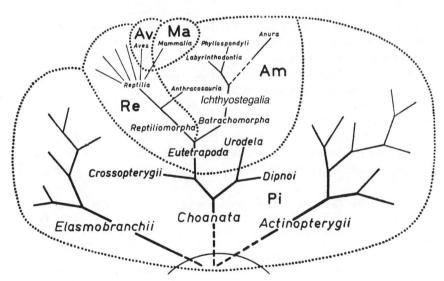

Fig. 4.1 Diagram of vertebrate (gnathostomes) genealogy and its system based on the phylogeny as proposed by Säve-Söderbergh. *Pi* = Pisces; *Am* = Amphibia; *Re* = Reptilia; *Av* = Aves; *Ma* = Mammalia in the usual sense and with the usual subdivisions of class.

area" hypothesis of Willis, there is no parallel between the geologic age and the distribution of individual groups of plants and animals.

Groups of animals of the same categorical rank do not, therefore, have to be of the same age at all. What are decisive, because the system is intended to express morphological circumstances, are the disparities among forms in the individual categories and, further, their richness in individual forms and special variants governed by the characters in question. But it must be acknowledged that mammals, in issuing from reptiles, have acquired such profound morphological differences that they differ more from reptiles than the orders of reptiles do from one another. The disparity of form between the actinopterygians and the crossopterygians is certainly considerably less than that separating them from the tetrapods.

Indeed, birds show scarcely greater structural independence than many orders of reptiles, yet they are assigned the rank of class because of the enormous number and diversity of forms that their structural type embraces. If the form had become extinct immediately after its appearance with *Archaeopteryx* and *Archaeornis,* it would only be added on to the reptiles as an aberrant, feather-bearing order. And so, because of their broad scope,[3] a contributory factor in the evaluation of characters, birds are regarded as a separate class, subdivided in turn into no fewer than forty-eight orders. If the pterosaurs had produced a similar number of species and abundance of forms, they also would have been granted a class of their own and not simply been assigned to the order of reptiles.

Under these circumstances, we do not see ourselves in a position to follow the thinking, interesting in itself, of Säve-Söderbergh and that behind similar views proposed by a few other authors. Moreover, a system built upon this foundation would be quite impractical: every new find can move the branching point of two lineages further back, which would upset the evaluations of the previous categories and, in certain cases, of the entire system.

Furthermore, it is impossible to use the time factor as such to segregate taxonomic categories. Many paleontologists have proposed species and genera that are morphologically identical with some already known but separated from them by lesser or greater periods of time. When this difference in age is considerable, it should be considered that the forms in question are in fact not the same and do

3. According to Ernst Mayr (1944, p. 286), Recent birds alone comprise nearly 2,600 genera with 8,500 species and 27,000 subspecies! Clearly, however, taxonomic splitting in this favorite field of study has been carried to extremes. This is shown by the fact that, according to the same author, for those 8,500 species there are more than 10,000 generic names! About 2,600 of these are commonly recognized, which, however, in view of the many monotypical genera, is still a very high number—even though in extremely broad groups it appears indicated and justified, in the interests of clarity of classification, to assign relatively higher rank to lesser disparities of form than is done in smaller groups of forms.

not belong together taxonomically. Careful examination will then always reveal morphological differences as well.

In every case, a morphological indicator that would allow the form to be identified independently of its stratigraphic age must be required; purely temporal grounds are inadmissible. Also, the practical consequences of evaluating fossils by chronology would be disastrous. We would be identifying fossils on the basis of their geological age and then turning around and using these identifications to establish age!

* * *

The other great significance of paleontology for the system is that only when fossil material is taken into consideration does a complete panorama of the overall diversity of forms emerge. Only a knowledge of the entire abundance of forms allows the plant and animal types of the modern world to be arranged correctly within the context of all those forms, and only then do we really come to understand them.

The Recent organic world is the result of an infinitely long evolution. It consists of the provisional terminal stages of series of forms, each of which extends back into the past for varying lengths of time. *The true relationships among forms do not lie, therefore, across the horizontal, as represented by the present, but down along the vertical of the past.* The modern, isolated coexistence of individual forms and types only comes together as we trace them back through geologic time to their natural succession and divergence.

The present world of plants and animals and with it, naturally, its system, is contingent upon history. It is thus a gross error of material and method when, as happens repeatedly, Recent organisms are naively and nonchalantly set up as the ancestral forms of some morphological stepped series [*Stufenreihe*].[4]

Only by taking the fossil material completely into account is a relatively coherent general idea of the modification of forms and of the individual taxon possible. The additional information supplied especially by *those groups of animals* that are represented today by only a few meager, extremely isolated *relicts* of a long-gone, enormous abundance of forms, such as is the case with the cephalopods, brachiopods, crinoids, reptiles, and so on, is indispensable. The relationships among Recent forms can only be interpreted and arranged on the basis of and in closest connection with fossil forms.

MORPHOLOGICAL AND PHYLOGENETIC SYSTEMATICS

Many readers who have followed the preceding discussion will have noticed, perhaps with astonishment or even displeasure, that thus far not a word has been

4. [See chapter 3, note 10.—Ed.]

said about the *role of phylogeny in biological systematics.* We have stressed consultation with paleontology and acknowledged its great importance for understanding the course of evolution. This would seem to call for our giving paleontologically revealed phylogeny a preeminent place in the system, in keeping with the generally held notion. Instead, we consciously chose to discuss first only the *temporal aspect* of the paleontological contribution to the system and, indeed, only as a yardstick for establishing sequences and not for the ranking of categories and their phylogenetic interpretation.

After Ernst Haeckel, in his phylogenetic exuberance, declared that systematics was a subproblem of phylogeny and that the natural system was only another way of expressing the evolutionary tree, this view gained general acceptance and, apart from a few exceptions, remains the prevailing view even today. In the field of botany, it was probably Alexander Braun who was the first advocate of this view, which was later endorsed primarily by R. von Wettstein. It is furthermore very typical of this notion that J. P. Lotsy entitled his well-known work *Vorträge über botanische Stammesgeschichte: Ein Lehrbuch der Pflanzensystematik* [Reports on botanical phylogeny: A textbook of plant systematics], thereby directly equating phylogeny and taxonomy.

According to this view, the task of the natural system is to explicate the historical kinship relations to lineages and the phylogenetic links of individual groups of organisms, hence, to illustrate existence as a process of becoming. The categories are no longer to be defined primarily on the basis of morphology, as concepts expressing likeness of characters, but as groups sharing a common ancestry. The relationship of *form,* which in the morphological system bound the representatives of a category together, is construed as a relationship of *consanguinity,* Morphological series [*Formenreihen*] are seen as lineages [*Stammreihen*], and so on.[5] Evolutionary considerations, then, become the foremost criterion; that is why, the reasoning goes, phylogeny is the logical preliminary to a sort of *phylogenetic systematics.*[6] But in this very matter, paleontology would have to have something important to say.

For various reasons, we cannot support this interpretation. In itself, from a purely theoretical point of view, such a genetic system would be entirely possible. Classifications may be established on the basis of the most varied aspects and, as far as that goes, even on the *courses of events* that led to the phenomena at hand.

However, one should not give oneself over to the widespread belief that such a phylogenetic system actually functions as an explanation of causes and gives

---

5. [*Stammreihe* as used here is the same as Abel's *Ahnenreihe* (chap. 3, n. 10).—Ed.]

6. [It should be noted that Schindewolf's concept of "phylogenetic systematics" is not the same as Willi Hennig's. Incidentally Hennig's book *Phylogenetische Systematik* was also published in 1950.—Ed.]

an account of the origin of the conceptualized objects. It would only express our historical-evolutionary notions without saying a thing about the causation and conditionality of the forms surveyed. The true etiology lies in the complex of causes, directly comprehensible neither morphologically nor phylogenetically, that shape and modify the forms and, hence, for their own part, bring about evolution, which, like all historical processes, does not constitute an actual "cause." The individual steps in the history of a lineage have no immediacy as stages but are only the results and expressions of evolutionary processes.

Practically speaking, a phylogenetic systematics would be possible if there were at least *one* independent phylogenetic method upon which it could be based. This, however, is not the case and as an assumption must be dismissed. No one yet has been able to show how phylogeny can be revealed other than through an evaluation and dynamic reinterpretation of the static data gained from morphology, taking into consideration the temporal aspect that is the contribution of paleontology. As the real, objective basis for evaluation, we have only forms and morphological likenesses. "Comparative morphology, checked against the positive facts of paleontology must still be the chief foundation on which to base phyletic conclusions" (F. O. Bower). This is not only "still" the case but will in all probability always remain so.

Indeed, people make great demands on the system, talking of divergence morphology, which is supposed to be something different from purely comparative morphology, or in treatises on the "phylogenetic system" of individual groups of plants and animals, attempting to explain to what extent the refinement of our phylogenetic notions are supposed to have changed the system. Anyone who looks more closely at such remarks must recognize that the advance in interpretation is always brought about by one of the following two circumstances: (1) the discovery of new forms that disclose heretofore unknown morphological peculiarities and throw light on other, long insufficiently explainable or incorrectly interpreted species; (2) a different evaluation of morphological peculiarities already known or the observation of new, previously unnoticed characters that lead to an altered arrangement and classification. Never, however, do we arrive at new insights through, for example, the direct observation of any phylogenetic process or connection, an approach that, in view of the static character of the fossils, lies outside the realm of possibility.

*   *   *

One important, recent contribution to knowledge in vertebrate phylogeny is the determination that the Paleozoic armored fishes of the group of the *ostracoderms* (figs. 4.2, 4.3) should be combined with the living *cyclostomes* (lampreys and hagfishes, figs. 4.4, 4.5) in the class of the jawless *agnathans*. This assessment was arrived at because the ostracoderms, as very careful investigations have shown, closely resemble Recent cyclostomes in the structure of the endocra-

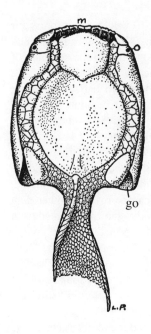

Fig. 4.2. *Drepanaspis gemuendenensis* Schlüt. as an example of the Heterostraci (Pteraspidomorphi), an order of agnathans in which, among other things, the armor is subdivided into several individual elements and, in places, into a mosaic; the eyes are lateral, the nasal pouches are situated forward, and on each side there is only a single external gill aperture. The species illustrated comes from the Lower Devonian roofslate of Bundenbach, in Hunsrück, and is shown dorsally. About ¼ ×. *m* = mouth aperture; *o* = orbit; *go* = gill aperture at the posterior end of the armor, which in this form spreads out like that of a skate. (After Traquair, Kiaer, and Heintz, from A. S. Romer 1933.)

nium, the brain, and the nervous and vascular systems (pl. 10B; pl. 11) and because these like characters have a higher value, that is, are considered to be more typical and constant, than the very divergent shape of the body and the dermal armor of the Paleozoic forms. Because formerly, only those contrasting characters were known, the idea of a phylogenetic relationship with the cyclostomes naturally did not come up. Thus, purely morphological circumstances, and not some possible directly observable relationship of descent, are decisive for our modern interpretation.

This case is an especially extreme example inasmuch as between the Devonian, when the ostracoderms made their last observed appearance, and now, we

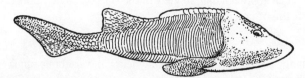

Fig. 4.3. *Cephalaspis,* a typical representative of the Osteostraci (Cephalaspidomorphi), from the Gothlandian and Devonian, with solid, rigid armored shell enclosing the head and often also the forward part of the body, dorsal eyes and nasal apertures, and many (10) ventral gill apertures. Furthermore, powerful electrical organs have developed, and behind the head, there are two scale-covered structures that correspond to the paired pectoral fins of fish. (From E. S. Goodrich.)

do not have a single intermediate form. Here, for once, the postulation of a gap in the fossil record might be well founded, for it may be assumed that if there is a relationship, the prerequisite intermediate forms lost their armor sometime after the Devonian and, like the Recent cyclostomes, did not have skeletal elements susceptible to preservation. (Insofar as such elements were present, we categorically deny, as already laid out in detail, the incompleteness of the fossil record in the usual sense, in particular in such a case as this one, in which the gap would inexplicably extend over such an enormous period of time.) The conceptual joining of the ostracoderms with the cyclostomes in the category of the agnathans is thus based only on *observed peculiarities of form,* and not on the undocumented, only assumed evolutionary connections.

On the details of the relation between the ostracoderms and the Recent cyclostomes, the interpretations of the various authors are sometimes at odds. Because of the different positions of the sensory organs, of the nasal and gill openings, because of the different histological structure of the elements of the armor, the ostracoderms are divided into two groups, the *Heterostraci,* (figs. 2.25, 4.2) and the *Osteostraci* (fig. 4.3). Some researchers (E. A. Stensiö, in particular) are now of the opinion that the Heterostraci are directly connected to the modern *myxinoids* (fig. 4.5), whereas the Osteostraci should be joined with the *petromyzonts* (lampreys) (fig. 4.4). Accordingly, the two modern orders could be traced separately back to the Devonian. Other investigators (E. I. White, among them), however, hold the view that the ostracoderms form a homogeneous lineage and the splitting into the Recent orders did not take place until recently.

This, too, is a matter of the evaluation of the characters determining the evolutionary explanations. *In each instance, purely morphological grounds are decisive for both the arrangement in the natural system and for the phylogenetic concepts.* Both spring from the same source; there is, however, *no special phylogenetic method of any kind* that would be independent of morphological data and that would replace or even just supplement it.

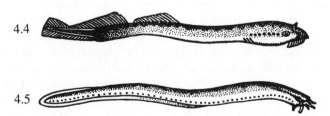

Fig. 4.4. *Petromyzon,* the Recent lamprey, according to E. A. Stensiö a direct descendant of the Osteostraci. The skeleton of living cyclostomes is purely cartilaginous; the loss of the armoring and also the lack of paired fins is construed as a secondary degeneration. (After B. Dean.)

Fig. 4.5. *Myxine,* the Recent hagfish, according to Stensiö a descendant of the Heterostraci. Reduced. (After B. Dean.)

The fact that, unlike Cuvier, we no longer lump the protozoans, echinoderms, and corals together as *Radiata* has also been held up as a result of our improved phylogenetic understanding. Certainly, we are convinced today that there is no direct evolutionary relationship between the groups cited, but this conviction is not based on the fact that we have proven that phylogenetic links do not exist— a proof that would hardly be convincing—but rests on a *more detailed investigation of the structural designs in question* and their morphological features. Only this has led to the conclusion that a major discontinuity separates the protozoans from the metazoans and, furthermore, that there are profound differences between the corals and the echinoderms (possession of a secondary body cavity, an ambulacral vascular system, and so on), which exclude both the possibility of combining them in the natural system and the assumption of direct phylogenetic connections.

Nowadays, we no longer divide the birds into the two major groups of the ratites and the carinates, the flightless and flying birds, as Merrem did. It has been shown that the ratites are not a homogeneous group but rather derive through loss of the ability to fly from probably five different groups, which means that they are only convergent with regard to their negative character of flightlessness. Therefore, they are divided into five orders, which are contrasted with the orders of flying birds on an equal basis. But here, too, there are obviously no directly observable findings with regard to the phylogeny; rather, these results are based on more penetrating morphological investigations, which have discovered the typical resemblances of form and the true structural relationship.

In the same way, it is not a primary phylogenetic realization that first separated the amphibians from the reptiles and that today causes them to be understood as being polyphyletic. Here, too, it is more a matter of morphological observations, or better, of nonphylogenetic systematics, for which a phylogenetic ground was supplied after the fact.

*   *   *

In view of the claims for phylogeny, we repeat the point of view we have already spelled out, that the construction of a system, including that of taxonomy, is a logical operation, whose task is to arrange objects conceptually and not to theorize and interpret. Furthermore, the view that a natural system *must* necessarily be genetic is, we believe, unjustified. The periodic table of chemical elements is unquestionably a natural system. Its order is based on the empirically determined atomic weight of the elements as well as on periodic similarities and dissimilarities of their chemical and physical properties. It yields in a natural way series and groups of elements, which, however, have by no means evolved from one another historically and genetically. At least there is no documentary evidence for it that would have played any part at all in the structuring of the system.

The methodology of taxonomy is autonomous and completely independent of ideas about descent. What is real and what we must order is the multiplicity of organisms. Their structures and their differences exist, regardless of whether we think of the forms as having been created or as having evolved from one another.

This independence of the theory of evolutionary descent is already evident in just the historical fact that the beginnings of natural systems as we understand them appeared at a time (de Jussieu, de Candolle, Cuvier, Geoffroy–St. Hilaire, Agassiz, and so on) when the theory of descent was not yet known or before it was raised by Darwin to be the dominant concept in biology. And the fundamentals of these morphological systems, which were uninfluenced by the theory of descent at the time they were set up, have retained their value on into the present. As our knowledge of forms has grown, the systems have naturally been considerably refined, but phylogenetic interpretation and reinterpretation of classification has not been able to add anything essentially new. Even without the introduction of evolutionary theory, appropriate changes would have been carried out, and the system would essentially be the same as it is today.

And it could not possibly be otherwise. What we observe directly is not lineages but varying degrees of morphological similarity in the temporal succession of organisms, which suggest to us assumptions of evolutionary connections. Only morphology, including comparative embryology, decides whether an evolutionary relationship should be presumed, and the temporal component explains the direction of the presumed connections. *Morphology and its expression in the natural system are the foundations of the general theory of evolutionary descent and of the particular history of the evolutionary process; it is not the other way around, that phylogeny is the basis of the system.*

The phylogenetic approach contributes only *secondarily* a historical-genetic, dynamic factor to the already existing system (which in form and content is static) and interprets the ideal-conceptual relationship of forms as the expression of actual consanguinity.[7] This is thoroughly justified and even essential unless we just want to dispense with understanding the forms at all. But it should not be forgotten that it is only an *interpretation of the existing system,* that one has surely not created new knowledge using any particular phylogenetic method. Those authors who advocate a view to the contrary appear not to be sufficiently aware of the logic of these circumstances.

* * *

Let us now take another brief look at what kind of "phylogenetic" methods and criteria people try to assert through taxonomy.

  1. A respected scientist (H. F. Osborn 1923), who regards taxonomy as a form

---

7. The system can only be considered static, however, in contrast to the dynamic process of evolution. Its nature is not purely static, inasmuch as it also includes particular states of variability and of ontogeny.

of phylogenetic expression, says that phylogeny can be revealed through an analysis of *mode of life,* of changes in feeding and locomotion. These factors determine how the structure of the teeth, the skull, the limbs, and ultimately the entire body is shaped and changed (a Lamarckian view, which we reject, but that has nothing to do with the issue here). Thereupon, however, it is proposed that these notions about the mode of life can be deduced from the structures of the teeth, the skull, the feet, and so on, that we have before us—hence, from purely morphological data!

Further, in that paper, which makes a comparison between the Linnaean and the so-called phylogenetic system for one particular group of animals, Osborn goes on to observe in conclusion that the former—and he means the natural morphological system for which Linnaeus had already laid the groundwork in his later years—can be translated into a phylogenetic system without more ado and will faithfully reflect the varying degrees of consanguinity of the individual lineages.

If matters are such that (a) phylogeny can be revealed indirectly by mode of life simply as it is inferred from morphology; that (b) the "phylogenetic system" has not grown out of its own independent base of knowledge but is only an interpretation of the morphologically structured natural system; and that (c) with regard to form, the "phylogenetic system" is to a large extent identical with the natural system, why do we even delude ourselves into believing in methods that do not really exist?

2. Other authors set up the realm, the *geographic distribution* of animals and plants, as an important aid in discovering phylogenetic connections. But this biogeographical principle also adds nothing critical to purely morphological inferences. If the range of a particular group of organisms is very large—say there are representatives of the same lineage in both Europe and America that are identical in all essential features—then the criterion of realm carries with it no considerable restriction of the possibilities for phylogenetic associations.

Or if the lineages replacing one another in various parts of the world are not identical but only similar to a certain degree, making it seem impossible to join phylogenetically the living forms of one continent with the Diluvian or Tertiary forms of another, then once again it is only morphology that explains the existing differences. In every instance of superficial similarity, there are particular characters that demonstrate the differences between the forms in question and keep us, also without going by way of geographic-phylogenetic constructions, from arranging them incorrectly within the carefully worked out natural system.

It is certainly true, as George Gaylord Simpson (1945) pointed out, that similar animals that live or lived in neighboring areas are in all probability more closely related to one another than to other similar forms that are widely separated. But as we have already said, it is only morphology that provides information about the similarities among animals and the possibility of their association with supposed ancestral stocks. Morphology shows that in spite of their broadly disjunct distribution, the tapirs of South America and the Indo-Malaysian realm are related forms and derive from common ancestors in the Northern Hemisphere; that the South American llama is much more closely related in form and genealogy to the Asiatic camel, which lives far from it, than to, say, the South American white-lipped peccary or the pampas deer.

In taxonomy itself, however, geographic factors have no place as criteria or as elements of definitions. The system can only be based on what the organisms themselves are and the characters they display. The system's job is to put us in the position of being able to recognize and name forms even without knowing their origin and distribution. We first have to be able to identify organisms before we can establish their geographic range.

Biogeography is thus rooted in morphological findings that have already been analyzed to the fullest in the natural system, and moreover, it can add nothing to the system. It provides only data on the centers of origin, migration routes, and distributions of organisms *after* their phylogeny has been revealed by the methods of morphology taking temporal factors into account.

3. We have already spoken briefly about the importance of *ontogeny*. Since the natural system considers not only the adult, mature forms but the entire life cycle of organisms, which includes their individual development, all information that ontogeny can provide with regard to morphological connections has already been worked into the system. From the phylogenetic point of view, in contrast, ontogeny, when regarded as "recapitulation," provides only certain general and sometimes ambiguous clues, which are borne out solely by the evidence of the actual existence of corresponding forms; it is the inventory of the system that provides the information. Consequently, inferring phylogeny from ontogeny will result in nothing new for taxonomy, and for phylogeny itself nothing different turns up that a phylogenetic interpretation of the natural system would not also provide. Ontogeny is a valuable aid in discovering homologues. It allows us to determine that certain parts, the homologous ones, which in different organisms may have different shapes and functions, arise from the same embryonic tissue and at relatively the same places within the structure of the basic design. But at first, these are only conceptual relationships, which find their expression in the system and in themselves say nothing about the real evolutionary context of the forms and their causation. If we draw such conclusions, they are not direct observations but interpretations of observations.

4. Further, *monstrosities* and deformities are often used for phylogenetic deductions. All they provide, however, is a picture of presently existing possibilities for modification. Whether this picture refers to the past, however, and corresponds to ancestral states, as the popular interpretation of these phenomena as being *atavisms* usually assumes, is in every case open to question. It is true for the teratological polydactyly of Recent horses (fig. 4.6) and similarly also for

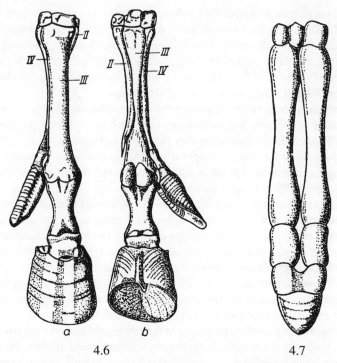

4.6   4.7

Fig. 4.6. Right forefoot of a polydactylous domestic horse. *a.* Front view. *b.* Back view. Toe IV, with the long pointed hoof, has developed abnormally as an atavistic reversion to the three-toed *Hipparion* stage. However, the return to that phylogenetic stage is incomplete, for only this one digit has been reactivated. (After Wood-Mason and J. E. V. Boas.)

Fig. 4.7. Skeleton of the foot of a one-toed pig (mutant of the domestic pig *Sus scrofa* L.), with digits III and IV fused distally. (The lateral toes, displaced backward and no longer functional, are not shown.)

the same condition occasionally observed in cattle, as paleontological evidence shows, even though it is usually not a complete reversion to the foot structure of actual ancestral forms.

In contrast, however, the one-toed pig (fig. 4.7), with toes III and IV fused distally, which is an abnormal mutant of the domestic pig, is absolutely not to be evaluated as meaning that the suids descended from the ungulates! In monstrosities such as these, it must always be taken into account that they can also be a matter of a developmental inhibition or of a neomorph with far-reaching, future significance, having, therefore, nothing to do with previous phylogeny.

5. Finally, *serodiagnosis* is often glorified as a tool in phylogenetic discovery and often even as experimental *proof* of the theory of evolutionary descent itself. Particularly in botany (C. Mez and his school) has it been proposed as the only reliable method for developing the evolutionary tree that is supposed to be far

superior to the morphological one. Obviously, however, this process only gives a picture of the chemical similarity in the composition of the proteins, thus, at the most, of a physiological-chemical relationship among living plants and animals, but it can never show the evolutionary tree, the true circumstances of past evolution.

Purely with regard to method and the kind of knowledge obtained, serodiagnosis is on the very same level as morphology; in both cases, it is chiefly a matter of determining resemblances, likenesses and differences, one working with form and the other with the chemical behavior of the proteins; neither perceives genetic lineages directly. These data are purely static and not dynamic in nature. Further, since the serodiagnostic method must be limited to Recent organisms and is not applicable to the only true evidence of phylogeny, it is at an undisputed disadvantage compared with morphology, although it should not be overlooked that serodiagnosis can provide valuable support for morphology.

Consequently, there can be no question of serodiagnosis providing experimental proof for consanguinity in the evolutionary sense, or for the justification of the concept of evolutionary descent. Here, it is easy to confuse the actual and the figurative meanings of the term "consanguinity." The consanguinity of serological experiments describes only ideal, conceptual relations based on similarities in the serum and has no more importance basically than the morphological relatedness of skulls, skeletons, and so on. All these provisionally ideal kinds of relatedness only become evidence of real, phylogenetic consanguinity in lineages that have already been established according to the theory of descent.

*   *   *

Therefore, autonomous phylogenetic methods that would be called upon to play a role in taxonomy and to influence it do not exist. People who profess to be using one or another are hiding the fact that their methods are *purely morphological;* they are not engaged in true phylogeny but only constructing conceptual connections of the kind that will be expressed anyway by basic, unpretentious taxonomy. Phylogenetic notions, however, which for the most part are derived from the natural morphological system, have nothing to do with the essence of taxonomy and should not be introduced into it.

It has been repeatedly emphasized, and rightly so, that the natural system as a reflection of objective, existing morphological patterns and laws, or of the possibility of establishing these, based as it is on the graded, stair-stepped diversity of forms, constitutes a major proof of the theory of evolutionary descent. But then, if we do not want to end up in a vicious circle and rob ourselves even of this strong support for concepts of descent, we should not credit phylogenetic interpretations with any influence at all on the development of the system. The natural system would become invalid as evidence for theoretical arguments of descent if we were first to put into it what we want later to read out of it.

Phylogenetic approaches and forms of expression have no place in taxonomy; because of their often hypothetical character, they would lead to a collapse of the system, which, in its morphological design, is sound and, because of its purely classificatory nature, is unassailable. Even George Gaylord Simpson (1945, p. 21), who takes up the cause of a phylogenetic basis, has to admit: "Phyletic theory is the most subjective element in taxonomy, the most influenced by differences of opinion, and the most liable to radical change with the advance of knowledge." But since the system has an important practical mission to fulfill, it must guarantee a certain constancy and will not go undamaged if it is altered by that subjective "difference of opinion."

There is also no actual reason for and logical possibility of introducing phylogenetic aspects. The foundation of both diagnostic taxonomy and interpretive phylogeny is morphology. The two are different ways of looking at the same set of facts; they diverge from the same root and are therefore coordinated with one another. Usually, however, phylogeny is based on the classificátory context clearly expressed in the system and in that respect is subordinate to taxonomy. But then it cannot also be at the same time the logical preliminary stage of taxonomy. The two ways of viewing matters are essentially different; *the way to express assumed real relationships is the evolutionary tree or bush, and not a system of any kind.*

With regard to the practical application of taxonomy, it would also be impossible to use the *criterion of common descent* in the diagnoses of the categories, to require proof of genetic connections. The result would be that we could only identify forms of known descent, and since we are usually lacking an objective knowledge of this, we would *not* be able to identify the overwhelming majority of forms. It seems to be particularly absurd that even zoologists and botanists continue to call for a phylogenetic system; absurd, for they, with their own methods, are not in a position to establish phylogeny, that is, ancient evolutionary connections. They would kill themselves trying to set up a system for living animals and plants!

The only element that we allow entry into the system aside from morphology is that of *time.* One might object that this is just a euphemism for the phylogenetic principle and, when all is said and done, is identical with it. This, however, is not so. *The factor of time is completely nontheoretical; it, like morphology, is based purely on observation and experience* and does not include any kind of uncontrollable concepts. It only shows the objective historical sequence of individual morphological states, which exist as such no matter whether we support the theoretical view of descent or not.

We can then interpret this sequence of forms in keeping with the theory of descent as the history of forms, or *the phylogeny of characters,* and represent it by an evolutionary tree of characters (O. Lorenz). Such a phylogeny of characters, which only expresses the relationships among features, is still not yet the same as the actual *phylogeny of kinship,* which asserts very particular existential relationships and is concerned with the real unfolding of the most highly specialized lineages and offshoots, about which, nevertheless, we are often able to express only subjective assumptions. It is these very relationships of descent, however, with their attendant uncertainty, especially with regard to the lower categories, that are supposed to form the criteria for the phylogenetic system!

Furthermore, we concede to the temporal factor no influence of any kind on the development, formulation, and evaluation of the individual categories themselves, but rather we regard it, as already stated, only as the guidepost pointing out the direction in which the series of forms revealed by comparative morphology should be arranged within the system. The system expresses the *sequence of morphological steps* and that alone, as, for example, A. Engler—in contrast to the earlier spokesmen for phylogeny in the field of botany—consciously tried to achieve in his *Natürlichen Pflanzenfamilien* [Natural families of plants] and his *Syllabus.*

Let us illustrate what has been said with a practical example. The family of the *horse* (Equidae) is represented today by only the single genus *Equus.* To be able to understand the anatomical structure of horses, it is necessary to take up the fossil material and to work within the system. Now, as is well known, we have from the Tertiary and the Diluvian remains of four-, three-, two-, and one-toed horses.

If these fossil forms had all been known in pre-Darwinian times and if, without any understanding of phylogenetic relationships and without consideration of the temporal factor, they had been arranged within a natural system purely on the basis of morphology, the separate categories of this system would be exactly like those obtained by taking the phylogenetic approach. Even the various successions of forms within the scope of foot and dentition modifications, which show some morphological difference and which are construed today as parallel lineages, would have been recognized back then and segregated.

The only thing that would have been left in doubt is the direction in which to rank the forms, for the series established by comparative morphology are reversible. This is where the time factor comes into play and shows that the modification proceeded from four toes to one; hence, the individual genera should be grouped in that sequence.

The observation that today there are no horses with four toes, only those with one, would not by itself be sufficient to substantiate this result firmly. We have numerous examples in paleontology of the fact that more primitively organized beings represent long-lasting, conservative lines and survive all their collateral lines of more highly differentiated forms. Thus, it would be entirely possible that the one-toed horse as a primitive type would extend on into the present, whereas the forms with more than one toe would represent derivative forms that arose early and became extinct (somewhat analogous to the manticocerates, presented earlier, fig. 3.72). To be absolutely certain, more detailed inquiry is necessary pertaining to which forms of the horse line are the oldest and in what sequence the rest of the types appeared. These investigations have led to the conclusions cited above and already discussed in detail (fig. 3.93).

* * *

When this has been done—when the relationships of form have been explained morphologically and those of time clarified chronologically for each morphological complex, we can transfer these insights as expressed in the system to the concept of an evolutionary tree, in an effort to understand their nature. However, the view of many authors (for example, Ernst Mayr 1944), that only through such phylogenetic interpretation does the system become natural, lose its supposedly arbitrary character, and reflect the objective state of affairs, seems unfounded to us.

After the introduction of the theory of evolutionary descent, phylogenetic re-evaluation was simply grafted onto the existing, completely morphologically based and conceived system; in structure, scope, boundaries, and nomenclature, the categories remained the same; no extensive changes were made, nor were any necessary. The character of the system, however, cannot be altered merely through a change of concept—interpretation of the relationships among forms as being relationships among lineages, of the morphological type as being the general picture of the phyletic form. Was the periodic table of the elements by any chance less natural and objective before acquiring its modern interpretation attributing chemical and physical properties to atomic structure?

In general, there is probably a consensus today that the system can never portray phylogeny [the course of evolution] completely and give it adequate expression. In the system, the various categories must be arranged in linear fashion; the multidimensional parallelisms, the separations and divergences, the root positions of common ancestral groups, the different rates of evolution in time and over space—none of this can be represented. The illustration of these relationships is, however, not at all the meaning and purpose of the system but rather the task of the evolutionary tree.

Nowadays, therefore, the view of the essence of the "phylogenetic system" is often more carefully phrased in terms of its having a *phylogenetic foundation*. Its goal is held to consist of constructing only those categories that represent phylogenetic units, have common origins, and are based on actual consanguinity. And yet, since these features, as we have already seen, can be revealed only morphologically, and the categories can always only be established with the aid of the kind and number of common characters, the "phylogenetic system" called for is in this sense, too, nothing other than the old, embattled natural system with the time factor taken into consideration in establishing the sequence of individual morphological steps.

Furthermore, since even with a phylogenetic basis, which means presuming a phylogeny that is consistent and completely sorted out, there are still various possibilities for structuring systems or categories, the claim of a "phylogenetic system" lacks any positive justification. An example of one of the various ways taxonomy can be structured lies with the problem of vertical or horizontal classification, which we shall discuss in a subsequent chapter.

Looked at objectively, the logical demand that in the natural system the constituents of a category should be more closely related in form to one another than to the members of another coordinate unit of equal rank is correct. This closer affinity of form is based on the fact that there is a distinct separation between the categories in question, between two genera, for example, and the species belonging to the categories are alike in the diagnostically supraordinate generic characters. However, the existing closer relationship is not valid in the genealogical sense. As George Gaylord Simpson has already pointed out (1945, p. 19), if our two genera are two phyletic units that have a relationship of descent, the last species of the ancestral genus is more closely related generationally to the first species of the daughter genus than to the first species of its own genus, from which it is separated by many generations.

To sum up the findings of this section, we observe that only morphological circumstances arranged conceptually within the system decide the possibility of phylogenetic affinities; it is not the other way around—that phylogenetic concepts determine the given morphological facts. Phylogeny, as a product of taxonomy augmented by spatial and temporal data cannot, therefore, at the same time be its foundation. *The call for "phylogenetic systems" is misguided and cannot be responded to either theoretically or practically.* It is another matter when we carry out a secondary phylogenetic interpretation of the natural morphological system in order to arrive at a theoretical understanding of the relationships among forms expressed by it.

ON IDEALISTIC MORPHOLOGY

A completely morphologically based, nonphylogenetic systematics, as we advocate here, is usually referred to—or better, branded as—the brainchild of idealistic morphology, because for many biologists there is scarcely a more reprehensible notion than that one. In any event, most modern biologists probably regard idealistic morphology as an outdated, anachronistic, undesirable line of thinking, or even as being unscientific, and firmly reject it.

Idealistic morphology is said to arise from a subjective mental state, to disregard objective facts, to take a purely ideational-constructional approach to the organic world, in contrast to phylogenetic morphology, which, leaving aside all human speculation and conceptualization, is the result of completely objective cognition. The type and the homologue of "formal" morphology are said to be subjective, abstract concepts; only processes of descent have any objective character. Phylogeny, correspondingly, is held up as an "empirical-rational science," and idealistic morphology as an "irrational train of thought."

Now, the position we take here is morphologically idealistic inasmuch as it consciously sets up as the basis for its system only the morphological relationships among organisms, and indeed, in keeping with the nature of the system, establishes a purely logical-conceptual order, without getting into interpretations of the origin of forms and types. But our position is not that of Goethe and his pre-Darwinian successors, who saw in idealistic morphology the ultimate ideal of biological knowledge and linked it to all kinds of now outdated notions of *Naturphilosophie.*

Nowadays, obviously, we should disregard these dated ingredients, which have nothing to do with the actual core of idealistic morphology. The central point of idealistic, or nonphylogenetic, morphology consists, rather, in the *comparison of organic forms, in the investigation of the graded levels of similarity and the determination of the relationships among forms;* it is therefore nothing but *comparative or systematic morphology, or typology,* as the method has also often been called.

Such a primary morphological way of looking at things, of course, by no means excludes the acceptance of the theory of evolutionary descent. On the contrary, by far the majority of morphologists are fully convinced that the morphological circumstances they observe can only be meaningfully understood when an evolutionary connection among the individual morphological steps is assumed. We disassociate ourselves emphatically from those few today who still support orthodox idealistic morphology, who deny the theory of descent—which, considering the foregoing discussion of evolution, hardly needs saying.

In other words, we do not see morphology and its role in constructing the system *as an end in itself and an ultimate goal for biology.* We do not think of it as a substitute for phylogenetics but only as a transitional stage on the way to a goal we support—knowledge of evolution, which alone makes the graded diversity of forms comprehensible as being the result of a historical, natural unfolding. First, however, this diversity of forms must be described, sorted out, and arranged, and this is the task of morphology and morphological taxonomy.

* * *

Looked at in this way and purged of all nonessentials and frills, idealistic morphology is anything but a subjective or irrational approach (in so saying, we are obviously disregarding the phenomenalistic theory of the subjective, purely sensory nature of reality). That assertion, in fact, can be completely turned around to read just the opposite:

Morphology and its natural system is purely empirical scientific research. It proceeds with rigorous objectivity from the real, natural data, from the existing forms and the graded, successive steps of their diversity, and arranges them according to logical principles in a graduated conceptual system.

On the other hand, it is phylogenetic concepts that are interpretations and subjective readings. Indeed, we hold the *general* theory of evolutionary descent to be objectively verified fact; but our ideas about the *courses of events and processes of phylogeny* are certainly not based on the empirical. In particular, in the interpretation of *highly specialized* lineages there is often considerable room for subjectivity; we have already given one example of this (ostracoderms and cyclostomes), and in the next section there is another illustration analyzed in more detail.

The possibility of there being differing interpretations of evolution is based on the circumstance that nature primarily presents us only with relationships of form and not of evolutionary descent. On the other hand, interpretations are also often based on morphology that has not yet been sufficiently explained; the phylogenetic approach, however, as we have seen, is not in a position to add a thing to morphological data. Why, then, do we want to introduce these subjective interpretations and reinterpretations of morphology into the system, and how is this supposed to be an improvement?

We reject, therefore, any influencing of taxonomy through phylogenetic interpretations, and acknowledge only the temporal aspect of forms. And this is also purely objective, based as it is on observation and experience. When we observe any sort of morphological modification at a particular geological horizon, it is a binding fact about which there can be no differences of opinion.

In this point, however, we again part company with idealistic morphology, which does not take the time factor into consideration and attempts to investigate the "archetype," from which all the other morphological stages derive, by purely intuitive means. In contrast, we stay with the objective natural data and strive to arrange the morphological steps in the system in their natural sequence. This can then be interpreted without hesitation as phylogeny. It is not, however, phylogeny in and of itself but only becomes it within the framework of the theory of evolutionary descent.

It is difficult to understand, furthermore, why a purely comparative morphology should be unscientific. Nature does not consist just of processes and forces, but (in the world of material phenomena) also of states, of static manifestations and objects, and it is important to describe and classify these. Evolutionary processes as such are inaccessible to us; only their stationary morphological manifestations are there for us to assess. The comparative study of these phenomena is based on real natural objects and can certainly not be said to be subjective, merely "man-made."

All other processes, courses of events, and expressions of life are also manifested materially; the vehicles are the forms and their structures. Clear-cut knowledge of them is therefore an indispensable prerequisite for all approaches to biology. In this regard, comparative morphology and its distillation in the nomenclature of taxonomy is also certainly not superfluous; rather, it is an essential, compulsory discipline of great practical significance. It can even be said without exaggeration that morphology and taxonomy are the *backbone of all of biology,* the foundation of all of its specialized disciplines. Faunistics and floristics, ecology, chorology, chronology, physiology, phylogenetics—all causal and functional approaches presume a mastery of the diversity of forms. In addition to its primary basic character, however, taxonomy also acts as the comprehensive *superstructure of all of biological knowledge.* In a certain sense, it is the file cabinet for the state of biological knowledge at a given time; in it, gaps and discrepancies become immediately apparent, calling for renewed, more thorough investigation. As it develops, then, the system is continuously contributing to the advancement of biology as a whole, while at the same time it synthesizes everything known about the nature of organisms and makes the knowledge useful.

* * *

Now, one could, of course, support the view that pure morphology becomes outdated and superfluous the moment evolutionary thought provides us with insight into the real relationships among forms and the inner context of morphological modification. This view holds that *morphological* series no longer have to be investigated once we are able to establish *evolutionary* series. That as a practical matter we are *not* in a position to do that directly, owing to the lack of independent phylogenetic methods, has already been brought out well enough. At the most, the objection might theoretically be justified for the time when a final, consistent morphology has been completely reinterpreted by phylogeny and has merged with it. But even this is debatable.

Let us make one thing clear, that the individual biological approaches operate on very different levels and proceed, each with a different kind of goal, from the surface of the visible phenomenon—of the organic form—deeper and deeper to the actual inner core of the expression of life and its circumstances. Morphology analyzes the diversity of organisms, classifies them, and situates them within a clearly arranged system. Phylogenetics goes a step further and traces the systematized, graded succession of forms back to a historical transformation and the fixed pattern it followed. Then genetics and developmental physiology come along and, through investigation of the genes and their mechanisms, attempt to understand the causes underlying morphological transformation.

However, the chain of causality does not end there. The genes and mutations are in turn conditioned by molecular, atomic, and even subatomic processes, to which all material phenomena can ultimately be attributed. The first attempts to trace the development of forms back to molecular processes and their physical causes are just now beginning to emerge, and we may expect that further, informative results will be achieved in this field.

But even if the effort to express the manifestations of life ultimately through chemical and physical formulas is successful, if the explanations come from the inorganic realm of physics and chemistry, would biology then be superfluous? Could we then dispense with genetics, with its explanation of the processes of heredity, or with phylogeny as a representation of the historical modification of forms? That is absolutely and unquestionably not the case; research into *all* of these questions, which lie on various horizons and represent individual stages along the path toward an understanding of ultimate causes of organic structure, its being and becoming, is and remains objectively justified and essential if we do not want to forgo a comprehensive overview of the organic world.

No more than genetics can be completely replaced by physics and chemistry, or phylogenetics by genetics, will morphology be replaced and made superfluous by phylogeny. *Comparative morphology remains an indispensable and logical autonomous discipline, independent of any causal interpretation;* its role will never cease as long as biology is pursued.

*   *   *

In another connection, we have already acknowledged the great importance of the natural system as validating evidence for the theory of evolutionary descent, and we have designated this as a major reason that phylogenetic thought should not be admitted into morphology and taxonomy. Far from wanting to reject the phylogenetic approach as such, we are only divesting it of a role that has been assigned to it incorrectly, one it has no call to play and in which it can only cause harm; the only thing we reject is a task assigned to it in error, which it cannot possibly fulfill with the means at its disposal.

A completely different matter is the *subsequent* phylogenetic reading and reinterpretation of the natural system, which rises considerably above the ideal knowledge and capacity for knowing of the former idealistic morphology. Whereas that concept viewed the types as archetypical phenomena, ideal forms not linked physically with one another and not further reducible—forms beyond which no questions were asked—types now acquire through the introduction of evolutionary thought a very particular meaning within the context of natural law. For the understanding of the types and above all of the natural system, however, it seems to us that it is precisely *the interpretation of the course of evolution* that we have developed in previous chapters based on fossil material that yields new insights and information.

It has already been pointed out that many authors have held up the species unit as a reality, in contrast to the higher taxonomic categories, which are said to be pure abstractions. We have seen that such a contrast is untenable, for all the units of the system, without exception, are developed along the same lines of conceptual abstraction. Nevertheless, a certain justification for this difference might be seen in the circumstance that the species, an interbreeding community, is subject to particular natural laws.

Such a grounding in reality would be lacking for the higher categories if one holds the view that the entire course of phylogeny consists of an endless chain of successive speciation. In a continuous series of species succeeding one another endlessly, all more significant divisions in the system would of course appear artificial and as pure abstractions.

But we have shown with our paleontological material that this reading of evolution cannot be correct, that it is not just a matter of forms being remodeled through smoothly flowing, progressive speciation but that—and this is much more important—at the beginning of every phylum and individual lineage there are discontinuities and leaps. The species at the root of new evolutionary trends display not only a simple recasting of the specific characters but at the same time a more or less far-reaching transformation of prior structural elements, with the result that these root species become the first representatives of a new genus, family, or order. Complex major mutations must have brought new types, or structural designs, into being directly and without transition.

Here, however, we understand the types as the conceptualized characteristics of the basic structure of any taxonomic category whatsoever, not just the higher ones, as did Blainville and Cuvier. The type, of course, can never be embodied in a single individual, for an individual always has, in addition to the general features of types of various rank, special and individual characteristics, which are disregarded in exposing the type itself. Hence, we are by no means so naive, as has been charged, as to accept the concepts of types as realities and to use them to track down illusory scientific problems. This charge reflects back to those who make it, who themselves, against all logic, interpret the species concept and certain monotypical higher categories as realities.

On the other hand, however, we saw that all the higher types, like the species concept, are based on the objective, factual circumstances found in nature. These consist in part of the natural divisions between the groups of forms encompassed by the individual basic organic structures, brought about by sudden, type-building developmental steps interrupting the flow of evolution; they also consist of all of that portion of the genome, of a particular combination of genes, that produces the basic organizational features common to that group of forms.

In the light of our interpretation, then, concrete evolutionary processes and the genetic factors responsible for them also constitute the objective ground for all higher types and taxonomic categories. And although these, like the species concept, have no material tangibility, we still do not consider them, in contrast to idealistic morphology, to be insubstantial, metaphysical, archetypical phenomena that elude scientific interpretation. Nor are they, contrary to the conventional evolutionary wisdom, empty abstractions and artificial creations of the mind applied arbitrarily to the organic world.

* * *

Looked at the other way around, the characteristic gradations of the system and its discontinuities may well confirm *our ideas of the saltational nature of phylogeny*. Since, further, the organizational traits of a higher category are common to all the categories subordinate to it, we must conclude *that the type features of the higher units arose historically-temporally before those of the lower ones, that the general features arose before the progressively more specialized sets of characters.*

For example, the essential feature of the type of the *proboscids* consists of the pronounced enlargement of the second *incisors*.[8] It is diagnostic for all members of the order and already appears in its typical manifestation in the oldest known representatives (*Moeritherium,* fig. 3.111; *Phiomia,* fig. 3.112), even though it intensifies considerably in later descendants (figs. 3.113–19; pl. 22B). This general type feature of the comprehensive category was thus emplaced first, before all highly specialized characters. The lengthened incisors subsequently developed in several different evolutionary directions, and on that basis the proboscids are divided into the subtypes of the Moeritherioidea, Dinotherioidea, Mastodontoidea, and Elephantoidea.

Within the framework of these suborders, that is, *after* their basic structure had been formed, there followed a diverse remodeling of the *cheek teeth,* which led to the divisions of families and subfamilies. In continuing descending order, the genera in turn differentiated from the general basic form of their corresponding subfamilies, and the species from that of the genera; it was not the other way around, as is usually assumed, that the generic characters developed through gradual accumulation by way of speciation, that family traits are continuations of the generic transformations, and that finally, as differences continued to increase, the combination of characters of the most all-inclusive categories supposedly developed, crowning the whole.

We have already seen other examples of temporal succession in the recasting of types in descending grades: in the bony fishes, the ammonoids, and the amphibians. The problem for taxonomic classification is to sort out cleanly the different classes or superimposed layers of differences among features and to present them. With the interpretation of the evolutionary processes of phylogeny advocated here, *the essence and principle of the hierarchy of types is finally comprehensible.*

---

8. The trunk is a correlative consequence of the lengthened incisors. It is not an obligatory character of the proboscids but appears as a projecting, free structure only after the incisors have reached a certain length.

* * *

For another thing, in the light of our interpretations the organic system gains *a certain rationale,* that is, it gives an account of the intrinsic, orderly relationships, of the mode of origin and development of the forms included within the categories. With organisms, however, we should not expect a rational system as rigorous as geometry, which represents the prototype of this kind of system in the sense of H. Driesch (1921), or as in crystallography, where the system has acquired an infallible, nomothetic foundation through investigation of the structure of crystals.

We can, however, compare the natural system with, for example, the approximately rational system of chemical compounds, which indeed says nothing about the absolute number of possible compounds but does give information about their composition and physical properties, since the individual compounds obey the law expressed in the supraordinate conceptual group. This is the same with organic forms, which, as we have already demonstrated, are no more purely accidental, disorderly aggregates of characters and combinations than chemical compounds are.

In the latter case, there would of course be no principles of order governing the diversity of forms, and there is no question of the possibility of a rational system. We have seen, however, that the transformation of organisms, the production of their differences, proceeds along very particular paths according to very particular rules, that the number and kind of potentially possible evolutionary trends are restricted and confined right from the start. The system reflects these natural circumstances, which in turn lend the system a deeper meaning, raising it above the level of being simply a superficial catalog of forms.

THE SYSTEMATICS OF PARALLEL LINEAGES

A confusing taxonomic problem comes up for paleontology in the very widespread phenomenon of parallel evolution, which has already been discussed in another context. We single out two examples here, the analysis of which offers the opportunity to use concrete examples to illustrate the relationship between morphological systematizing and phylogenetic interpretation.

A diagnostic and extremely common index form of the early Upper Devonian is the goniatite genus *Cheiloceras,* whose shell has a narrow umbilicus, convex growth rings, and a very simple suture line (figs. 2.87, 3.39*a–c,* 4.8). The lateral lobe, found in other goniatite groups at the center of the sides, here lies on the umbilicus. It is bisected by the seam—shown in our figure 4.8 (left column), as usual, by a short vertical line—so that only one side of it is visible on the outer portion of the whorl. To it is added, toward the outside, an adventitious lobe (A_1), which arose during ontogeny through the splitting of the originally uniform outer saddle, and at the center of the external side there is the external lobe.

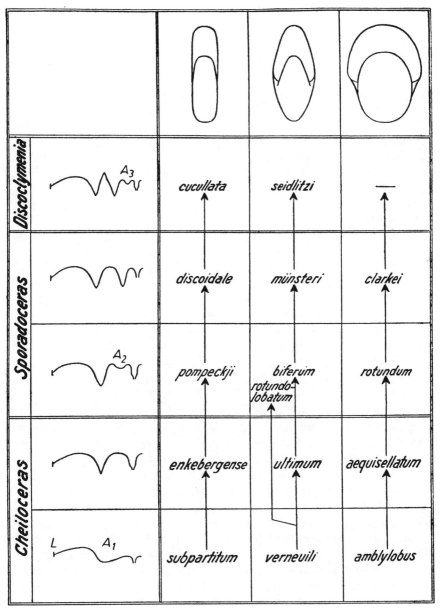

Fig. 4.8. Parallel lineages within the Upper Devonian goniatite genera *Cheiloceras, Sporadoceras,* and *Discoclymenia,* whose boundaries appear to be polyphyletically exceeded. (From Schindewolf 1928.)

In the oldest representatives of the genus *Cheiloceras,* the adventitious lobe is at first very smooth and broadly rounded, and in the younger species it deepens, narrows, and becomes pointed.

Cheiloceras is followed morphologically and stratigraphically by the genus *Sporadoceras* (figs. 2.21, 3.39*d,* 4.8). Here, through further splitting of the external saddle, a second adventitious lobe (A_2) appears, which is shallow at first and then increases in depth in later forms. This evolutionary path of the suture line, initiated by *Cheiloceras* and *Sporadoceras,* continues in the late Upper Devonian in the genus *Discoclymenia* (figs. 3.39*e,* 4.8; pl. 32A), which acquires yet another adventitious lobe (A_3). What we have here seems to be a closed, homogeneous stepped series of forms [*Stufenreihe*], the major grades of which segregated as genera. Because, further, the corresponding temporal and spatial prerequisites are met, we might well interpret this series as an evolutionary line [*Ahnenreihe*].[9]

Upon closer examination, however, the matter does not look so simple. A survey of the diversity of forms within the genera *Cheiloceras, Sporadoceras,* and *Discoclymenia* reveals three different, sharply contrasting shell types. The differences are especially pronounced in the cross sections of the shell (fig. 4.8, top row). We have here (1) shallow disk-shaped shells with flattened, parallel flanks; (2) fatter shells with sides vaulting as they converge with the external shell; and (3) more or less round shell types with pronounced bulging of the sides, which curve broadly as they merge with the external side.

On the basis of this different shaping of the shell, the individual species come together to make three parallel vertical series. Since the shape of the shell apparently does not change and is adhered to tenaciously, it seems that there is a correspondingly sharp distinction and separation between the three rows and that they evolved independently, in parallel, producing the transformational steps of the suture line already described more or less at the same time, each thereby exceeding the boundaries of the individual genera separately.

We must dispense with a discussion of the details here. We would only like to point out that the reshaping of the suture line in each of the series took place very gradually and continuously. In our illustration only the main steps of the developmental process are shown, and the names of some few species are inserted that correspond to the particular degree of evolution. In reality, the evolutionary process involved a much greater number of individual elements.

For example, between the two developmental stages of the suture line illustrated under *Cheiloceras,* several intermediate forms fit in, in which the adventitious lobe deepens step by step or the saddle situated between it and the external lobe rises higher, forming a deep adventitious lobe, rounded on the bottom and enclosed by two saddles of even height. As the adventitious lobe narrows and

9. [See chapter 3, note 10.—Ed.]

comes to a point, one of the suture lines so structured finally gives rise to the upper one shown for *Cheiloceras*. In a similar fashion, between the two illustrated stages of the suture line of *Sporadoceras*, there are in each of the three series a number of intervening species, in which, analogous to the previous example, the *second* adventitious lobe drops progressively lower and finally comes to a point.

Owing to recent investigations, some of the details in the table reproduced here, which I published twenty years ago (1928), should be altered. Some of the specific names should be replaced by others, but this is unimportant for the present discussion.

*   *   *

Thus we have the circumstance in which the boundaries of the genera *Cheiloceras*, *Sporadoceras*, and *Discoclymenia*, defined morphologically based on the level of differentiation of the suture line, cut right across our three parallel evolutionary series. Each of the genera as a horizontal unit combines segments of different vertical lines showing the same developmental level of the suture line.

Is such a classification permissible from the viewpoint of natural systematics? Those investigators who support the view that the system should express phylogenetic concepts and should be a reflection of them will not consider it justified. They will demand that in strict observance of their point of view the genera in question be termed *polyphyletic* (consisting of several lineages) and broken up.

In the vocabulary of the nonphylogenetic taxonomy that we emphatically advocate, the term "polyphyletic" does not exist; it could only refer to artificial or heterogeneous groups in a purely morphological and historical-developmental sense. Here, however, that is not the case. Detailed investigations of the comparative morphology and the ontogeny of the larger groups of forms to which our goniatite genera belong have shown that the suture lines, and further, the form of the growth lines or the margins of the aperture that correspond to them, are the most important criteria for classification and systematization.

The special shell forms, in contrast, are of extremely subordinate importance; they are directly connected to mode of life, are therefore, in general, not very constant, and consequently should be evaluated at a lower level than the other characters cited, which to a large extent are independent of mode of life. The same types of shells are repeated in quite distant groups of goniatites; they appear in combination with the most varied of designs of suture line and growth rings and are thus not very diagnostic and are unsuitable for understanding the larger units of form. As specific characters, they stand in the same relation to the generic features cited as certain racial or varietal peculiarities within individual species do to the concept of the species.

When, after thoroughly weighing the available characters, we have defined the genus *Sporadoceras* as a group of forms having a suture line of particular structure and very particular mode of development, and have excluded everything else—forms with suture lines that are superficially similar but have a different developmental mode, and anything having to do with other features not recognized as essential for this group of forms, such as growth lines, siphuncles, and so on—the species arranged horizontally under *Sporadoceras* in the chart unquestionably fall within this generic concept and form a *natural group, a homogeneous category.* The demands of natural systematics are thereby satisfied.

* * *

Now we shall ask the counterquestion: Is it even possible to classify in another way, and to what practical consequence would it lead? From the standpoint of "phylogenetic systematics," the morphologically conceived genera would have to be broken up into as many units as there are genetic lines, the descendants of different root forms, in them.

Against this, it should be stressed that we know nothing certain about the origin of our series of forms and that it is impossible to pursue it beyond the boundaries of *Cheiloceras.* For the time being, it is not known whether the three species of *Cheiloceras* at the base of our series had already separated and arisen in parallel from *different* roots, or whether and how they derive from one another.

Furthermore, there is no way to determine with the required degree of certainty whether the series are really so sharply distinct throughout their entire extent, that is, whether the differences in shell shape, fairly unimportant in themselves, upon which we have established the series, persist in as completely consistent a manner as it at first seems. There is undoubtedly a certain variability in shape, which might, in time, even have become considerable. In any event, we cannot exclude the possibility of perhaps an enhanced variability of shell shape taking hold in one of the constituents of the center series and its extreme variants coming to resemble the shell shapes of the first and third series.

Then, from one such form others could issue that crossed over into the realm of the other two series, just as, in the reverse, there might be remodeling in the first or third series in the direction of the shell shape of the second series. Instead of the separate, parallel series we are assuming, there would then be a complicated network of intersecting lineages. No solution for this hypothetical model will be found, since the individual species offer no clues to their particular genealogical relationships.

The establishment of the three evolutionary series cannot, then, be substantiated by verified, objective circumstances; it is a purely *subjective assumption,* based only on unverifiable and perhaps deceptive grounds. If we took the three proposed series as the basis for classifying genera, we would end up with extremely uncertain groups loaded down with hypotheses, and it would not be at all certain that they were monophyletic. As a practical consequence, there would be the unsatisfactory necessity of dividing each of the former generic designations, *Cheiloceras, Sporadoceras,* and *Discoclymenia,* into three parts.

But the matter does not rest there. Years ago, I described a species of *Sporadoceras* (*S. rotundolobatum,* pl. 32B), from the middle Upper Devonian of eastern Thuringia, which has since been found with great frequency in the Upper Devonian of North Africa. It is unique among known forms of *Sporadoceras,* inasmuch as in it the second adventitious lobe is connected with a still very primitive, broadly rounded first adventitious lobe. Because of the vaulted sides of the shell, this forms belongs in the middle series of our chart and, because of the shape of the first adventitious lobe, should be linked with *Cheiloceras ovatum,* a species not shown in the chart, which has a deeper rounded adventitious lobe and higher external saddle than *C. verneuili.*

The more primitive developmental grade of A_1 makes it impossible, however, to arrange *S. rotundolobatum* in *the line of evolution* leading from *C. ovatum* through *C. ultimum* to *S. biferum,* since in the last two species the first adventitious lobe is narrow and pointed, hence more highly differentiated, than in *rotundolobatum.* Consequently, this form must be construed as an element of a collateral line that split off from *C. ovatum before* lobe A_1 began to come to a point and then developed parallel to the main center series. It would in turn have to be given a new name.

But what happened in one series could well have happened in the others, so that if all of these possibilities materialized—perhaps they have just not yet been observed—instead of the *one* genus, *Sporadoceras,* there would be *six.* If such a division were undertaken, serious difficulties would arise in the exact determination of genera for the extremely highly differentiated species of *Sporadoceras.* It absolutely cannot be maintained with incontrovertible certainty that, for example, *S. münsteri* issued from *S. biferum.* It would also be conceivable that *münsteri* evolved from *rotundolobatum* or from a later element of this offshoot through the concerted, simultaneous narrowing to a point of *both* adventitious lobes. In this case, the resultant generic classification would be different.

When we take these possible collateral lines into consideration, the origin of the discoclymenians could theoretically be sixfold (regardless of the fact that representatives of the third series are for the time being not even known). It is conceivable, furthermore, that in a manner analogous to the way that *S. rotundolobatum* issued directly from a round-lobed *Cheiloceras,* discoclymenians perhaps also derived directly from species of *Sporadoceras* with rounded A_2's to form other collateral lines. Then we would have *nine* genera just for our group of discoclymenians but would no longer be in a position to determine the generic affinity for any of the given forms.

*   *   *

All in all, from the standpoint of a "phylogenetic" classification and assuming the reality of all of the theoretical possibilities pointed out, the *three* good, unambiguous genera *Cheiloceras, Sporadoceras,* and *Discoclymenia* would be broken up into *eighteen* (!) genetic units. This consequence would make any economically minded systematicist uneasy. Purely practical considerations should not, however, keep us from proceeding in this way if there were compelling reasons to do so and we gained some assured knowledge in exchange.

But we believe to have shown that this is not the case, that, on the contrary, it would lead to complete chaos, and we would no longer be able to integrate the species in question taxonomically. The phylogenetic approach therefore offers no way to classify that is different from the method already used in this example. *No taxonomic procedure other than that based on the fundamental principles of comparative, nonphylogenetic morphology is possible.*

This impossibility is grounded in the fact that phylogeny is not the same in essence as genealogy. We cannot determine the genealogical origin for individuals of either our fossil material or Recent material (unless it has been raised under human control). This is a profound difference, one not to be overlooked, between what we are doing and genetics, which works with particular chains of generations created experimentally and, under certain circumstances is able to follow evolution from the individual to the race. Furthermore, in most instances, as has been shown, we are also never able to document the exact context of individual species with certainty.

The only thing granted to fossil material to establish unequivocally is *the temporal and phylogenetic sequence of steps of certain higher-ranking transformations of characters,* within which there are usually several possibilities for linking individual species with one another. In general, then, phylogeny has exhausted its possibilities when it has established stepped series of forms [*Stufenreihen*] and gives no account of the detailed genealogy of individuals and species in the more minor evolutionary branches that compose them. Only form series can be established with certainty, but these are certain beyond doubt. They constitute the objective morphological and temporal-historical evidence of which we spoke, *whereas all phylogeny of detailed kinship is subjective interpretation.*

* * *

This allows us to comment on another example: The coral genus *Plerophyllum,* already mentioned several times, belongs to the zaphrentoid stock and derives from the genus *Zaphrentoides,* as morphology, ontogeny, and temporal occurrence show. Within this genus *Zaphrentoides,* we encounter remarkable differences in the position of the cardinal septum. In one of the groups of forms (subgenus *Zaphrentoides* s. str.) the cardinal septum, embedded in a pronounced fossula, lies on the *convex* side of the corallum, and in the others (subgenus *Hapsiphyllum*), on the *concave* side (fig. 4.9, bottom). In both cases, the entire

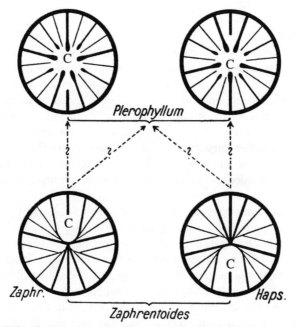

Fig. 4.9. Possibilities for derivation of the different groups of forms of *Plerophyllum* from the genus *Zaphrentoides*. In all illustrations, the convex side of the corallum is uppermost, and so the cardinal septum (C) appears alternately at the top or at the bottom.

septal apparatus shows a correspondingly opposite orientation with relation to the curvature of the coral. These differences in positional relationship are also found within the genus *Plerophyllum*, but there they are less constant and therefore not a suitable basis on which to found subgenera (fig. 4.9, top).

Now, in which direction do the genetic links between the individual species of *Plerophyllum* and *Zaphrentoides* run? There are various possibilities. The plerophylls with the cardinal septum either on the convex or on the concave sides of the corallum could each have evolved separately, in parallel, from the corresponding subgenera of *Zaphrentoides*. It is also conceivable, however, that starting from either *Zaphrentoides* s. str. or *Hapsiphyllum* there was first a general changeover of the septal apparatus to that of *Plerophyllum*, and that subsequently there was another oscillation (as already said, in this genus linked by transitional forms) in the position of the cardinal septum and the orientation of the septal apparatus.

What really happened we do not know and will perhaps never be able to explain unequivocally. We must be satisfied with being able to demonstrate the overall general relationship of form between *Plerophyllum* and the *Zaphrentoides* group. Only this assured fact of the morphological-temporal succession of steps can be expressed in the system.

* * *

From a theoretical point of view, the described limitations to the informative capabilities of phylogeny are not a substantial drawback. Only individuals are the vehicles of true evolution. Whether we abstract from the individual to the species or to the evolution of a character at the categorical level of, say, the genus, is only a *difference of degree*. If we take the genera as the basic units of phylogeny we arrive, in a manner of speaking, at lineages of higher rank, which inherently have the advantage of being more securely founded and less subject to attack. These higher categories are portrayed in the natural system. The phylogenetic approach can add nothing to this picture; its introduction into the system would only produce uncertainties and pretend an exactitude that does not in reality exist.

The fact that in our goniatites, and probably almost universally, the genera combine several minor evolutionary lines, or various series of species, is no argument against evaluating them as natural groups. Under the heading of *species,* too, numerous parallel reproductive lines, such as varieties and geographic races often widely separated in space and sometimes with considerable differences, are also customarily combined, the species concept being abstracted from them. Furthermore, even in species, dividing lines are established without hesitation, with no one taking umbrage, right across individual reproductive lines and strains, whenever the characters of a new species have developed. For the development of new species undoubtedly takes place not only as a unique event, in a single pair of parents, but often in a lesser or greater number of genealogical lines of a breeding community.

Nevertheless, the species is not described as polyphyletic and unnatural but as a natural unit having a common root. The decisive criterion is that the representatives of a species arose from a homogeneous breeding community, from a *single category of equal rank,* even if, in addition, they derive by way of numerous parallel evolutionary lines. In contrast, only those erroneously conceived species composed of descendants that evolved convergently from two or more different species are polyphyletic, or better, polygenetic.[10]

In a completely analogous manner, a *genus* must be regarded as homogeneous and monogenetic if it derives from a single category of the same rank preliminary to it morphologically and temporally, hence from another closed genus,

---

10. Because of the ambiguity of the term "polyphyletic," I recommended earlier (1928) that groups of forms going back to several roots be called "polygenetic" and that the use of the term "polyphyletic" be reserved only for those categories that indeed, like our genera of goniatites, are composed of several lines—hence, which include various subordinate evolutionary lines but which in totality derive from the same root—from a group of types of equal rank. In the latter sense (but to some extent also in the broader sense), H. J. Lam (1936) has spoken of *polyreithrie* (polylinearity).

regardless of whether one or several individual genealogical lines intervene between the two genera or their individual constituents. This stipulation is fulfilled in the examples discussed here. The genus *Sporadoceras,* in which we conceptually combine the putative series of forms of *S. pompeckji, bigerum,* and *rotundum* for morphological reasons, is derived in totality from the equal-ranking *Cheiloceras* group of forms and should therefore not be termed polygenetic. Our three genera, *Cheiloceras, Sporadoceras,* and *Discoclymenia,* like the genera *Zaphrentoides* and *Plerophyllus* in the example of the corals, are *natural categories, which only in this form, because of their sharply defined, unambiguous boundaries, are of practical use and theoretically sound.*

Remarkably enough, many researchers, George Gaylord Simpson among them, who advocate the viewpoint of a phylogenetically based system also endorse this interpretation. Simpson (1945, p. 17) considers it likely that the horse genus *Merychippus* arose from more than one species of *Parahippus.* Nonetheless, *Merychippus* should not be broken up into different genera, which would be very difficult or even impossible to tell apart with certainty. Such a subdivision would not express any assured phylogenetic facts.

Similarly, as Simpson theorized, the totality of mammals did not arise from a single species, genus, or perhaps not even from a single family of reptiles. They are, rather, a compound category, which, however, should not be broken up, because the forms it contains are rooted in one homogeneous group of reptiles at about the rank of (family or) superfamily. The demand for monophyletism is sufficiently met when there is descent from one ancestral group that is of a rank lower than or equal to the derived group.

According to Simpson, even from the phylogenetic viewpoint, horizontal and vertical classification should be considered as equally justifiable, for neither of the two phylogenies can really express the situation correctly. Any system is based on a combination of the two methods, and of the two, Simpson, as we do, regards the horizontal classification as the simpler and more objective, and also the more stable because it is less easily influenced by later discoveries. Similarly, Ernst Mayr (1944, p. 279) considers it possible that in parallel lineages both horizontal and vertical classification is possible, although, with his penchant for placing more emphasis on the phylogenetic, he at least theoretically prefers the vertical. The systematicist of Recent organisms struggles less with these kinds of difficulties of classification. This is not, however, due, as Mayr believes, to the fact that the biologist has complete material at his disposal and access to a greater number of characters than the paleontologist does. The reason is rather that it is a much easier task to classify material from a single unit of time, whereas the paleontologist must come to terms systematically with stages of phylogenetic series of differing times and from different places, with a broader range of evidence.

Truly polygenetic, nonhomogeneous categories, that is, those in which like does not come from like, but in which, rather, only apparent similarities have issued from different sources, must naturally be broken up and eliminated from the natural system. But here, it is *not an autonomous phylogenetic method* that clears up such incorrectly constituted systematic groups but *sophisticated morphological data,* which demonstrate the heterogeneous nature of such unnatural categories.

Bibliography

Other than the major textbooks, reference works, and periodicals, only a small selection of comprehensive works are included here, most of them containing detailed bibliographies of their own. Many specialized studies have been omitted, since these are somewhat inaccessible to general readers and are in any case known to those scholars who might need to refer to them.

I. INTRODUCTORY TEXTS, GENERAL WORKS, AND REFERENCE WORKS

Abel, O. *Paläontologie und Paläozoologie.* Kultur d. Gegenw., part 3, sec. 4, vol. 4, 303 – 95; 8 ills. Leipzig and Berlin: Teubner, 1914.

———. *Lehrbuch der Paläozoologie.* 2d ed.; xiv and 523 pp., 700 ills. Jena: Fischer, 1924.

Beurlen, K. *Erd- und Lebensgeschichte. Eine Einführung in die historische Geologie.* Viii and 462 pp., 227 ills., 29 tabs. Leipzig: Quelle and Meyer, 1939.

Boule, M., and J. Piveteau. *Les Fossiles. Eléments de Paléontologie.* Vii and 899 pp., 1,330 ills., 6 pls. Paris: Masson, 1935.

Camp, C. L., and G. D. Hanna. *Methods in Paleontology.* Xxiii and 153 pp., 58 ills., 1 pl. Berkeley: University of California Press, 1937.

Dacqué, E. *Das fossile Lebewesen. Eine Einführung in die Versteinerungskunde.* Verständl. Wissensch., vol. 4; vii and 184 pp., 93 ills. Berlin: Springer, 1928.

———. *Die Erdzeitalter.* Xi and 565 pp., 396 ills., 1 pl. Munich and Berlin: Oldenbourg, 1930.

Handbuch der Paläozoologie. Ed. O. H. Schindewolf. Berlin: Borntraeger, 1938–44. Nine publications to date, including the prosobranchiate gastropods, by W. Wenz; the Graptolithina, by O. M. B. Bulman; the Hydrozoa, by O. Kühn; the Scyphozoa, by A. Kieslinger; and the Conularida, by B. Bouček.

Hartmann, M. *Allgemeine Biologie. Eine Einführung in die Lehre vom Leben.* 3d ed.; xii and 869 pp., 714 ills., 1 pl. Jena: Fischer, 1947.

Hennig, E. *Wessen und Wege der Paläontologie. Eine Einführung in die Versteinerungslehre als Wissenschaft.* Iv and 512 pp., 198 ills. Berlin: Borntraeger, 1932.

Hirmer, M. *Handbuch der Paläobotanik,* vol. 1; 724 pp., 817 ills. Munich and Berlin: Oldenbourg, 1927. The only volume to date.

Moret, L. *Manuel de Paléontologie animale.* Vii and 675 pp., 241 ills., 12 tabs. Paris: Masson, 1940.

Neumayr, M. *Die Stämme des Thierreiches. Wirbellose Thiere,* vol. 1; vi and 603 pp.,

192 ills. Vienna and Prague: Tempsky, 1889. Somewhat outdated, but still very worthwhile presentation; unfortunately, out of print.

Potonić, H. *Lehrbuch der Paläobotanik.* 2d ed. with W. Gothan; vii and 538 pp., 326 ills. Berlin: Borntraeger, 1921. Soon to appear in a completely revised third edition.

Potonié, H., and W. Gothan. *Paläobotanisches Praktikum.* Bibl. naturwiss. Praxis, vol. 6; viii and 152 pp., 14 ills. Berlin: Borntraeger, 1913.

Romer, A. S. *Vertebrate Paleontology.* 3d ed.; xi and 687 pp., 377 ills. Chicago: University of Chicago Press, 1947.

Schindewolf, O. H. *Wesen und Geschichte der Paläontologie.* Probl. Wiss. i. Vergangenheit u. Gegenwart, vol. 9; 108 pp., 17 ills. Berlin: Wissenschaftliche Editionsgesellschaft, 1948.

Schmidt, H. *Einführung in die Palaeontologie.* Iii and 253 pp., 465 ills. Stuttgart: Enke, 1935.

Seitz, O., and W. Gothan. *Paläontologisches Praktikum.* Biol. Studienbücher, vol. 8; iv and 173 pp., 48 ills. Berlin: Springer, 1928.

Shimer, H. W., and R. R. Shrock. *Index Fossils of North America.* 4th ed.; ix and 837 pp., 303 pls. New York: Wiley, 1949.

Stromer von Reichenbach, E. *Lehrbuch der Paläozoologie.* Vol. 1: *Invertebrates;* x and 342 pp., 398 ills. Vol. 2: *Vertebrates;* ix and 325 pp., 234 ills. Leipzig and Berlin: Teubner, 1909, 1912.

———. *Paläozoologisches Praktikum.* Vii and 104 pp., 6 ills. Berlin: Borntraeger, 1920.

Swinnerton, H. H. *Outlines of Palaeontology.* 2d ed., xii and 420 pp., 368 ills. London: Arnold, 1930.

Walther, J. *Geschichte der Erde und des Lebens.* Iv and 571 pp., 353 ills. Leipzig: Veit, 1908.

———. *Allgemeine Palaeontologie. Geologische Fragen in biologischer Betrachtung.* X and 809 pp., 5 ills., 2 pls. Berlin: Borntraeger, 1919–27.

Wedekind, R. *Einführung in die Grundlagen der Historischen Geologie.* Vol. 1: *Die Ammoniten-, Trilobiten-, und Brachiopodenzeit;* vii and 109 pp., 19 ills., 27 text pls. Vol. 2: *Mikrobiostratigraphie. Die Korallen- und Foraminiferenzeit;* viii and 136 pp., 35 ills., 16 text pls. Stuttgart: Enke, 1935, 1937.

Zittel, K. A. *Handbuch der Palaeontologie.* 5 vols. Munich and Leipzig; Oldenbourg, 1876–93. Volume 1 of this basic but very outdated guide treats Protozoa, Coelenterata, Echinodermata, and Molluscoidea; vol. 2, Mollusca and Arthropoda; vol. 3, Vertebrata excluding the Mammalia; vol. 4, Mammalia; and vol. 5, edited by W. Ph. Schimper and A. Schenk, Paleobotany.

———. *Grundzüge der Paläontologie (Paläozoologie).* Vol. 1: *Invertebrata,* 6th ed., edited by F. Broili; viii and 733 pp., 1,467 ills. Vol. 2: *Vertebrata,* 4th ed., edited by F. Broili and M. Schlosser; v and 706 pp., 800 ills. Munich and Berlin, Oldenbourg, 1924, 1923.

II. SPECIALIZED STUDIES OF MAJOR INTEREST

Abel, O. *Grundzüge der Palaeobiologie der Wirbeltiere.* Xv and 708 pp., 470 ills., 1 pl. Stuttgart: Schweizerbart, 1912.

———. *Lebensbilder aus der Tierwelt der Vorzeit.* 2d ed.; viii and 714 pp., 551 ills., 2 pls. Jena: Fischer, 1927.

————. *Vorzeitliche Lebensspuren.* Xv and 644 pp., 530 ills. Jena: Fischer, 1935.

————. *Vorzeitliche Tierreste im Deutschen Mythus, Brauchtum und Volksglauben.* Xv and 304 pp., 186 ills. Jena: Fischer, 1939.

————. *Tiere der Vorzeit in ihrem Lebensraum.* 336 pp., 273 ills., 16 pls. Berlin: Deutsch. Verl., 1939.

Dacqué, E. *Vergleichende biologische Formenkunde der fossilen niederen Tiere.* Viii and 777 pp., 345 ills. Berlin: Borntraeger, 1921.

Deecke, W. *Die Fossilisation.* Vi and 216 pp. Berlin: Borntraeger, 1923.

Gothan, W. *Palaeobiologische Betrachtungen über die fossile Pflanzenwelt.* Fortschr. Geol. u. Palaeontol., no. 8; iii and 178 pp., title page, 26 ills. Berlin, 1924.

Joleaud, L. *Atlas de Paléobiogéographie.* 39 pp., 99 pls., 1 portr. Paris: Lechevalier, 1939.

Mägdefrau, K. *Paläobiologie der Pflanzen.* Vii and 396 pp., 305 ills. Jena: Fischer, 1942.

Stromer, E. *Gesicherte Ergebnisse der Paläozoologie.* Abh. bayer. Akad. Wiss., n.s. 54; 114 pp., 2 tabs. Munich, 1944.

III. PERIODICALS, COMPILATIONS, AND CATALOGUES

Palaeontographica. Beiträge zur Naturgeschichte der Vorzeit. Founded by W. Dunker and H. von Meyer. Sec. A (Paläozoologie—Stratigraphie), ed. O. H. Schindewolf. Sec. B (Paläophytologie), ed. M. Hirmer and H. Weyland. Stuttgart: Schweizerbart. Series of long monographs, published since 1864; the oldest scientific paleontological journal.

Monographs of the Palaeontographical Society. London. Comprehensive monographs on the Paleontology of Great Britain, published since 1847.

Annales de Paléontologie. Ed. J. Piveteau. Paris.

Palaeontographica Americana. Illustrated contributions to invertebrate paleontology. Ithaca, N. Y.: Paleontological Research Institution.

Palaeontographica Bohemiae. Prague: Czechoslovakian Academy of Sciences.

Paleontographica Italica. Pisa: Memorie di Paleontologia.

Palaeontologia Indica. Calcutta: Geological Survey of India.

Palaeontologia Pononica. Warsaw.

Palaeontologia Sinica. Peiping: National Geological Survey of China.

Paleontology of the USSR., Monographs. Leningrad-Moscow.

Schweizerische Paläontologische Abhandlungen. Formerly, *Abhandlungen der Schweizerischen Paläontologischen Gesellschaft.* Zurich.

Beiträge zur Paläontologie und Geologie Österreich-Ungarns und des Orients. Vienna: Braumüller.

Geologische und paläontologische Abhandlungen. Founded by W. Dames and E. Kayser, most recently edited by F. von Huene. Jena: Fischer. Begun in 1933.

Monographien zur Geologie und Palaeontologie. Ed. W. Soergel. Berlin: Borntraeger. One series in quarto and a second in octavo.

Geologica Hungarica. Series Palaeontologica. Budapest: Royal Hungarian Geological Institution. Formerly, *Palaeontologia Hungarica.* Long monographs on the paleontology of Hungary.

Palaeontologische Zeitschrift. Journal of the Palaeontologischen Gesellschaft. Ed. H. Schmidt. Frankfurt am Main.

Palaeobiologica. Archiv für die Erforschung des Lebens der Vorzeit. Journal of the Paläobiologischen Gesellschaft Wien. Ed. K. Ehrenberg. Vienna and Leipzig: Haim.

Journal of Paleontology. Journal of the Society of Economic Paleontologists and Mineralogists and the Paleontological Society of America, Tulsa, Oklahoma.

Bulletins of American Paleontology. Ithaca, N.Y.: Paleontological Research Institution.

Transactions and Proceedings of the Palaeontological Society of Japan. Tokyo. Appear in Journal of the Geological Society of Japan.

Neues Jahrbuch für Geologie und Paläontologie. Ed. F. Lotze, O. H. Schindewolf, and M. Schwarzbach. *Abhandlungen* (contains long studies) and *Monatshefte* (notes). Stuttgart: Schweizerbart.

Forschritte der Geologie und Paläontologie. Ed. W. Soergel. Berlin: Borntraeger. Individual works, primarily of a general nature.

Zentralblatt für Geologie und Paläontologie. Part II: *Historische Geologie und Paläontologie.* Ed. O. H. Schindewolf and M. Schwarzbach. Stuttgart: Schweizerbart. Journal of abstracts, with individual discussion of the literature.

Revue critique de Paléozoologie. Ed. M. Crossmann. Paris. Begun in 1924.

Rivista Italiana di Paleontologia. Ed. A. Desio. Pavia. Discussion of the paleontological literature of Italy; also short and, in the supplements, longer original studies.

Zoologischer Bericht. Ed. C. Apstein. Jena: Fischer. Contains abstracts, zoological studies, and some of the more important paleontological articles.

Biological Abstracts. Ed. J. E. Flynn. Philadelphia. Abstracts covering the entire biological literature, including a large part of paleontological literature.

Fortschritte der Paläontologie. Ed. O. H. Schindewolf. Berlin: Borntraeger. Summary of two years of reports on paleontological research.

Leitfossilien. An aid to identifying fossils in the laboratory and in the field. Founded by G. Gürich, published by E. Dacqué. Berlin: Borntraeger, 1908–42. Eight issues have appeared, Cambrian to Cretaceous, edited by E. Dacqué, C. Diener, W. Gothan, G. Gürich, and H. Schmidt.

Atlas of the Leading Forms of the Fossil Faunas of the USSR. Publ. by the Central geol. and prospect. Inst. Moscow-Leningrad. Individual volumes of the index fossils of the Soviet Union, arranged according to formation with many new descriptions.

Type Invertebrate Fossils of North America. Ed. E. M. Kindle. Philadelphia. Map catalog arranged by formation with original illustrations and descriptions of all North American fossils.

Palaeontologia Universalis. Paris: International Geological Congress. Map catalogue intended to illustrate and describe fossil type specimens published before 1850, but with inadequate information for modern needs.

Fossilium Catalogus. I: *Animalia,* ed. W. Quenstedt. II: *Plantae,* ed. W. Jongmans. Neubrandenburg: Feller. About 120 issues have appeared to date, containing critical synonyms of fossil genera and species.

Catalogues of the British Museum (Natural History). Department of Geology, London. This catalogue describes fossils from the collection of the British Museum and also provides comprehensive descriptions of the groups of animals and plants in question.

In addition to the studies contained in these specialized periodicals, numerous relevant contributions appear in the organs of the various geological societies, institutes, academies, and museums, as well as in other scientific journals.

IV. WORKS ON GEOLOGICAL CHRONOLOGY

Diener, C. *Grundzüge der Biostratigraphie.* Viii and 304 pp., 40 ills. Leipzig and Vienna: Deuticke, 1925.

Grabau, A. W. *Principles of Stratigraphy.* 2d ed.; xxxii and 1,185 pp., 264 ills. New York: Seiler, 1924.

Neaverson, E. *Stratigraphical Palaeontology. A Manual for Students and Field Geologists.* Xiii and 525 pp., 70 ills. London: Macmillan, 1928.

Oppel, A. *Die Juraformation Englands, Frankreichs und des südwestlichen Deutschlands.* Iv and 857 pp., 64 tabs., 1 map. Stuttgart: Ebner and Seubert, 1856–58.

Pia, J. von. *Grundbegriffe der Stratigraphie mit ausführlicher Anwendung auf die europäische Mitteltrias.* Iv and 252 pp., 3 ills. Leipzig and Vienna: Deuticke, 1930.

Schindewolf, O. H. *Grundlagen und Methoden der paläontologischen Chronologie.* 3d ed.; viii and 153 pp., 5 portr., 47 ills. Berlin: Borntraeger, 1950.

Wedekind, R. *Über die Grundlagen und Methoden der Biostratigraphie.* Iv and 60 pp., 18 ills. Berlin: Borntraeger, 1916.

V. WORKS ON ORGANIC PHYLOGENETIC DEVELOPMENT

Abel, O. *Paläobiologie und Stammesgeschichte.* X and 423 pp., 224 ills. Jena: Fischer, 1929.

Berg, L. S. *Nomogenesis or Evolution Determined by Law.* Xviii and 477 pp., 33 ills. London: Constable, 1926.

Beurlen, K. *Die stammesgeschichtlichen Grundlagen der Abstammungslehre.* Viii and 264 pp. Jena: Fischer, 1937.

Dacqué, E. *Organische Morphologie und Paläontologie.* Viii and 476 pp., 27 ills. Berlin: Borntraeger, 1935.

Darwin, Ch. *Über die Entstehung der Arten durch natürliche Zuchtwahl oder die Erhaltung der begünstigten Rassen im Kampfe ums Dasein.* Trans. H. G. Bronn; reviewed and corrected after the 6th English ed. by J. V. Carus; viii and 592 pp., portr. Stuttgart: Schweizerbart, 1876.

Davies, A. M. *Evolution and Its Modern Critics.* Xii and 277 pp., 30 ills. London: Murby, 1937.

Decugis, H. *Le Vieillissement du Monde vivant.* Vi and 357 pp., 147 ills. Paris: Plon, Masson, 1941.

Depéret, Ch. *Die Umbildung der Tierwelt. Eine Einführung in die Entwicklungsgeschichte auf palaeontologischer Grundlage.* Trans. R. N. Wegner; vi and 331 pp. Stuttgart: Schweizerbart, 1909.

Die Evolution der Organismen. Ergebnisse und Probleme der Abstammungslehre. Ed. G. Heberer, x and 774 pp., 323 ills. Jena: Fischer, 1943.

Diener, K. *Paläontologie und Abstammungslehre.* Samml. Göschen, 2d ed.,; 137 pp., 9 ills. Berlin and Leipzig: de Gruyter, 1920.

Dobzhansky, Th. *Die genetischen Grundlagen der Artbildung.* Trans. W. Lerche; viii and 252 pp., 22 ills. Jena: Fischer, 1939.

Dürken, R. *Entwicklungsbiologie und Ganzheit. Ein Beitrag zur Neugestaltung des Weltbildes.* Vi and 207 pp., 56 ills. Leipzig and Berlin: Teubner, 1936.

Goldschmidt, R. *The Material Basis of Evolution.* Xi and 436 pp., 83 ills. New Haven, Conn.: Yale University Press, 1940.

Heintz, A., and L. Störmer. *Dyrelivets Utvikling* [The evolution of the Animal Kingdom]. Wall chart printed in color, 1.20 × 1.80 m, with explanation, "Nökkel til Planchen om Dryrelivets Utvikling"; 24 pp., 1 fig., 1 tab. Oslo: Norli, 1938.

Hesse, R. *Abstammungslehre und Darwinismus.* 7th ed.; iv and 108 pp., 64 ills. Leipzig and Berlin: Teubner, 1936.

Karny, H. H. *Die Methoden der phylogenetischen (stammesgeschichtlichen) Forschung.* Handb. biol. Arbeitsmethod., ed. E. Abderhalden. Sec. 9, part 3, no. 2, 211–500, 40 ills. Vienna and Berlin: Urban and Schwarzenberg, 1925.

Kühn, A. *Grundriß der Vererbungslehre.* Viii and 164 pp., 115 ills. Leipzig: Quelle and Meyer, 1939.

Osborn, H. F. *The Titanotheres of Ancient Wyoming, Dakota, and Nebraska.* Monogr. U.S. Geol. Survey 55, 2 vols; xxxv and 953 pp., 797 ills., 236 pls. Washington, D.C., 1929.

―――. *Ursprung und Entwicklung des Lebens, auf Grund einer Theorie von der Wirkung, Gegenwirkung und Zwischenwirkung der Energie dargestellt.* Trans. A. Meyer; xxxviii and 328 pp., 1 portr., 135 ills. Stuttgart: Schweizerbart, 1930.

Philiptschenko, J. *Variabilität and Variation.* Vii and 101 pp., 4 ills. Berlin: Borntraeger, 1927.

Rabaud, E. *Transformisme et adaptation.* Bibl. Philos. Scient.; 265 pp., 51 ills. Paris: Flammarion, 1942.

Reinöhl, F. *Abstammungslehre.* Schr. deutsch Naturkundever., n.s. 11; 176 pp., 190 ills. Öhringen: Rau, 1940.

Rensch, B. *Neuere Probleme der Abstammungslehre. Die transspezifische Evolution.* Vii and 407 pp., 102 ills. Stuttgart: Enke, 1947.

Schaxel, J. *Grundzüge der Theorienbildung in der Biologie.* 2d ed., viii and 367 pp. Jena: Fischer, 1922.

Schindewolf, O. H. "Das Problem der Menschwerdung, ein paläontologischer Lösungsversuch." *Jb. preuß. geol. Landesanst.* 49, II (1928): 716–66; 30 ills. Berlin, 1929.

―――. *Vergleichende Morphologie und Phylogenie der Anfangskammern tetrabranchiater Cephalopoden. Eine Studie über Herkunft, Stammesentwicklung und System der niederen Ammoneen.* Abh. preuß. geol. Landesanst., n.s. 148; 115 pp., 34 ills., 4 pls. Berlin, 1933.

―――. *Paläontologie, Entwicklungslehre und Genetik. Kritik und Synthese.* Viii and 108 pp., 34 ills. Berlin: Borntraeger, 1936.

―――. *Zur Stratigraphie und Paläontologie der Wocklumer Schichten (Oberdevon).* Abh. preuß. geol. Landesanst., n.s. 178; 132 pp., 27 ills., 4 pls. Berlin, 1937.

―――. " 'Konvergenzen' bei Korallen und bei Ammoneen." *Fortschr. Geol. u. Palaeontol.* 12, no. 41, vii and 389–492, 33 ills., 1 pl. Berlin, 1940.

―――. *Zur Kenntnis der Polycoelien und Plerophyllen. Eine Studie über den Bau der "Tetrakorallen" und ihre Beziehungen zu den Madreporarien.* Abh. Reichsamts Bodenforsch., n.s. 204; 324 pp., 155 ills., 36 pls. Berlin, 1942.

Sewertzoff, A. N. *Morphologische Gesetzmäßigkeiten der Evolution.* Vi and 371 pp., 131 ills., 24 diagr. Jena: Fischer, 1931.

Simpson, G. G. *Tempo and Mode in Evolution*. Columbia Biol. Ser. 15, 2d ed.; xviii and 237 pp., 36 ills. New York: Columbia University Press, 1947.

Tschulok, S. *Deszendenzlehre (Entwicklungslehre). Ein Lehrbuch auf historischkritischer Grundlage*. Viii and 324 pp., 63 ills., 1 tab. Jena: Fischer, 1922.

Voight, E. "Die Erhaltung von Epithelzellen mit Zellkernen, von Chromatophoren und Corium in fossiler Froschhaut aus der mitteleozänen Braunkohle des Geiseltales." *Nova Acta leopold.*, n.s. 3, no. 14, 339–360, pls. 13–17. Halle am Salle, 1935.

———. "Weichteile an Fischen, Amphibien und Reptilien aus der eozänen Braunkohle des Geiseltales." *Nova Acta leopold.*, n.s. 6, no. 34, 1–38, 3 ills., pls. 1–7. Halle am Saale, 1938.

———. "Weichteile an fossilen Insekten aus der eozänen Braunkohle des Geiseltales bei Halle (Saale)." *Nova Acta leopold.*, n.s. 6, no. 34, 1–38, 3 ills., pls. 1–7. Halle am Saale, 1938.

Zimmermann, W. *Grundfragen der Deszendenzlehre*. Sammlg. Mod. Naturwissenschaft; 32 pp., 7 ills. Stuttgart: Kohlhammer, 1934.

———. *Vererbung "erworbener Eigenschaften" und Auslese*. Xii and 347 pp., 80 ills. Jena: Fischer, 1938.

———. *Grundfragen der Evolution*. 221 pp., 6 ills. Frankfurt am Main: Klostermann, 1948.

VI. BIOLOGICAL SYSTEMATICS

Driesch, H. *Philosophie des Organischen. Gifford-Vorlesungen 1907–1908*. 2d ed.; xvi and 608 pp., 14 ills. Leipzig: Engelmann, 1921.

Kühnelt, W. *Prinzipien der Systematik*. Handb. Biol., ed. L. von Bertalanffy. Vol. 6, pp. 1–16, ills. 1–12, pl. 1. Potsdam: Athenaion, 1942.

Mayr, E. *Systematics and the Origin of Species from the Viewpoint of a Zoologist*. Columbia Biol. Ser. 13, 2d ed.; xiv and 334 pp., 29 ills. New York: Columbia University Press, 1944.

Meyer, A. *Logik der Morphologie im Rahmen einer Logik der gesamten Biologie*. Vi and 290 pp. Berlin: Springer, 1926.

Naef, A. *Idealistische Morphologie und Phylogenetik. (Zur Methodik der systematischen Morphologie)*. Vi and 77 pp., 4 ills. Jena: Fischer, 1919.

Osborn, H. F. "Linnaean classification and phylogenetic classification of the Proboscidea." *Palaeontol. hungar.* 1 (1921–23): 35–54; 5 ills. Budapest, 1923.

Pia, J. von. *Untersuchungen über die Gattung Oxynoticeras und einige damit zusammenhängende allgemeine Fragen*. Abh k. k. geol. Reichsanst. 23, no. 1; iv and 179 pp., 5 ills., 13 pls. Vienna, 1914.

Plate, L. *Prinzipien der Systematik mit besonderer Berücksichtigung des Systems der Tiere*. Kultur d. Gegenw. part 3, sec. 4, vol. 4, pp. 92–164; 16 ills. Leipzig and Berlin: Teubner, 1914.

Richter, R. *Einführung in die Zoologische Normenklatur durch Erläuterung der Internationalen Regeln*. 2d ed.; 252 pp. Frankfurt am Main: Kramer, 1948.

Schindewolf, O. H. "Prinzipienfragen der biologischen Systematik." *Palaeontol. Z.* (Berlin) 9 (1928): 122–169. 2 ills., 4 tabs.

Sigwart, Chr. *Logik.* Vol. 2: *Die Methodenlehre.* 5th ed., ed. H. Maier; x and 887 pp. Tübingen: Mohr, 1924.

Simpson, G. G. *The principles of classification and a classification of mammals.* Bull. Amer. Mus. Nat. Hist. 85, i–xvi and 1–350. New York, 1945.

Troll, W. *Gestalt und Urbild. Gesammelte Aufsätze zu Grundfragen der organischen Morphologie. Gestalt, Abh. allg. Morphol.,* vol. 2. 2d ed.; vii and 182 pp., 30 ills. Halle am Saale: Niemeyer, 1942.

Wettstein, R. von. *Das System der Pflanzen.* Kultur d. Gegenw., part 3, sec. 4, vol. 4, pp. 165–75. Leipzig and Berlin: Teubner, 1914.

Wundt, W. *Logik. Eine Untersuchung der Prinzipien der Erkenntnis und der Methoden wissenschaftlicher Forschung.* Vol. 2: *Logik der exakten Wissenschaften.* 4th ed.; xv and 671 pp., 16 ills. Stuttgart: Enke, 1920.

Afterword

In 1950, three years after the zoologist Bernhard Rensch published his *Neuere Probleme der Abstammungslehre: Die transspezifische Evolution*, Otto Heinrich Schindewolf brought out his magnum opus: *Grundfragen der Paläontologie: Geologische Zeitmessung, Organische Stammesentwicklung, Biologische Systematik* (References, parts 1 and 2, below).

Both books were written during the difficult times of World War II and the immediate postwar period. Their appearance was delayed until publishing companies in Germany began to work normally again, which required resources such as paper and a more general economic recovery. The two books are the only major summaries by single authors that influenced the discussion of evolution in Germany after the war, but they have had a very different history. Rensch's book won international fame and was published in English in 1960, whereas the influence of Schindewolf's remained restricted to German paleontology.

Rensch's theoretical account shows that transspecific evolution (= macroevolution, as it used to be defined) can largely be explained by extrapolation of the microevolution in living populations. This book was his major contribution to the "Modern Synthetic Theory of Evolution" (References, part 3). Apart from the work of Rensch and his less well known colleagues, the zoologist Gerhard Heberer and the botanist Walter Zimmermann, this synthesis was an almost exclusively Anglo-American enterprise developed in the major publications of its founders Theodosius Dobzhansky (genetics, 1937), Julian Huxley (zoology, 1942), Ernst Mayr (systematics, 1942), George Gaylord Simpson (paleontology, 1944), and Ledyard Stebbins (botany, 1950).

The purpose of the largest of the four parts of Schindewolf's book was also to develop a theory of macroevolution, but one of an anti-Darwinian kind. This book thus continued a strong anti-Darwinian movement that existed in the paleontology of the German-speaking countries from the decades before World War II well into the 1970s. Indeed, it still today has not been completely overcome (References, part 4). After the war, this tradition isolated German paleontology from biology and prevented it from participating in the enthusiasm and the research program of the Modern Synthesis. Schindewolf's book was criti-

I thank Roger D. K. Thomas, of Lancaster, for critically reading an earlier version of this essay.

cized from time to time by Anglo-American authors, who understood its significance, though very few were able to read it right through. However, it was not possible for most biologists and paleontologists outside the central European tradition to understand the different historical trends in evolution and morphology that produced this discrepancy between German and Anglo-American evolutionary paleontology. Additionally, German paleontologists had problems keeping up with the vastly increasing English literature on evolution, as their own main interests were in systematics, stratigraphy, and regional geology. In what follows, I will explain the historical developments that led to Schindewolf's book and the strong influence it had for more than two decades after its publication.

PRECURSORS

After Darwin's publication of *On the Origin of Species* (1859), some German-speaking paleontologists were quicker than many of their colleagues in other countries to accept the fact of evolution. The first phylogenetic tree based on empirical evidence appears in a Tübingen dissertation of 1863 on the evolution of fossil snails from a Miocene lake at Steinheim in eastern Württemberg. The author, Franz Hilgendorf, and a few other paleontologists, most notably the Viennese professor of geology and paleontology Melchior Neumayr, accepted not only the fact of evolution, but also Darwin's mechanism, natural selection.

Over the next thirty years, virtually all German paleontologists accepted the reality of descent from common ancestors and used it as a basis for their own concepts and practical work. However, few of them had much enthusiasm for natural selection. Neumayr (1845–90) died young, leaving behind the best treatise on Darwinism that one could imagine given the state of knowledge at that time. However, Neumayr's work (1889) was little read and even less understood. The few other Darwinians who remained active in paleontology were of little influence.

For German paleontologists of the two decades before the turn of the century (and for practically all their colleagues in Western Europe and America), the fossil record did not show the pattern of gradual improvement they deduced from Darwin's theory. In addition to that disappointment, a strong self-consciousness had grown among the German scientists, who claimed that even if microevolution observed by biologists in living populations was controlled by natural selection, macroevolution documented in the fossil record could only be explained by specific macroevolutionary processes.

An outstanding problem faced by both biologists and paleontologists for a long time was the Lamarckian mechanism of the inheritance of acquired characters. This was accepted by Darwin himself but clearly rejected by August Weismann in 1886. This date marks the origin of neo-Darwinian theory, Darwinism purged of Lamarckism (References, part 3). Lamarckism nevertheless remained an issue in German biology into the 1920s. The Austrian paleontologist Franz Baron Nopcsa claimed until his death, in 1933, that it meant nothing that experiments seemed to have refuted the inheritance of acquired characters. He argued that an extremely small heritability (not detectable in the experiments) and the long geological time available to evolution allowed Lamarckian mechanisms an important part in the concert of factors involved in evolution (References, part 4).

Discussion of evolutionary theory has always played a small role in German paleontology. We find statements of the authors often buried in textbooks or in articles that address only the specialist. Nonetheless, a large number of possible mechanisms for macroevolution have been drawn from the biological literature or deduced from interpretations of the fossil record. Almost all authors proposed their own theories, each with his own use of terms and a different combination of evolutionary factors.

One of the more important of these models was proposed in 1928 by the Viennese founder of the research program of "paleobiology," Othenio Abel. To a paleobiologist and functional morphologist like Abel, adaptations were more important than they were to stratigraphers. In Abel's theory, adaptations were brought about mainly by Lamarckian mechanisms, but they were nonetheless only minor corrections in the flow of the evolutionary stream, which had its own internal driving force and hence a momentum independent of environmental factors. Thus a lineage could evolve beyond the adaptive realm and acquire hypertrophied organs and other deleterious characters that led inexorably to extinction. Abel argued that this "biological law of inertia" embraced four subsidiary generalizations that had previously been proposed: (1) Haacke's law of orthogenesis (1893), the internal evolutionary driving force; (2) Rosa's law of continuous reduction in variability (1899), and hence adaptive versatility, in the evolution of a higher taxon; (3) Cope's law of the unspecialized stem-group (1893), which meant for Abel that a new higher taxon arrives on the evolutionary stage with a new reservoir of variability; and (4) Dollo's law of the irreversibility of evolution, which is self-evident, in that the high momentum of evolution does not allow any reversals.

By the 1930s, a major divergence in thought had developed between biologists and paleontologists. As the topic of evolution played almost no role in the education of paleontological taxonomists and biostratigraphers, it was hardly discussed at scientific meetings. Moreover, the great differences of opinion and terminology among authors must have been quite frustrating for readers of the paleontological literature. This situation, which paralleled that in England and America, called for an Alexander to split the Gordian knot. In the development of synthetic theory, it was the vertebrate paleontologist George Gaylord Simpson who reconciled the evolutionary underpinnings of the disciplines in his *Tempo and Mode in Evolution* (1944). He put the interpretation of the fossil record on a firm neo-Darwinian basis.

History went quite differently in Germany. As early as 1921 the developmental biologist and geneticist Bernhard Dürken and the ammonite stratigrapher Hans Salfeld, both of Göttingen University, published a small book on the mechanisms of evolution. Their separate contributions were not integrated. Salfeld emphasized that evolution was driven by internal mechanisms (orthogenesis) and often made significant jumps (saltations). He also favoured Rosa's law and Lamarckian mechanisms. Dürken saw it as his task to propose genetic mechanisms that could account for the paleontological observations. His central hypothesis was that genes are synthesized in the cytoplasm and then incorporated into the genome in the nucleus. Hence there was a weak kind of inheritance (= Lamarckian inheritance) by information transfer from the environment to the genes in the cytoplasm, and a strong kind of inheritance controlled by the genes in the genome. Autonomous changes of genes in the genome were the main driving force of evolution.

An attempt to harmonize the theoretical claims of biologists and paleontologists was made at a joint meeting of the Paläontologische Gesellschaft and the Deutsche Gesellschaft für Vererbungsforschung on September 8, 1929, in Tübingen. The attempt failed—not only because of the different languages of the two disciplines but also because the spook of Lamarckism had not yet been exorcized in either field.

SOURCES OF THE TYPOSTROPHIC THEORY

In the late 1920s, Schindewolf and Karl Beurlen, two of the leading invertebrate paleontologists and biostratigraphers in Germany despite their youth, independently deduced macroevolutionary patterns from detailed studies of the stratigraphic distribution and morphological trends observed in ammonites, crustaceans, and corals. They never published together, and it is doubtful whether they ever had much discussion because their political views and their scientific philosophies were very different.

The first generalization that is clearly the basis for what would later be called the "theory of typostrophism" was Beurlen's hypothesis, laid out in his "Funktion und Form" (1932). This essay, which appeared in *Naturwissenschaften,* was based on a detailed study of all aspects of the evolution of crustaceans:

> It is a very general rule that the pathway of evolution within a taxon—irrespective of whether it is a unit of higher or lower rank—is cyclic. Evolution starts with a first phase of rich saltation and explosive creation of forms. It goes on to a phase of orthogenetic continuity that is directional and purposive and does not produce new types of forms. Finally, a phase of degeneration and disintegration of forms leads to extinction. (1932, p. 76)

Beurlen's ultimate mechanisms for the creation of new types (*Baupläne* of higher taxa) and control of evolutionary cycles were vitalistic forces ("creative life reality," "will to existence," "will to power"). Schindewolf would ultimately adopt Beurlen's cyclic pattern of evolution, but he vigorously opposed the invocation of these vitalistic mechanisms. (For Beurlen's book-length treatment of his theory, see References, part 4.)

In his own early work, Schindewolf was strongly influenced by his teacher Rudolf Wedekind (1883–1961), who in 1935–37 tried to reduce the morphological evolution of ammonites to a few internally controlled (i.e., orthogenetic, rather than adaptive) rules. Schindewolf's strong stand against Lamarckism—in *Grundfragen der Paläontologie,* but also in earlier publications—stemmed from Wedekind, who opposed it as early as 1916 in his first book, citing the works of the geneticists Johannsen and Goldschmidt. This book was devoted to the methods of biostratigraphy (a subject that was to become central in Schindewolf's later writings) and especially to the species concept in paleontology. Wedekind used biometric methods and tried to find Gaussian distributions of Mendelian ratios to show that species can be objectively delineated in paleontology. In Schindewolf's work, biometry never played a large role.

In 1936 Schindewolf published a comparatively small book, *Paläontologie, Entwicklungslehre, und Genetik* (108 pp.), in which he tried to bridge the gap between paleontology and genetics. (The war prevented the English version of the book from being printed.) The preface mounts a strong attack against mysticism:

> Wherever the bases of firmly established knowledge are missing or seem to shake, mystical lines of thought expand. In the case of evolution a wrongly conceptualized holism has seduced authors to abandon causal research and strict scientific analysis. Irrational circumlocutions have replaced clear terms and approaches, without providing any heuristic value.

In this book cyclic patterns of evolution played a smaller role than in the contemporary writings of Beurlen. Rather, Schindewolf concentrated on paleontological evidence for the origin of new types in evolution and the expression of character complexes in morphogenesis of animal skeletons. Schindewolf regarded classical Darwinism as incomplete and Lamarckism as invalid. This led him to scan the literature of genetics for results bearing on the then very important question as to whether there are micromutations, which account for the fine-tuning of adaptation, and macromutations (or systemic mutations), which give rise to new character complexes.

The most important author for Schindewolf was Richard Goldschmidt, who was, as we have seen, already well-known to Wedekind and to Beurlen. Goldschmidt is largely remembered today critically, for his ideas on the saltational origin of species, by means of (now disproven) macromutations and "hopeful monsters." However, it is usually forgotten that Goldschmidt was one of the first people to be interested not only in Mendelian genetics, that is, the transmission of characters from one generation to the next, but also in the stage and manner in which genes and new mutations are expressed in ontogeny. This approach to what we now call developmental genetics perfectly complemented Schindewolf's interest in the origin of new characters and their subsequent evolutionary expression and modification by way of allometry and heterochrony. Goldschmidt was a director at a Kaiser Wilhelm–Institut in Berlin from 1913 to his emigration in 1936, and Schindewolf was a director at the German Geological Survey in Berlin from 1933 to 1947. However, it is not known whether they ever met and discussed their ideas. In any case, Schindewolf felt fully justified, both in 1936 and still in 1950, in assuming that mutations can have vastly different effects on the ontogeny of an organism, given the work of Goldschmidt and other geneticists.

The last paragraph of Schindewolf's 1936 book is quite revealing:

> For the time being, scientific analysis of organic evolution is limited by the assumption that mutations (of the genes) are ultimate causes. The irrational remainder cannot be resolved any further with the epistemological means available to natural science.

This shows that in 1936 he regarded evolutionary theory (the combination of genetics as he understood it and the evidence of evolution from the fossil record) as incomplete. He distinguished scientific theory sharply from the "inexplicable" remainder. The notion of a possible incompleteness of evolutionary theory no longer appears in *Grundfragen der Paläontologie*.

Between 1936 and 1942, Schindewolf consolidated his system of ideas. He began to take a strongly cyclic, antiadaptationist, and antiselectionist stance. He incorporated the three modes of macroevolution (explosive origin, orthogenesis, and decline of a taxon) more explicitly than before. The phases are the same as Beurlen's, but Schindewolf did not refer to his work at that time because of his antagonism to Beurlen's vitalism. (However, in the *Grundfragen* Schindewolf mentioned that Beurlen was a major proponent of cyclism, and that he had developed the idea of a rejuvenation of a taxon after racial senescence).

Idealistic morphology had been a new development after the First World War. One of its main proponents was the zoologist Adolf Naef (*Idealistische Morphologie und Phylogenetik,* 1917). Zoologists, botanists, and paleontologists of Schindewolf's generation were all faced with these new ideas. Each author reacted differently, from extreme mysticism to extreme soberness. The influence of idealistic morphology survived in all three areas after the Second World War, but it faded more rapidly in zoology and botany than in paleontology. Idealistic morphology had its roots in the writings of Goethe and the *Naturphilosophen,* and it is probably no coincidence that it was rediscovered after the defeat of

World War I, together with other ideas and values of the Romantic period. Some authors claimed that the mystical kind of idealistic morphology was an expression of the typical German way of thinking.

The complexity of the debate and its expression in the literature between the wars and Schindewolf's reluctance to deal with philosophical issues explains his rather pragmatic position. Schindewolf equated "idealistic morphology," which interpreted types in a Platonic way and which regarded morphological laws as a research goal, with "comparative morphology." He was convinced that this was the only productive method of morphology and that it had *long* been freed from any metaphysical connotations. Schindewolf's ideas can only be understood if one keeps in mind that he regarded himself as a natural scientist who strictly avoided recourse to mysticism and to misleading analogies, such as Abel's analogy between the inertia of physical mechanics and the hypothetical "inertia" of his evolutionary theory.

The year 1943 was marked by an important event in the development of the Modern Synthesis in Germany and also for Schindewolf's development of his own theory: A book on evolution edited by Gerhard Heberer was published. It included chapters by nineteen authors—all convinced selectionists and not one a Lamarckist. They came from all fields of biology except paleontology. In "Die paläontologischen Evolutionsregeln," an article published the same year in *Biologia Generalis,* macroevolution was dealt with by Rensch. Rather than rejecting the patterns described by the paleontologists, Rensch tried to show that they

> can be explained by directionless mutation and natural selection without the help of inner evolutionary forces of unfolding or trends of formation. Only change in the environment is important for evolutionary novelties. Hence evolution presents itself altogether, not as autogenesis [driven by inner forces = orthogenesis], but rather as ectogenesis [driven by outer forces]. The possibility of autogenetic processes cannot be denied in principle. However, as evidence of their effects, we require a different set of facts from those provided by paleontology so far. (p. 53)

This is moderate language, not that of a revolutionary founding father of the new synthetic theory. It was deliberately intended to invite paleontologists convinced of autogenesis to continue the discussion. Rensch took a much stricter ectogenetic stance in his *Neuere Probleme* (1947).

Heberer's book shows clearly that at the beginning of World War II international discussion of synthetic theory was well under way. Dobzhansky's book of 1937 had appeared in German translation in 1939 and German and Anglo-American authors were not completely isolated from each other during the first years of the war.

Heberer's own contribution to the book, entitled "The Type Problem in Macroevolution," was important for Schindewolf. Heberer redefined the ambiguous term "type," which in idealistic morphology had often been loaded with mystical connotations, equating it simply with "taxon." He was an outspoken selectionist and gradualist, regarding the boundaries between types as artificial. Like Rensch, Heberer was interested in continuation of the discussion with the paleontologists. He referred to the cyclic concept of Beurlen, Schindewolf, and others as "two-phase paleontology." He called the explosive phase "typogenesis," and the phase of orthogenetic continuation "adaptogenesis." The final phase of degeneration and disintegration, not taken seriously by Heberer, was not discussed. Heberer showed clearly that there is no difference whatever between the origin of taxa (typogenesis) and their adaptation (adaptogenesis). Types (= taxa) do not evolve in jumps but, rather, emerge gradually. Heberer called this the "hypothesis of the additive origin of types." Evolutionary phases of rapid diversification, like all other events in micro- and macroevolution, can be explained by the mechanisms of natural selection. They are real phenomena, but they do not require special evolutionary mechanisms such as those advocated by proponents of autogenesis.

In November 1943, Schindewolf gave a paper in Hungary that was not published until 1945. He defended his own theory against Rensch but adopted Heberer's term "typogenesis" for the explosive phase of evolution. He introduced "typostasis" for the orthogenetic phase, and "typolysis" for the degenerative phase. One complete cycle was called a "typostrophe."

Before his *Grundfragen* was published, Schindewolf had the opportunity to present his "typostrophic theory" at the Senckenberg-Museum in Frankfurt am Main. It appeared as the first publication of the now famous series *Aufsätze und Reden*. Robert Mertens, the herpetologist at the museum, criticized Schindewolf's ideas from a Darwinian point of view in the next issue in the series, arguing as Rensch and Heberer had done that mutation and selection are sufficient to explain all aspects of evolution.

THE GRUNDFRAGEN

Schindewolf's *Grundfragen der Paläontologie* is a unique book. It has influenced generations of German-speaking paleontologists in their evolutionary thinking. It was used as a textbook and a reference for approaches of general paleontology. Schindewolf's preface informs us that the book was commissioned by a publisher, probably Borntraeger in Berlin, with whom he had long cooperated. It was ultimately published by Schweizerbart, Stuttgart, as the first publisher was unable to print it and other attempts had failed. The original suggestion by the first publisher, which Schindewolf eagerly took up, was to provide an account of the basic concepts of paleontology for the educated layman in order to increase general knowledge of the field and to counteract misinformation in newspapers and popular magazines. Schindewolf focused on the three fields he regarded as most important—stratigraphy, evolution, and systematics—but he kept the chapter on stratigraphy short because he had already written another book on that subject (*Grundlagen und Methoden der paläontologischen Chronologie,* 1943).

Schindewolf's intent to address a wide readership is evident in several sections that provide little information for a professional, by his attempt to include as many illustrations as possible despite the technical difficulties of those years, and by ending the book with general lists of publications the layman could easily obtain. Furthermore, Schindewolf provides the names of authors but not dates or citations of works on which he has drawn. However, there is no doubt that the book was much too technical to be intelligible to a general reader. Even if we grant that there was a large demand after the war for uncensored literature free of political *Weltanschauung,* both in the arts and in the sciences, public interest in the theoretical concepts and foundations of paleontology must have been very small.

We have to assume instead that the *Grundfragen* was recognized from the date of its publication as the complete and authoritative, but at the same time rather easily accessible, postwar text on macroevolution for German paleontologists. Simpson's *Tempo and Mode in Evolution* (1944) was translated by Heberer in 1951, but it never had a significant circulation. His *Major Features of Evolution* (1953) was never translated. Evolution never played a large role in the education of German geologists and paleontologists, so all interested students had to use Schindewolf's book.

Chapter 1 is a general introduction. Chapter 2, on stratigraphy, was rather uncontroversial at that time. An active stratigrapher for decades, Schindewolf had strong opinions on the methods of stratigraphy. Between 1928 and 1967 he wrote numerous papers on stratigraphical problems of the Paleozoic, the most important of which are listed in his book *Grundlagen und Methoden* and in part 1 of the References at the end of this Afterword. Controversy arose later when Schindewolf's discussion with American stratigraphers became intense. In 1960, Hollis D. Hedberg gave a report to the Twenty-first International Geological Congress in Norden and edited the so-called Copenhagen Guidelines (*Stratigraphic Classification and Terminology,* Report of the Twenty-first International Geological Congress, Norden [Copenhagen, 1961]). Schindewolf's long and elaborate response came in the last publication (1970b, n. 1) that he saw through press. However, it was not published in an international journal but in the *Proceedings* of the Academy in Mainz, which has a small circulation. We will not discuss here to what degree he was influential in shaping the now more or less universally accepted international codification of stratigraphic terminology.

Chapter 4, on systematics, was also not very controversial in 1950. Obviously, there were close connections between Schindewolf's thinking on evolution and systematics and the taxonomic methods that he applied. Coincidentally, Willi Hennig's *Phylogenetische Systematik* was published in the same year as the *Grundfragen*. However, this book had little influence until it was translated into English in 1968. It is clear that Schindewolf was aware of Hennig's predecessors who strongly demanded that taxonomy accept only real lineages as natural taxa (see pp. 419–20 of this volume). In other words, they advocated monophyly in the strict, rather than in the loose sense of Schindewolf, and also of Simpson. Schindewolf found this methodological prescription impractical. In discussions in the 1960s, Schindewolf opposed Hennig's emphasis on monophyly strictly defined and on branching and cladograms rather than lineage and phylogenetic trees. However, I have not found any reference to this in his later writings.

The main part of the book, Chapter 3, is devoted to Schindewolf's evolutionary theory. He made this account as simple as possible, but I doubt that many paleontologists have read the almost 340 pages carefully. First we are informed about the quality of the fossil record. Many fossil assemblages include remarkably well preserved invertebrates, vertebrates (even with cell nuclei preserved! pl. 14A), and plants. The gaps in the fossil record are real, indicating that there never were transitional forms between discrete *Baupläne*. Schindewolf leaves no doubt that this contradicts Darwinian expectations. Paleontologists must study patterns directly and develop specific explanations which best fit the fossil record. (It should be noted here that Schindewolf's key term is *Stammesgeschichte,* or *Stammesentwicklung.* This term, "phylogeny," often means more than it says. In most cases, it encompasses all aspects of evolution that are deduced from the fossil record, including lawlike patterns. We have used the translations "evolution" or "phylogeny" where appropriate.)

The lawlike patterns (*Gesetzmäßigkeiten*) that were to form the basis for his macroevolutionary theory are deduced from the two taxa that Schindewolf knew best, cephalopods and corals (chap. 3, pp. 111–64). In each group, he traced the evolution of basic morphological elements (morphogeny of shell shape and ornamentation, of the siphuncular neck, of the bulging of the septal plane and of the suture line in the cephalopods and morphogenetic patterns of the septal plan in the corals) through time. Schindewolf's preference for these two groups stemmed not only from the influence of his teacher Wedekind but also from the fact that the morphogenesis of all these features could be studied in any well-preserved specimen. Growth forms from the youngest stage to the adult could be reduced to simple line drawings that allowed easy comparisons. When I first met Schindewolf as a student, I was interested in sponge spicules and sponge skeletons. Schindewolf told me that he preferred taxa with distinct morphology to those with fuzzy morphology like sponges, in which no morphogenetic studies could be carried out.

Schindewolf generalizes the results of these two analyses and draws further conclusions under nine general themes. "Evolution or Creation" was really a rhetorical question, which served to emphasize that the gaps between the types are real, contrary to the expectations of the Darwinians. It was also a response to the influential invertebrate paleontologists Edgar Dacqué and Oskar Kuhn, who in the 1940s (but not before and not afterward) had denied the reality of evolution altogether. They had carried typological thinking so far that they could not imagine any transition between the types of the higher taxa but only a kind of unfolding within them, a clear consequence of the strict application of the ideas of idealistic morphology. Kuhn's statement of 1943, "The theory of descent has collapsed," was a serious challenge for Schindewolf, because he shared the view that intermediate stages between types never had existed.

Consequently, Schindewolf's goal was to demonstrate that the gaps between types are not unbridgeable, as they appear in adult stages of organisms. Every organism during its ontogeny first develops the typical characters of the phylum, then of the class, then of the order, and so on. The specific characters of the individual develop last (figs. 3.88, 3.89, 3.90, 3.156). The younger stages are more plastic than the older ones. Hence, the earlier a mutation is expressed in ontogeny, the more significant are its consequences, especially if it has pleiotropic effects. At an early growth stage, new characters and inherited ones can more easily be integrated into a harmonious organism. A mutation that affects the "order stage" of ontogeny will lead the organism to cross the boundary to a new order, because it will express characters typical for that new order (fig. 3.90). Despite this suddenness, the elaboration of the structure of this new order and its diversification at lower taxonomic levels (families, genera, etc.) will take some time after the new order first appears. The transition from one species to the next (speciation) within the character complex of one genus is only the simplest case of such a transition from one type to the next.

Because, in his theory, ontogeny could be changed at any stage, Schindewolf was strongly opposed to Haeckel's "biogenetic law of recapitulation," with its terminal addition of new characters and the subsequent condensation of the inherited stages. Rather, Schindewolf defined an important heterochronous mode as "proterogenesis" (pp. 215–24) in which a new character complex is introduced at an early ontogenetic stage and then expands to later and later stages in descendant taxa of the lineage.

In a saltationist theory like that of Schindewolf, new character complexes do not evolve gradually by mutation and recombination within populations. They arise spontaneously within an individual. Schindewolf overcame Goldschmidt's problem of how "hopeful monsters" would find mating partners by assuming that the boundaries between types were crossed several times in parallel.

The possibility that the boundary between taxon A and taxon B was crossed several times could in theory cause problems for stratigraphy, because the new taxon would not have one center of origin, arising in different places or at different times. This was obvious to Schindewolf. He assumed (1) that if newly founded groups differ too much with respect to their times and places of origin they would not become one integrated taxon but rather would diverge and hence be easily distinguished; and (2) that stratigraphic units extend over enough time for rapid migration to form an integrated new taxon that would still constitute a useful index fossil.

Schindewolf repeatedly wrote about parallel evolution, emphasizing that it is the general phenomenon of the typostatic phase (see especially fig. 3.73; also Schindewolf 1936, p. 99). However, it is not easy to be sure exactly what he meant by this, particularly as in the text he seems not always to be fully consistent with the rather clear diagrams of figures 3.77–79, 3.91, and 4.8.

Figures 3.77–79 set out a classical case of parallel evolution. Within a family of clymenian ammonoids there is first a typogenetic splitting into three lineages, then a typostatic elaboration of the characters and evolution of new low-level taxa (genera and species). In this typostatic phase, a triangular shell shape evolved independently in the three related lineages. According to Schindewolf, this had no adaptive value whatsoever but, rather, arose by orthogenetic unfolding of genetic information inherited from the common ancestor of the three lineages.

Schindewolf cites numerous examples of parallel trends of morphological change in higher taxa of invertebrates, vertebrates, and plants—all of which have their own orthogenetic momentum, because they occurred in different environments and at different times. One such example is the independent modification of the skull in the three orders of amphibians—labyrinthodonts, phyllospondyles, and lepospondyles. The important parallel changes are flattening of the skull roof, reduction in ossification, disappearance of the interorbital septum, shifting of the brain and brain nerves, and so on. The independence of the trends, occurring at different times and in different environments in the various lineages, demonstrated to Schindewolf that they had no adaptive significance but were orthogenetic trends.

From this extensive development of supposedly nonadaptive parallel trends, Schindewolf inferred that new character combinations capable of giving rise to new taxa of any rank can arise spontaneously and independently in several individuals.

In figure 3.91 several individuals of species a, which belongs to an unnamed genus and family of order A, cross the border and found the new order B, with the new family I, the new genus 1, and the new species b. Species b then gives rise to more new species, which found new genera and families. All genera and families in the diagram are clearly monophyletic (or, at most, paraphyletic in the Hennigian sense).

In figure 4.8, several individuals of *Sporadoceras discoidale* give rise to *Discoclymenia cucullata,* and parallel to this process several individuals of *Sporadoceras münsteri* found the other new species *Discoclymenia seidlitzi.*

This is also monophyly in the traditional sense defended by Schindewolf and Simpson, although not in a cladistic sense.

Saltationism, the assumption that there were no gradual transitions from ancestor to descendant, is the first essential element of Schindewolf's theory. In the *Grundfragen,* as in his 1936 book, he cited much work in genetics to show that the occurrence of macromutations was plausible. The emergence of a new type by parallel evolution was quite acceptable because geneticists often find several copies of the same mutation in their culture.

The second basic element of the theory is *orthogenesis.* Schindewolf used a narrow definition of orthogenesis. He regarded typogenesis, the spontaneous origin of taxa, as being in accord with the experimental finding that mutation is random with respect to adaptive directions. In his theory, orthogenesis occurs in the constrained unfolding of typostasis. Being suspicious of mysticism and vitalism, Schindewolf argued strongly against a teleological, finalistic interpretation of orthogenesis. In his view, orthogenesis is not directed *toward* a goal but is rather constrained *from* its particular starting point. From a Darwinian perspective, it is clear that Schindewolf's macroevolutionary theory was orthogenetic ("autogenetic" in Rensch's terminology) in all its phases, because it is not based on Darwinian mechanisms such as mutation, selection, isolation, and genetic drift.

The third, and the central coordinating element of Schindewolf's theory, is *cyclism.* All three elements had a long history in anti-Darwinian theories, but the particular combination of the three and especially the strong emphasis on cyclism made the typostrophic theory unique. Like the cycle of individual development from youth to old age, taxa on all levels arise by spontaneous saltations. They have a life cycle of (1) spontaneous diversification; (2) elaboration and stabilization, which leads to specialization; and finally (3) degeneration, which leads to extinction. The last phase (fig. 3.73) is characterized by deleterious overspecialization, as orthogenesis takes lineages far into the inadaptive realm. At this point, only a few individuals have the capacity to avoid extinction and to found new taxa by saltation. These cycles are hierarchically stacked at successive taxonomic levels. The cycle of an order includes several cycles of families, which in turn embrace several cycles of genera, and so on. The exact taxonomic ranks given to each taxon are to some degree arbitrary. Its autonomous cyclic behavior gives each taxon a high degree of reality.

Schindewolf's argument that species and all higher taxa have only a "metaphysical reality" seems to contradict the ontological status of taxa derived from autonomous cyclic behavior. However, this was obviously no problem for Schindewolf, who stated that *Baupläne* and taxa are at the same time real things *and* mental constructs derived from natural objects by objective ways of reasoning. The topic can be found in several contexts of the book (e.g., pp. 203–6, 415). We can interpret it from a modern point of view as Schindewolf's synthesis between the methods of comparative morphology on the one hand and idealistic morphology cleaned of mysticism on the other.

The framework of the typostrophic theory has several implications:

1. Dollo's rule of the irreversibility of evolution poses no problems.
2. The typostrophic cycles are autonomous. They are not influenced by important events in earth history such as transgressions, regressions, or orogenic cycles.
3. Phyletic increase in size is not caused by the environment. It is a truly macroevolutionary phenomenon and can be regarded as a special case of orthogenesis.
4. The essential processes of evolution are not the formation of races or species adapted by micromutation and selection (i.e., microevolution) to local environments, but rather the development of new types as higher taxa by a process of genetic rejuvenation.
5. These new types are comparatively unspecialized (Cope's rule), providing a lineage with new variability which is then used up by one-sided adaptations during the life cycle of the type (Rosa's rule!). "One-sided adaptations" are dead-end roads that constrain the future evolutionary versatility of a taxon. In the end, all Darwinian adaptation is one-sided.
6. The processes of adaptation are inimical to evolution.
7. If no new types arise, a lineage dies out from racial senescence.
8. Only adaptive characters within the essential morphologies of higher taxa, not the essential character complexes themselves, are subject to Darwinian natural selection. Selection is thus irrelevant with respect to the origin of higher taxa.

SCHINDEWOLF'S THEORY AFTER THE WAR

In 1948, after many years at the Geological Survey in Berlin and a short time at the Humboldt University, Schindewolf was appointed to the chair for geology and paleontology at the University of Tübingen, which had been created in 1837 for Friedrich August Quenstedt (1809–89). With him Schindewolf shared an interest in ammonites and stratigraphy. (For political reasons, Schindewolf was not granted a full professorship until the end of the Third Reich.) His main work in Tübingen was devoted to the gigantic "Studies on the Evolution of Ammonites" (860 pp.), which was published between 1961 and 1968 in seven parts in the *Abhandlungen der Akademie für Wissenschaft und Literatur* in Mainz. At the same time, he continued to make individual contributions to all his fields of interest.

The *Grundfragen*'s enormous influence must be at least partly understood from the fact that Schindewolf had virtually no more competitors in evolutionary paleontology in Germany and that he was regarded as the authority in all the fields discussed in the *Grundfragen*. After the war, most earth scientists were involved in practical rather than theoretical problems. Also, Germany had since 1933 lost many scientists, who had either been killed or had been forced to emigrate. The last wave of emigration was that of Nazis after 1945. Among them was Karl Beurlen, an opportunist, as demonstrated in his book on evolution

(1937), who had attacked Schindewolf politically during the Third Reich. Beurlen emigrated to Brazil to become a mining geologist there. He returned home to Tübingen after his retirement (around 1970) and used the library of the Geological department for writing popular books on geology and paleontology. In these he based his evolutionary accounts on Schindewolf's postwar publications. (When I interviewed Beurlen in 1982 about the early development of his own ideas on evolution, it turned out that he had completely forgotten not only the 1937 book but also his earlier papers.)

I will refer briefly to four of the twenty-two papers relating to evolution that Schindewolf wrote after 1950. Mass extinctions, or *Faunenwenden* ("changes in faunas") as they have been called in German, play no role in the *Grundfragen*. (Not even the term appears in the index.) As it gradually became evident that the extinctions of many groups were often remarkably simultaneous (hence the term *Faunenwende*), nonbiological signals for this synchroneity were sought more intensively than in 1950. In 1956 (and in several other papers), Schindewolf argued that evolution as a whole is a largely autonomous process. As in the *Grundfragen,* he rejected repeated assertions of geologists that tectonic processes influenced major evolutionary events like the origin of higher taxa, extinction, and sudden changes in diversity. In 1963, Schindewolf extended his hypothesis that cosmic radiation caused the phases of rapid extinction and species turnover at the end of the Paleozoic and of the Mesozoic. He attributed the sudden increase in intensity of the cosmic radiation to the explosion of supernovas. Remarkably, this hypothesis of an extraterrestrial cause helped Schindewolf to protect his view of evolution as an autogenetic process against the ectogenetic challenges of the geologists. Cosmic radiation increases mutation rate. Hence it has an influence—if a rather unspecific one—on the typostrophic cycle. If the genome of certain taxa has the right susceptibility when the mutation rate is increased, it may be transferred from the typostatic to the typolytic phase. This could result in hyperspecializations, deleterious organs, and extinction.

In 1964, the year of his retirement, Schindewolf read a paper, published in the same year, called "Erdgeschichte und Weltgeschichte" [History of the earth and history of mankind] at the Academy in Mainz. Here, he defended his position that the earth, life, and mankind have histories, against claims of humanists who wanted to reserve the term history for cultures with a written tradition. Schindewolf showed that the three phenomena of history have fundamental properties in common, in addition to irreversibility. These properties do not require any special explanation; they are essential properties of the historical process. Cycles of rapid origination and diversification, followed by elaboration, stasis, and extinction, which he had found in the history of life, could also be found in the history of the earth and in the history of mankind. The key witness for earth history was the eminent German geologist Hans Stille (1876–1966). His orogenic cycles began with rapid mountain building, followed by long elaborations, stabilization, solidification, and rapid (revolutionary) rejuvenation of the crust. An important parallel with the role of Schindewolf's ideas is seen in the fact that Stille's strong and pervasive influence delayed acceptance of the theory of plate tectonics by more than a decade in Germany. To demonstrate the cycles in human history, Schindewolf referred to Oskar Spengler's *Der Untergang des Abendlandes: Umrisse einer Morphologie der Geschichte* (1919–22) and to the voluminous publications of A. J. Toynbee. Schindewolf's library no longer exists, but there is no doubt that he knew Spengler's work, which was regarded as one of the major contributions to the German worldview after World War I. Schindewolf showed in his *Grundfragen* that evolutionary theories of cyclism can be traced back to Ernst Haeckel. However, he and Beurlen may also have been strongly influenced by Spengler.

In the second to last paper that Schindewolf saw through press (1969), he defended the typological approach as that most appropriate and fruitful for tackling problems of morphology, systematics, and phylogenetic reconstruction. However, he did not use the terminology of his typostrophic theory any longer. He explicitly withdrew from his earlier attempts to bridge the gaps between morphology and genetics. He regarded the synthetic theory as still significantly incomplete and defended the idea that the evolution of the character complexes of higher taxa is not a process of gradual adaptive evolution. Rather, it can only be based on the rapid restructuring of conservative gene-complexes. He did not doubt that such "supergenes" would eventually be found. Consequently, the character complex of the placentals, for example, is selection-neutral because it arose by a selection-neutral saltation and has been expressed over millions of years by an extremely conservative gene-complex. It has nothing to do with the special adaptations of the various mammalians at lower taxonomic levels. In the 1969 paper, the typostatic phase was also still an important concept in Schindewolf's evolutionary theory, but he did not discuss overspecializations, deleterious characters, and other aspects of typolysis.

If we characterize typostrophic theory as a combination of cyclism, saltationism, and orthogenesis ("autogenesis" in Rensch's terminology) based on idealistic morphology, we see that Schindewolf retreated significantly from cyclism at the end of his career but not so much from the other factors. For him, synthetic theory had not solved the problem of the evolutionary origin of the discrete *Baupläne* of higher taxa. He emphasized that adaptive bridges between them could not be imagined and that they had a remarkable evolutionary stability. In conclusion, when we look at the long-term influence of Schindewolf's typostrophic theory, we find it had a variety of ramifications for German paleontology.

First, for a long time, the theory appeared in all textbooks on paleontology, and weak forms of typostrophism can still be found in contemporary books. Second, in research projects derived from typostrophism, the stratigraphic distribution of higher taxa and the development of their diversity was compiled. Ectogenetic causes for the patterns were rejected. Rather, as an explanation, autogenetic mechanisms were advocated. Indeed, because higher taxa were understood to arise by saltation, no attempts were made to reconstruct the adaptive pathways that led to new *Baupläne*. Only rather late was the obvious demonstrated, namely, that a fine resolution of the fossil record revealed the gradual evolution of the character complex of a higher taxon.

Many German paleontologists avoided any discussion of evolutionary theory or explicit opinions on typostrophism versus the Modern Synthesis either in writing or at scientific meetings. Zoologists were reluctant to discuss the origin of higher taxa with paleontologists. Furthermore, no attempts were made to give adaptive explanations for seeming overspecializations. Orthogenesis in a strict sense ("rectilinear evolution") was taken as a reliable model and as a measure for evolutionary progress. This went so far that a certain "evolutionary height" of a taxon was used as a stratigraphic marker. Finally, as late as the 1970s, young authors risked censure by their superiors if they discussed typostrophism critically. Under the influence of Schindewolf's authority, evolution was no topic for the would-be paleontologist.

REFERENCES

1. *WORKS BY SCHINDEWOLF*

Before and after the *Grundfragen,* Schindewolf wrote numerous theoretical accounts on the three main topics of that book, namely, stratigraphy, evolution, and systematics. The most important ones (excluding those that are mentioned in the references to this translation) are listed below.

Stratigraphy

1957. Comments on some stratigraphic terms. *Amer. J. Sci.* 255:394–99.

1960. Stratigraphische Methodik und Terminologie. *Geol. Rundschau* 49:1–35 (Stuttgart).

[1964] 1967. Logic and method of stratigraphy. *Trans. Proc. Geol. Soc. South Africa* 67:306–10 (Johannesburg).

1970a. Stratigraphical principles. *Newsletters on Stratigraphy* 1:17–24.

1970b. Stratigraphie und Stratotypus. *Abhandl. Akad. Wiss. u. Lit., Math.-naturwiss. Kl.* 1970(2):100–232 (Mainz).

Evolution and Systematics

1925. Entwurf einer Systematik der Perisphinkten. *N. Jb. Miner. etc., Beil.-Bd.* (B), 52:309–43 (Stuttgart).

1929. Ontogenie und Phylogenie. *Paläont. Zeitschr.* 11:54–67 (Stuttgart).

1937. Beobachtungen und Gedanken zur Deszendenzlehre. *Acta biotheoretica* 3:195–212 (Leiden).

1942. Evolution im Lichte der Paläontologie: Bilder aus der Stammesgeschichte der Cephalopoden. *Jenaische Z. Med. u. Naturwiss.* 75:324–86.

1944. Zum Kampf um die Gestaltung der Abstammungslehre. *Naturwissenschaften* 31: 269–82 (Berlin).

1945. Darwinismus oder Typostrophismus? *Kulönnyomat a Magyar Biol. Kutat. Munkaibol.* 16:107–77 (Tihany).

1947. Fragen der Abstammungslehre. *Aufsätze und Reden der senckenbergischen naturforschenden Gesellschaft* 1:23 pp. (Frankfurt am Main).

1948. *Wesen und Geschichte der Paläontologie.*, Wiss. Editionsges. (Berlin), 108 pp.

1954. Über die möglichen Ursachen der großen erdgeschichtlichen Faunenschnitte. *N. Jb. Geol. Paläont., Mh.,* pp. 457–65 (Stuttgart).

1956. Tektonische Triebkräfte der Lebensentwicklung. *Geolog. Rundschau* 45:1–17.

1962. Neue Systematik. *Paläont. Zeitschr.* 36:59–78 (Stuttgart).

1963. Neokatastrophismus? *Zeitschr. dt. geol. Ges.* 114 (1962):430–45 (Hannover).

1964. Erdgeschichte und Weltgeschichte. *Abh. Akad. Wiss. u. Lit., Math.-naturwiss. Kl.* 1964 (2):53–104 (Mainz).

1968. Homologie und Taxonomie: Morphologische Grundlegung und phylogenetische Auslegung. *Acta biotheoretica* 18:235–83 (Leiden).

1969. Über den "Typus" in morphologischer und phylogenetischer Biologie. *Abh. Akad. Wiss. u. Lit., Math.-naturwiss. Kl.* 1969 (4):55–131 (Mainz).

2. SCHINDEWOLF OBITUARIES AND BIBLIOGRAPHIES

Erben, H. K. 1971. Nachruf auf Otto Heinrich Schindewolf. *Jahrbuch 1971 der Akademie der Wissenschaften und der Literatur in Mainz,* pp. 75–86 (with a complete bibliography).

Kullmann, J., and J. Wiedmann. 1971. Otto Heinrich Schindewolf zum Gedächtnis, 7.6.1896–10.6.1971. *Atempto Tübingen* 39/40:120–21.

Seilacher, A. 1972. Otto H. Schindewolf, 7. Juni 1896–10. Juni 1971. *Neues Jahrbuch für Geologie und Paläontologie Monatshefte* 1972 (2):69–71 (Stuttgart). With additions to the bibliography published in *Neues Jahrbuch Geol. Paläont. Abhandlungen* 125 (Festband Schindewolf), pp. xvii–xxiv (Stuttgart, 1966).

3. ON MODERN EVOLUTIONARY SYNTHESIS

Lehmann, U. 1985. *Paläontologischen Wörterbuch.* 3d. ed. Stuttgart: Ferdinand Enke-Verlag. 440 pp.

Mayr, E., and W. B. Provine, eds. 1980. *The Evolutionary Synthesis: Perspectives on the Unification of Biology.* Cambridge, Mass.: Harvard University Press. 488 pp.

4. ON THE DEVELOPMENT OF EVOLUTIONARY THINKING IN GERMAN PALEONTOLOGY

Beurlen, K. *Die Stammesgeschictlichen Grundlagen der Abstammungslehre.* Jena: Gustav Fischer Verlag, 1937.

Reif, W.-E. 1975. Lenkende und limitierende Faktoren in der Evolution. *Acta biotheoretica* 24:136–62.

————. 1980. Paleobiology today and fifty years ago: A review of two journals. *N. Jb. Geol. Paläont. Mh.* 1980:361–72.

————. 1983a. Evolutionary theory in German paleontology. In *Dimensions of Darwinism,* ed. M. Grene, pp. 173–203. Cambridge and New York: Cambridge University Press.

————. 1983b. Hilgendorf's (1863) dissertation on the Steinheim planorbids (Gastropoda; Miocene): The development of a phylogenetic research program for paleontology. *Paläont. Zeitschr.* 57:7–20.

————. 1986. The search for a macroevolutionary theory in German paleontology. *Journal of the History of Biology* 19 (1):79–130.

Weishampel, D. B., and W.-E. Reif. 1984a. Paleoecology and evolution in the work of Friedrich von Huene. In *Third Symposium on Mesozoic Terrestrial Ecosystems,* ed. W.-E. Reif and F. Westphal, pp. 193–98. Tübingen: Attempto Verlag.

————. 1984b. The work of Franz Baron Nopcsa (1877–1933): Dinosaurs, evolution, and theoretical tectonics. *Journal der geologischen Bundesanstalt Wien* 127:187–203.

Index

Page numbers in bold type refer to illustrations.

Abel, Othenio, 176, 226n, 263, 283, 316, 322–23, 355, 357–61, 437
Abiologic events, in delimiting geologic time, 67
Abstraction, 378
Aceratherium, **294**
Acme, 193
Acromegaly, 307
Actinocerans, 116, **119,** 257
Actinopterygians, 32, 35, 275, 338
Actualism, 330
Adaptations, 326–28, 341, 349
Adaptogenesis, 343, 442
Agnathans
  appearance in Ordovician, 21, 27–28, **27**
  evolution of, 165, 197, **198,** 275, 337, 397, **398,** 399
Ahnenreihe, 226n
Alcyonarians, 106
Algonkians, 15
Allometric growth, 361–62, **362, 363,** 364–65, **366,** 367–69
Amia, 323
Ammonites
  cretaceous, 135, 298, **299**
  Devonian, 22–24, 365, 367
  and extinction, 319
  Jurassic, 135
  mesozoic, 37, 39, **44,** 45, 48, 140–41, **142,** 143–46
  paleozoic, 28, 31, 125–26, **127,** 128–29, 131
  and parallel evolution, **127,** 277, 334, **335**
  Permian, 132, **133**
  preservational state of fossils for, 93, **94,** 95–96

shell in, 208, 217–18, 269
  Triassic, 35, 132, **133,** 134, 135, 195–96, 197
Ammonitoceras, **142**
Ammonoidea, 257
Amphibians, evolution of, 197, **198,** 275–76, 337–41, **339**
Amphicyon geoffroyi, **78**
Amynodon, **294**
Analogous organs, 278, **279**
Analogues, 387
Anancus arvernensis, **280, 283**
Anaptychus, 126
Anarcestes, 69, **70, 127, 128,** 129, **130**
Anaximander, 75
Anchitherium, 247, **248**
Ancistroceras, **123**
Ancyloceras, **142**
Anemone pulsatilla, **241**
Angiosperms, 42
Animal kingdom, evolutionary bush of, **344**
Anoplotherium, 265
Anticipatory selection, 357–61
Antirrhinum majus, 253–54
Anura, 37
Aplacentals, 260, 262
Applied micropaleontology, 56
Aptychus, 126
Aquatic birds, 327
Archaeocyathids, 18, **19,** 147
Archaeopteris hibernica, **32**
Archaeopteryx, 103
Archaeornis siemensi, **44**
Archaic strata, 15
Archelon ischyros, **181**
Archetype, 411

455

Plates

Plate 1

Plate 1 The statue of the lindworm, in Klagenfurt (Carinthia, Austria). The head of this statue was created in 1590 by the sculptor Ulrich Vogelsang, who based his work on the "lindworm skull." According to old accounts, this skull was found in 1335; it can still be seen, preserved in good condition, in the museum of the city of Klagenfurt. It is the skull of an Ice Age wooly rhinocerous (*Coelodonta antiquitatis* [Blumenb.]), the features of which have been rendered unmistakably in the head of the statue; but the body that goes with it has been "reconstructed" in the likeness of the winged dragon of myth and fable. The true appearance of the animal can be seen in plate 13A. (From O. Abel 1939, *Tiere der Vorzeit* [Prehistoric animals]).*

*When the sources of the illustrations are indicated in the bibliography at the end of the book, the author's name is given along with the year of publication and, in the event that several publications of the same year are cited, the abbreviated title is included.

Plate 2

A

Plate 2 A. A *Wirfelstein* [*Wirfel* = the staggers; *Wirbel* = vortex], a pebble from the Upper Cretaceous (Gosau formation) of the Eastern Alps, with cross sections of large snails of the genus *Actaeonella,* which stand out clearly as white spirals against the dark background. The alpine farmers of the region put stones of this kind into the water troughs to protect their cattle, according to popular belief, from the staggers. The specimen in the picture came from the Stodertal (Upper Austria). (From O. Abel 1939, *Vorzeitliche Tierreste im deutschen Mythus* . . . [Prehistoric animal remains in German mythology, etc.].)

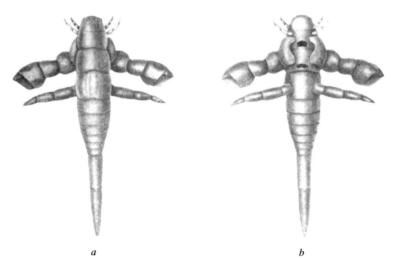

B

a   b

B. Reconstruction of the large arthropod *Protadelaidea howchini* Till., which occupies a class (Arthrocephala) and order (Protadelaidoidea) of its own because of the unfused segments of the head section and other peculiarities. Upper Proterozoic (lower division of the Adelaide strata) from Adelaide (South Australia). ⅓ natural size. *a.* Dorsal view. *b.* Ventral view. (After T. W. E. David and R. J. Tillyard.)

Plate 3

A

Plate 3 A. *Peytoia nathorsti* Walc., a scyphomedusa from the Middle Cambrian Burgess Shale of British Columbia. Impression of the underside with four wide sectors (*x*) radiating out from the four-sided oral aperture; each of the quadrants so defined contains seven narrow sectors. (From C. D. Walcott.)

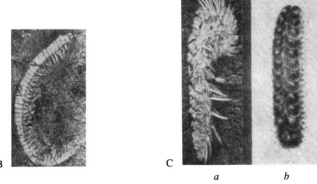

B   C

a   b

B. *Worthenella cambria* Walc., a polychaete worm from the Middle Cambrian Burgess Shale. The segmentation of the body, the two-lobed parapodia, and the tentacles on the head are some of the many features that have been beautifully preserved.

C. *Canadia spinosa* Walc., another polychaete worm from the same locality (*a*), with dorsal scales (elytrae) overlapping like roof tiles, similar to recent aphroditids, for example, *Polynoe squamata,* which is shown in 3-D for comparison (*b*). (After C. D. Walcott, from H. F. Osborn 1930.)

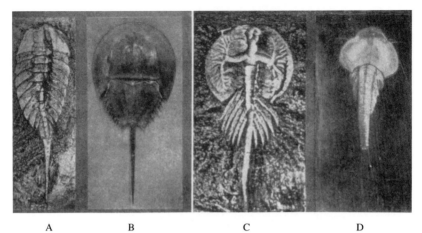

A   B   C   D

Plate 4 A. *Molaria spinifera* Walc., a merostome from the Middle Cambrian Burgess Shale of British Columbia. (The Merostomata and the Arachnoidea, or spiderlike animals, together form the class Chelicerata, mentioned in the text; the class is distinguished by having chelicerae instead of antennae.) Of the subclass Merostomata, order Xiphosura, the only genus still living is *Limulus,* represented by a few species, one of which, *L. polyphemus* L., is shown in figure B, much reduced, for comparison.

B. *Limulus polyphemus* L.

C. *Burgessia bella* Walc., a crustacean from the Phyllopoda group (branchiopods). Middle Cambrian Burgess Shale. About 2 ×. Noteworthy is the outstanding preservation of the legs, the gut, the highly developed liver sacs, and so on. The legs show an original segmentation and are not yet phyllopodia. This form also differs from Recent forms in its long tail spine (telson). However, the segmentation of the large dorsal shield and legless abdomen, characteristic of modern forms, is already complete.

D. *Triops lucasanus,* a Recent form very similar in shape to the form in figure C, belongs to the same order of crustaceans.

(After C. D. Walcott, from H. F. Osborn 1930.)

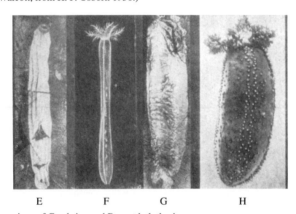

E   F   G   H

E–H. Comparison of Cambrian and Recent holothurians.

E. *Mackenzia costalis* Walc. from the Middle Cambrian Burgess Shale, a form that embodies the type of holothurians that live in cavities or in burrows in marine bottoms. The elongate muscle, a ring of tentacular nodes around the mouth, and many other details have been extremely well preserved. ⅔ ×.

F. For comparison, a corresponding Recent species, *Synapta girardii.*

G. *Louisella pedunculata* Walc., another sessile holothurian type from the Middle Cambrian Burgess Shale, with a double row of tube feet and a crown of tentacles around the mouth preserved.

H. A similar recent form: *Pentacta frondosa.*

(After C. D. Walcott, from H. F. Osborn 1930.)

Plate 5

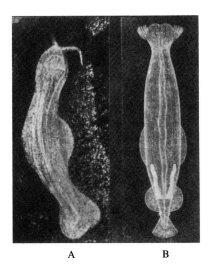

Plate 5 A. *Amiskwia sagittiformis* Walc., a fossil from the Middle Cambrian Burgess Shale, which, based on its outer form, must be interpreted as being a representative of the chaetognaths. 2.5 ×. The body is divided into a head, torso, and tail, just like the body of the Recent chaetognath *Sagitta gardineri*, shown in figure B for comparison. The gut and other digestive organs are seen clearly through the thin body wall. The pair of tentacles at the anterior end of the head is lacking in the living form. In many Recent forms, however, there are sensory papillae arranged on a stalk on each side of the head. (After C. D. Walcott, from H. F. Osborn 1930.)

   B. *Sagitta gardineri.*

A   B

C. *Phillipsastraea ananas* (Goldf.), a colonial reef coral from the lower Upper Devonian of the Rhenish Mountains. 1 ×.

C

D

D. *Cheriolepis trailli* Ag., a Devonian bony fish from the subclass Actinopterygii (order Chondrostei, represented today by the sturgeons). The surface of the body is covered with small, rhomboidal ganoid scales arranged in diagonal rows. About ⅓ ×. (After R. H. Traquair.)

Plate 6

Plate 6 *Hapalocrinus frechi* (Jkl.), a magnificently preserved group of crinoids from the Lower Devonian roof slate at Bundenbach, in Hunsrück. The largest individual, the one in the center, has attached itself to a foreign body; five younger animals have colonized the stem of the largest one. 1 ×. (After R. Opitz.) The Bundenbach Shale produces an extremely diverse fauna, in particular, sea lilies and sea stars. (A sea star, albeit not too well preserved, can be seen at the bottom left of the picture.) Owing to the fine grain of the Bundenbach Shale and to the special conditions of burial, this rock is noted for the outstanding preservational state of the most fragile structures and even of soft parts (cf. figs. 2.24, 2.27, 3.8–10, 3.12, 4.2). The Bundenbach Shale is for the Devonian what the Burgess Shale is for the Cambrian.

Plate 7

Plate 7 *Ceratites semipartitus* (Montf.), a typical representative of the Ceratitacea (ammonoid line) from the German Triassic (Upper Muschelkalk, Germany). ½ ×. The original is a steinkern without the shell and shows, therefore, the inner chambering of the spiral. The unchambered terminal portion of the shell, the last living-chamber, has not been preserved. In this form, it is typical for the sutures, that is, the outer edge of the walls separating the chambers, that the portions that are convex toward the front (the saddles) are entire, whereas those that curve in the opposite direction (the lobes) are finely saw-toothed.

Plate 8

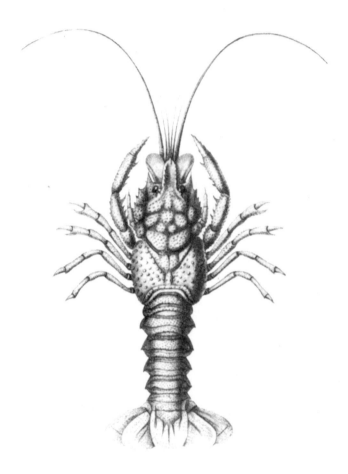

Plate 8 *Pemphix sueurii* (Desm.), a crustacean from the group of long-tailed decapods (Macrura), considered to be the ancestral form of the crabs (Brachyura). Upper Muschelkalk from Württemberg. ½ ×. (After P. Assmann. Based on more recent research, this reconstruction should be modified to show that the first pair of pereiopods have claws, just as the two following do.)

Plate 9

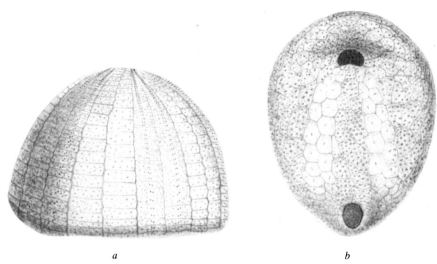

A

a   *b*

Plate 9 A. *Echinocorys ovatus* (Leske), another irregular sea urchin (cf. fig. 2.46) from the Upper Cretaceous of northern Germany, in which both the mouth (front) *and* the anal openings (back) are situated on the underside of the shell (*b*). ⅔ ×. (From E. Stromer von Reichenbach 1909.)

B. Leaf of *Credneria triacuminata* Hampe from the Upper Cretaceous of the piedmont of the Harz Mountains. The plant that produced this large leaf, very typical of the deposits of that time, is thought to be a precursor of our modern plane trees. ⅔ ×. (After H. Potonié.)

B

Plate 10

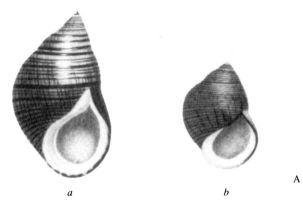

A

a   *b*

Plate 10 A. *Littorina* (*Algaroda*) *littorea* (L.), a snail found in modern northern seas, on the Atlantic coasts of Europe and North America, and in the Mediterranean. The two specimens illustrated (*a* and *b*) came from the North Sea. 1 ×. (After H. C. Küster, in W. Wenz.)

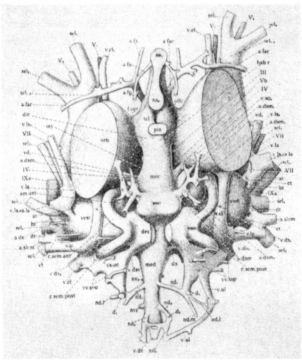

B

B. Brain and sensory organs (orbits, labyrinthine cavities, olfactory organ) with connecting nerve fibers, arteries, and veins, of *Cephalaspis hoeli* Stensiö, from the Upper Gothlandian of Spitzbergen. Dorsal view, drawn from a wax-plate model. About 7 ×. (After E. A. Stensiö.)

Plate 11

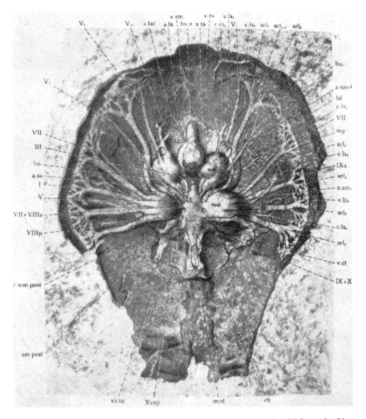

Plate 11 Head shield of *Kiaeraspis auchenaspidoides* Stensiö, a cephalaspid from the Upper Gothlandian of Spitzbergen. Endoskeleton in ventral view. About 6 ×. In spite of the overwhelming
abundance of features preserved here, the various compartments of the cranial cavity, the electrical
fields on the right and left lateral margins, and the nerves and blood vessels are clearly discernible
in every detail, right down to the finest branchings. The most important pairs of nerves in the heads
of vertebrates are already present in their characteristic form: I (olfactorius) provides nerves to the
olfactory organ; II (opticus), to the eyes; V (trigeminus), to the first three gill arches (in which
differentiation into mandibular and hyoid arches has not yet taken place); VII (facialis) leads to
the electrical fields (as in modern electric rays) and the third gill arch; VIII (acusticus) leads to the
labyrinthine canals; IX and X lead to gill arches 4–10. (After E. A. Stensiö.)

Plate 12

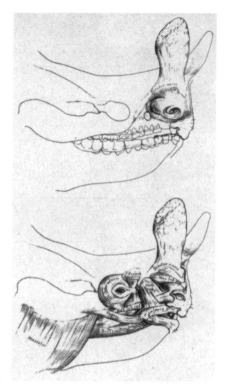

Plate 12 A. Facial musculature and nasal cartilage of the Lower Oligocene titanothere *Brontotherium platyceras* (Scott and Osb.). *a*. Superficial layer of muscle. *b*. Position of the nasal cavity and the muscles surrounding it, which attach to the skull. About ¹⁄₁₂ ×. (After H. F. Osborn 1929.)

B. Reconstruction of the heads of four Lower Oligocene titanotheres. About ¹⁄₁₂ ×. *a. Brontops robustus* Marsh. *b. Menodus giganteus* Pom. *c. Megaceros copei* (Osb.) *d. Brontotherium platyceras* (Scott and Osb.). (After H. F. Osborn 1929.)

A

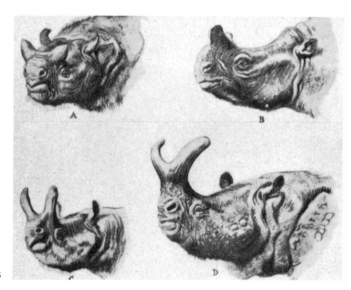

B

Plate 13

A

Plate 13 A. *Coelodonta antiquitatis* (Blumenb.), the Ice Age woolly rhinoceros excavated in 1929 from the ozokerite of Starunia (Polish Galicia), in the collection of the Physiographic Museum of the Polish Academy of Sciences, in Cracow. This specimen was missing its horns and hooves; these were supplied from an earlier find. (After a photograph supplied by the Polish Academy.)

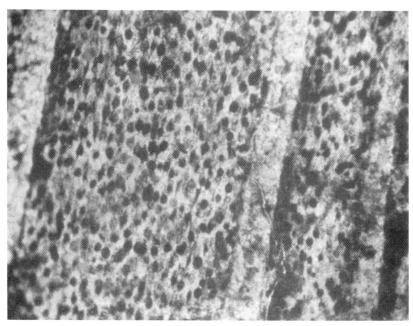

B

B. Melanophores (black pigment cells) on the dorsal fin (dorsalis) of the fish *Anthracoperca siebergi* Voigt, from the Eocene lignite of the Geiseltal, near Halle am Saale (Germany). 75 ×. The light, unpigmented longitudinal bands represent fin rays; on the fin membrane stretched between them, the pigment cells, elongate with pointed ends and oriented lengthwise, are arranged in rows, indicating a faint longitudinal striping. (This illustration and the following ones on pates 14, 15, and 16A are from original photographs kindly supplied by Professor E. Voigt.)

Plate 14

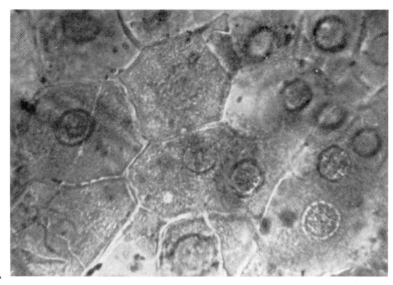

A

Plate 14 A. Beautifully preserved polygonal epithelial plate cells with nuclei, from the epidermis of an Eocene frog from the Geiseltal lignite. 1,000 ×. The cells are part of the stratum corneum, the outermost epithelial layer of the frog's skin; at the top right there are a few cells from the next deeper layer, the stratum germinativum. The structure of the epidermis of these fossil frogs is identical with that of Recent forms.

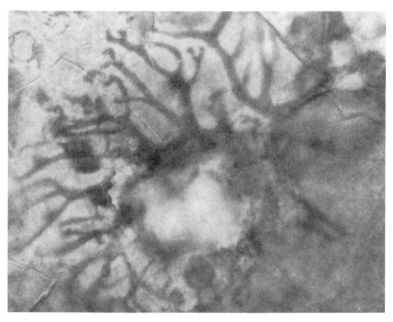

B

B. Explanation on facing page.

Plate 15

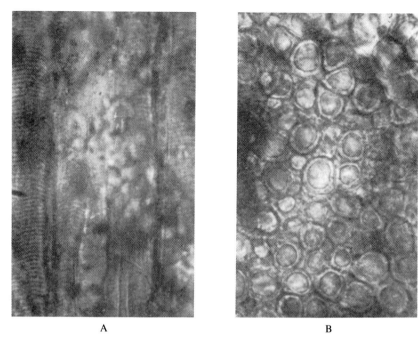

A   B

Plate 15 A. Transversely striated muscle from the femur of the bat *Cecilionycteris prisca* Hell., from the Eocene lignite of the Geiseltal. 450 ×.

B. Reticulate cartilage from the ear of the bat *Cecilionycteris prisca* Hell., from the Eocene lignite of the Geiseltal. 450 ×. The outstanding quality of this preparation, which, like the others shown here, was obtained by the lacquer film technique worked out by E. Voigt, is comparable in every way with that of microtome sections of Recent material.

---

14 B. Large, amoebalike melanophores (black pigment cells) with many extensions, from the skin (stratum vasculare, the layer separating the true skin, or derma, from the cuticle) of a frog from the Eocene lignite of the Geiseltal. 1,000 ×. The melanophores are in a state of pigment *expansion:* the long, branching extensions of the pigment cells are evenly filled with dark granules of melanin right to the very tips. In contrast, in Recent frogs, at the time of death and the cessation of circulation there is generally a *concentration* of pigment: the pigment granules withdraw from the extensions and collect at the center of the cell. There is only a single instance in which such a post mortem pigment agglomeration does not occur, and that is when death is caused by lack of oxygen. The conclusion to be drawn from this is that the fossilized frogs suffocated in small pools left behind when the rest of the habitat dried up, a conclusion completely consistent with observations on other aquatic fauna of the Geiseltal.

In addition to melanophores, these fossil frogs also exhibit xantholeucophores, which carry reddish-yellow pigments; further, granules of guanine were found in them. In living frogs, the blue iridescence of the guanine and the yellow pigment combine against the dark ground color created by the melanophores to produce green tones. Consequently, this fossil species of frog must have been some shade of green.

Plate 16

A

Plate 16 A. Bundle of muscle fibers from the beetle *Eopyrophorus,* from the Eocene lignite of the Geiseltal, near Halle, am Saale. 250 ×. The section illustrated shows a contractile wave with distinct enlargement at the belly and considerably tighter transverse striation than in the uncontracted part of the bundle. The dark A-bands, consisting of an anisotropic (doubly refractive) substance, alternate with the lighter, isotropic (singly refractive) I-bands, revealing clearly in several places dark transverse striations (Hensen's lines, or M-disks). The microstructure of the individual layers, the heights of the loculi, and other kinds of differentiation are completely the same as in Recent beetles; the absolutely outstanding state of structural preservation is astonishing. Moreover, the degree of shortening of the A-bands and the broadening of the I-bands in this fossil material allow the degree of the stimulus that led to the formation of the belly to be determined directly and beyond doubt.

Further, a delicate longitudinal striation of the bundle of muscle fibers is clearly discernible, which indicates the fine fibrils that make up the fascicle. On the sarcolemma, the sheath enclosing the fibrils, which is also preserved, unevenly coiling lines (T) can be seen; these represent the trachaeal tubules that supplied the muscle with oxygen.

B

a   *b*

B. Two ammonites that have been colonized by serpulid worms. *a. Ammonites (Arnioceratoides)* cf. *kridion* Ziet., from the Lower Liassic (ariete beds), from Völpke (northwest Germany). 1½ ×. *b. Schlotheimia (Scamnoceras) angulosa* Lange, from the Lower Liassic (angulati beds) from Eime (northwest Germany). 1½ ×. (From Schindewolf.)

C

C. *Argonauta sismondae* Bell. Pliocene, from northern Italy. 1 ×. (From E. Stromer von Reichenbach 1909.)

Plate 17

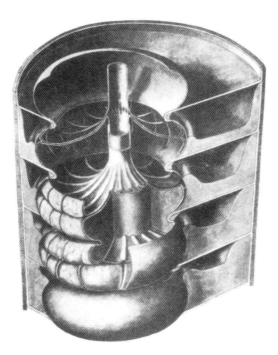

Plate 17 Reconstruction of the internal structure of an actinoceran, a representative of the Early Paleozoic nautiloids. (After C. Teichert.) The illustration shows a section of the straightened shell with the transverse partitions, or septa, which divide the interior space into a system of successive air chambers. The chambers are run through by a siphuncle, very thick in this form, which narrows at the points where it pierces the septa (septal necks) and bulges sharply outward in the chambers. The inside of this structure is filled with organic calcareous deposits, which leave free the tube running through the middle (the endosiphuncle) and the complex system of canals radiating out from it. In the most posterior chamber shown in this illustration (at the bottom), the siphuncle is shown with its closed wall (connecting ring). The next two chambers show the calcareous deposits and the radial canals, which open out into a ring canal, the perispatium. In the next-to-the-last segment, only the canal system is shown. At this point, the interior of the siphuncle is thought to be not yet calcified—the canals were embedded in endosiphonal tissue that was calcified later; this tissue and its ring membranes have been omitted from the illustration. The chambers also contain organic calcareous deposits that decrease in thickness and extent from the older to the younger. The multitude of forms these deposits take also provide diagnostic characters. The illustration is based on the examination of a large number of thin and serial sections; it reflects actual observations. The size of the illustration corresponds to a medium-sized to large actinoceran.

Plate 18

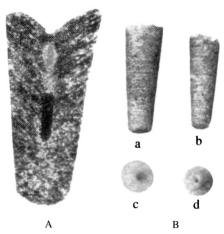

A   B

Plate 18 A. *Volborthella conica* Schdwf. Lower Cambrian, from Reval [Talinn] (Estonia). Medial longitudinal section. 20 ×.

B. *Volborthella tenuis* Schm. Lower Cambrian, from Reval. 10 ×. *a, b*. Side views of steinkerns. *c*. Upper, adoral fracture surface, looking down at the funnel-like depression in the filling of the chamber and the central perforation of the siphuncle. *d*. Lower, adapical fracture surface looking down at the chamber filling, here protruding in a conical shape, with the opening for the siphuncle. (From Schindewolf.)

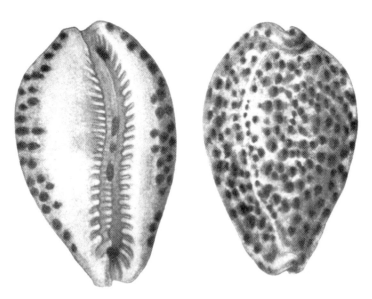

C

C. *Cypraea (Cypraea) tigris* L., the type of the superfamily Cypraeacea, studied by Schilder. Recent, Indo-Pacific. 1 ×. (After Sowerby, in W. Wenz.)

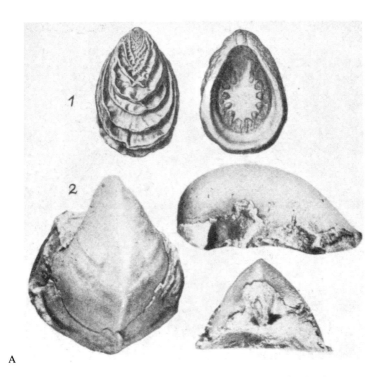

A

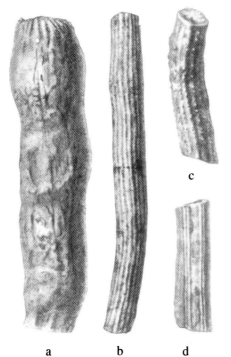

Plate 19 A. Shells of representatives of the Tryblidiacea and the Bellerophontacea, with impressions of muscle attachments. *a. Tryblidium reticulatum* Ldstr. Middle Gothlandian, from Gothland. Exterior and interior of the shell. 1 ×. (After Lindström.) *b. Cyrtonella mitella* Hall. Middle Devonian, from Michigan. Three views of the shell. 1 ×. (Both from W. Wenz.)

c

B. Lateral view of the corallites of various representatives of the Heterocorallia, from the Lower Carboniferous (Upper Visean) of Lower Silesia. *a. Heterophyllia (Heterophyllia) grandis* Mc-Coy. 1⅓ ×. *b. Heterophyllia (Heterophylloides) reducta* Schdwf. 1½ ×. *c. Heterophyllia (Heterophyllia) parva* Schdwf. 3 ×. *d. Hexaphyllia mirabilis* (Dunc.). 3 ×. (From Schindewolf.)

a   b   d

B

Plate 20

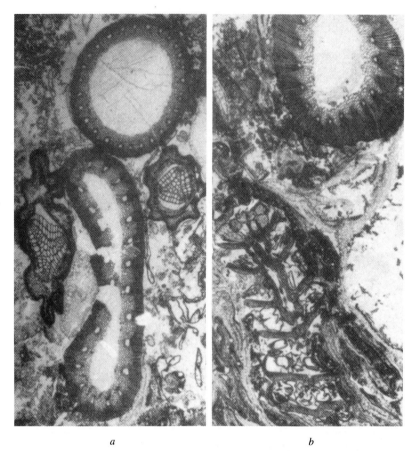

a   *b*

Plate 20 Two thin sections of dolomitized coal balls from the coal of the Upper Carboniferous of Rheinland-Westfalen. These nodular or spherical structures are fossilized pieces of ancient peat, the source material of the coal seams; they are characterized by the outstanding preservation of the enclosed plant remains. About 12 ×. *a.* Cross section through two calamite stems, the lower of which has been collapsed. These tree-high horsetail-like plants of the Carboniferous differ from our modern horsetails in the development of a true xylem, which can be seen in the cross section. To the right and left of the lower calamite are cross sections through the stems of *Sphenophyllum,* with the characteristic triangular phloem, its individual layers of tissue preserved with extreme clarity. *b.* Above: Cross section through another calamite stem with wedges of secondary xylem. Bottom left: Longitudinal section through a calamite cone, its upper portion with full spore sacks, the lower portion with empty ones; between the spore sacks are the sterile bracts. (After W. Gothan and W. Hartung.)

Plate 21

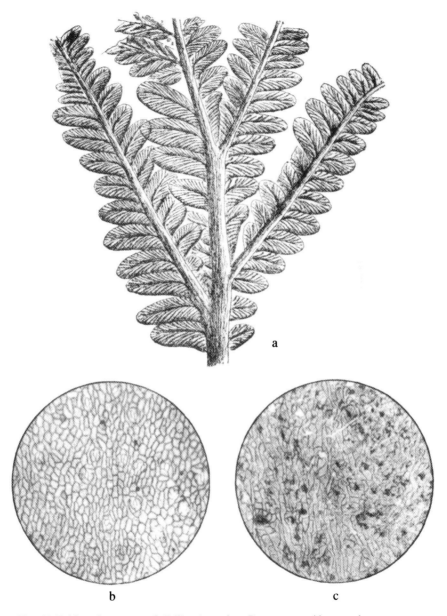

Plate 21 Epidermal structures of *Callipteris conferta* Brongn., a pteridosperm that serves as an index fossil of the Rothliegende Age (Lower Permian), from Crock, in Thuringia. *a.* Remains of a frond, natural size. *b.* Cuticle of a leaflet with extremely well preserved cells. The roundish cellular structures distributed unevenly over the surface are probably related to internal glands. 100 ×. *c.* Hypodermis with stomal openings (black spots). Left of center, the conspicuous area with more elongate cells and fewer stomata corresponds to a vein. 100 ×. (After W. Gothan.)

Plate 22

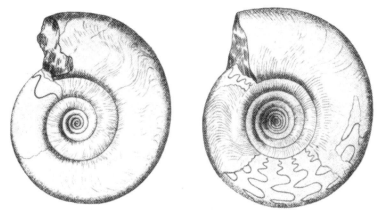

A

Plate 22 A. Two members of the goniatite family Manticoceratidae, from zone I*a*, the immediate base of the Upper Devonian, in the Dill region (Rhenish Mountains). 1 ×. *a. Ponticeras aequabile* (Beyr.), the most primitive representative with regard to the level of differentiation in the suture line. *b. Pharciceras tridens* (Sdbg.), one of the most highly developed, terminal members of the evolutionary lineage.

B. Skeleton of *Elephas* (*Mammonteus*) *columbi* Falc., from the Middle Diluvian of North America. The specimen illustrated (original in the American Museum of Natural History, New York) measures 5.43 meters in length and 3.20 meters at the shoulder, and the right tusk, measured on the convex outer side, is 3.47 meters long! (After H. F. Osborn, from O. Abel 1929.)

B

Plate 23

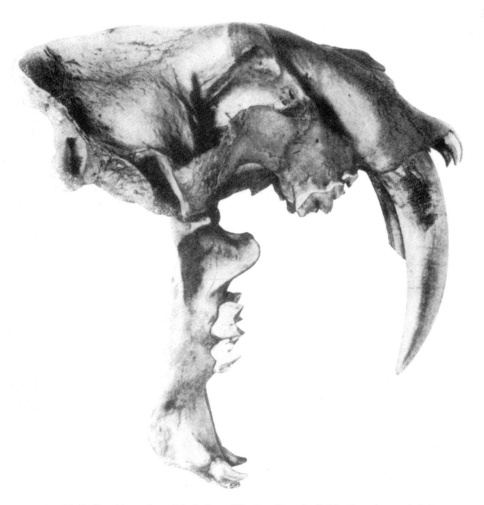

Plate 23 Skull and lower jaw of *Smilodon californicus* Bov., the Californian saber-toothed tiger, from the Diluvian [Pleistocene] asphalt deposits at Rancho La Brea (California). ⅓ ×. (From J. C. Merriam and C. Stock.)

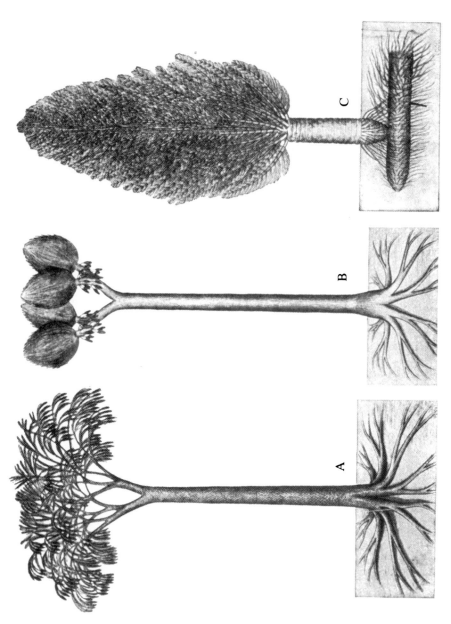

Plate 24 Reconstructions of arborescent pteriodophytes of the Carboniferous period.

A. Lepidodendron (*Lepidodendron*), a club moss about 16 meters tall with leaf cushions of an elongate rhomboidal shape, abundantly branched crown, and terminal cones.

B. Sigillaria (*Sigillaria*), another representative of the club mosses, about 8 meters tall, with longitudinal rows of hexagonal or elongate-ovate leaf scars, only a slightly branched crown, and cauliflorous cones.

C. Horsetail (*Calamites*), reaching a height of about 9 meters, with unbranched, spirally arranged leaves.

None of these giant forms are the immediate ancestors of modern club mosses or horsetails but are, rather, extinct offshoots of the lineages leading to the modern forms. The ancestors of Recent horsetails were small, herbaceous plants that lived during the Carboniferous period along with the large calamites. Thus, the giant plants of the Carboniferous do not in any way contradict the rule of phylogenetic increase in size. (After M. Hirmer, from F. Reinöhl 1940.)

Plate 25

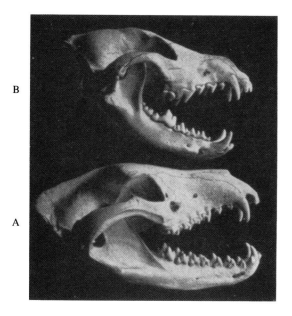

Plate 25 A. Skull of the marsupial wolf *Thylacinus cynocephalus* Wagn. Recent, Tasmania.

B. Skull of the wolf *Canis lupus* L. Recent, Europe and Asia.

A comparison of the illustrations shows the surprising similarity in the shape of the skulls and the jaws between the aplacental mammal and the analogous placental mammal. (From F. Reinöhl 1940.)

C. An example of external similarity of form from the plant world: The development of the "cactus" type—columnar, leafless stems with fleshy, succulent tissue—in xerophytes (plants adapted to dry habitats) from three different families of plants. *a. Stapelia grandiflora. b.* Cactus (*Cereus Pringlei*). *c.* Euphorbia (*Euphorbia erosa*). (After Fitting, from F. Reinöhl, 1940.)

Plate 26

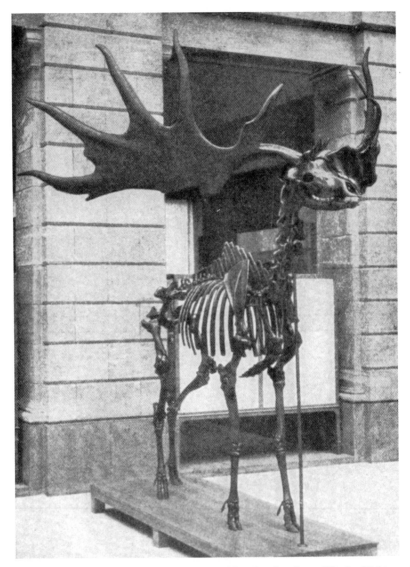

Plate 26 Skeleton of the gigantic Irish elk *Megaceros hibernicus* Ow., from a Diluvian [Pleistocene] peat bog in Ireland. Original in the Senckenberg Museum of Natural History, in Frankfurt am Main (After F. Drevermann, from O. Abel 1929.)

A

Plate 27 A. Group of mounted skeletons of the small, delicate precursor of the camel, *Stenomylus hitchcocki* Loom., from the Middle Miocene of western Nebraska; on exhibit at the American Museum of Natural History, New York. Reduced. (From F. Drevermann.)

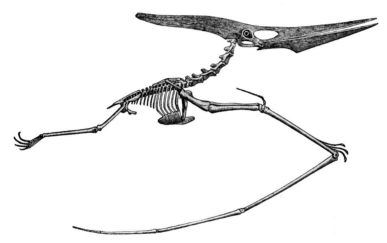

B

B. Skeleton of *Pteranodon ingens* Marsh, from the Upper Cretaceous of Kansas. For clarity, only the two extremities facing the viewer are shown. Much reduced. (After G. F. Eaton.)

C

C. The Komodo monitor, *Varanus komodensis*, the largest living lizard (3 meters long), from the island of Komodo, Dutch East Indies [Indonesia]. (After F. Reinöhl 1940.)

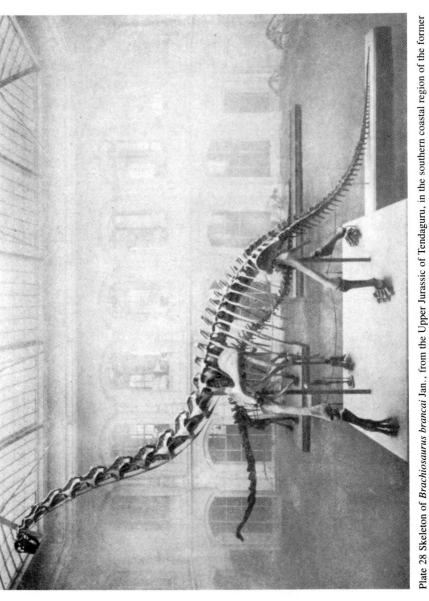

Plate 28 Skeleton of *Brachiosaurus brancai* Jan., from the Upper Jurassic of Tendaguru, in the southern coastal region of the former German East Africa [Tanzania]. (After an original photograph of the Berlin specimen, made available by Professor W. Janensch.)

Plate 29

Plate 29 The *Brachiosaurus* skeleton during its mounting in the Lichthof [skylight court] of the Berlin Museum of Natural History. (After W. Janensch.)

Plate 30

Plate 30 A. *Pleurotomaria* (*Entemnotrochus*) *adansoniana* Crosse and P. Fischer, one of the few living species of the genus *Pleurotomaria,* of which the typical subgenus (*Pleurotomaria* s. str.) is known only as a fossil. Recent. Lesser Antilles. ½ ×. (After Damon, from W. Wenz.)

A

B

B. *Dinornis maximus* Ow., a giant representative of the moas of New Zealand, which coexisted with humans in the recent past, but which is now extinct. It grew to a height of more than three meters; its eggs were three times as large as ostrich eggs. (From O. Abel 1939, *Tiere der Vorzeit* [Prehistoric animals].)

Plate 31

A

Plate 31 A. *Sphenodon punctatum* Gray, the Recent tuatara, found on a few islands of northern New Zealand.

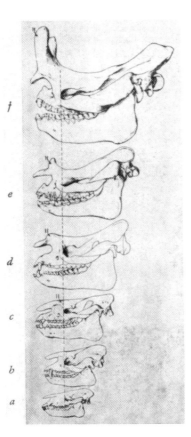

B

B. Evolution of the skull and horns in titanotheres; progressive increase in size and allometric growth of the various parts of the skull. In the initial form, *Eotitanops* (*a*), the face part of the skull is longer than the brain case; in the final form, *Brontotherium* (*f*), the relationship is the reverse. The dotted line running down through the forward edge of the eye sockets makes the shift in proportions stand out clearly. Small, hornlike swellings appear first in *Manteoceras* (*c*) and then increase rapidly in size in subsequent evolutionary stages in correlation with the increase in size of the entire skull. During the course of this development, the line of the forehead becomes deeply concave. *a. Eotitanops borealis* (Cope). Lower Eocene. *b. Limnohyops priscus* Osb. Early Middle Eocene. *c. Manteoceras manteoceras* Hay. Late Middle Eocene. *d. Protitanotherium emarginatum* Hatch. Upper Eocene. *e. Brontotherium leidyi* Osb. Early Lower Oligocene. *f. Brontotherium gigas* Marsh. Late Lower Oligocene. (After H. F. Osborn 1929, much reduced.)

Plate 32

A

Plate 32 A. *Discoclymenia kayseri* (Schdwf.). Late Upper Devonian, from Kowala, near Kielce (Polish Mittelgebirge [Holy Cross Mountains]). 1½ ×. (From Schindewolf 1944.)

B

a

B. *Sporadoceras rotundolobatum* Schdwf. Middle Upper Devonian, from Saalfeld, in eastern Thuringia. *a*. Lateral view. 1 ×. *b*. Suture line. (From Schindewolf 1928.)